THE SHAPE OF SPECTATORSHIP

FILM AND CULTURE
John Belton, Editor

FILM AND CULTURE

A series of Columbia University Press
Edited by John Belton

For the list of titles in this series, see page 373.

THE SHAPE OF SPECTATORSHIP

ART, SCIENCE, AND EARLY CINEMA IN GERMANY

SCOTT CURTIS

COLUMBIA UNIVERSITY PRESS
NEW YORK

Columbia University Press
Publishers Since 1893
New York   Chichester, West Sussex
cup.columbia.edu
Copyright © 2015 Columbia University Press
All rights reserved

Library of Congress Cataloging-in-Publication Data
Curtis, Scott.
The shape of spectatorship : art, science, and early cinema
in Germany / Scott Curtis.
   pages cm. — (Film and culture)
Includes bibliographical references and index.
ISBN 978-0-231-13402-6 (cloth : alk. paper) — ISBN 978-0-231-13403-3
(pbk. : alk. paper) — ISBN 978-0-231-50863-6 (ebook)
 1. Motion pictures—Germany—History—20th century. 2. Motion picture
audiences—Germany—History—20th century. 3. Motion pictures—Aesthetics.
4. Motion pictures in science—Germany. 5. Documentary films—Germany—
History—20th century. I. Title.
PN1993.5.G3C88 2015
791.430943—dc23
2015010546

Columbia University Press books are printed
on permanent and durable acid-free paper.

This book is printed on paper with recycled content.
Printed in the United States of America

Cover Design: Jordan Wannemacher
Cover Image: From Wilhelm Braune and Otto Fischer, "Versuche am unbelasteten
und belasteten Menschen," *Abhandlungen der Mathematisch-Physischen Klasse der
Königlich Sächsischen Gesellschaft der Wissenschaften* 21, no. 4 (1895): 151–322.

References to websites (URLs) were accurate at the time of writing.
Neither the author nor Columbia University Press is responsible for URLs
that may have expired or changed since the manuscript was prepared.

To my parents
◊ ◊ ◊

CONTENTS

List of Illustrations ix
Acknowledgments xi

INTRODUCTION 1

1. SCIENCE'S CINEMATIC METHOD: MOTION PICTURES AND SCIENTIFIC RESEARCH 19
    Early Scientific Filmmaking: An Overview 26
    Bergson, Cinema, and Science 32
    The Science of Work and the Work of Science 37
    Brownian Motion and "the Space Between" 62
    Nerve Fibers, Tissue Cultures, and Motion Pictures 76

2. BETWEEN OBSERVATION AND SPECTATORSHIP: MEDICINE, MOVIES, AND MASS CULTURE 90
    The Multiple Functions of the Medical Film 96
    Motion Pictures and Medical Observation 110
    Time, Spectatorship, and the Will 125

3. THE TASTE OF A NATION: EDUCATING THE SENSES AND SENSIBILITIES OF FILM SPECTATORS 142
   Cinema and the Spirit of Reform 147
   Children, Crowds, and the Education of Vision and Taste 162
   "Cinematic Lesson Plans" in Elementary and Adult Education 176

4. THE PROBLEM WITH PASSIVITY: AESTHETIC CONTEMPLATION AND FILM SPECTATORSHIP 193
   Agency and Temporality in the Aesthetic Experience of Cinema 202
   Einfühlung, Identity, and Embodied Vision 214
   The Politics of Contemplation 230

CONCLUSION: TOWARD A TACTILE HISTORIOGRAPHY 243

Notes 253

Bibliography 313

Index 355

LIST OF ILLUSTRATIONS

Figure 1.1. Braune and Fischer's military recruit in the experimental suit   47
Figure 1.2. The subject at rest with the grid superimposed   48
Figure 1.3. Braune and Fischer's camera placement   50
Figure 1.4. The resulting chronophotograph   52
Figure 1.5. Determination of the coordinates of a point P from the projections of P on two planes as seen from the two cameras   53
Figure 1.6. Side and top views of the instrument used to measure coordinates   54
Figure 1.7. Measurement of a coordinate   55
Figure 1.8. A table of the coordinates derived from experiment 1   57
Figure 1.9. The graph of the coordinates (view from the right side)   58
Figure 1.10. The graph of the coordinates (view from above of different body parts)   59
Figure 1.11. Left and back views of the tridimensional model representing the attitudes of the human body during walking   61

Figure 1.12. Seddig's cinematic apparatus for measuring Brownian motion   71
Figure 1.13. Seddig's photographic rendering of Brownian motion   71
Figure 1.14. The irregular paths of Brownian motion   72
Figure 1.15. Hermann Braus   78
Figure 1.16. Ross Granville Harrison   79
Figure 1.17. Harrison's sketches of the elongation of frog nerve fibers grown in culture   80
Figure 2.1. Frames from Ludwig Braun's film of a dog's beating heart   99
Figure 2.2. Groedel's serial cassette X-ray apparatus   101
Figure 2.3. Max Nordau   129
Figure 2.4. A hypnotist practicing his craft, France, circa 1900   138
Figure 3.1. A typical storefront movie theater (*Ladenkino*) from the pre–World War I era   153
Figure 3.2. *The Wanderings of Odysseus*, touted to be a "Reformfilm"   158
Figure 3.3. "The Cultural Work of the Film Theater: Thoughts from the Year 1784 by Friedrich von Schiller"   167
Figure 3.4. The geometry of taste   168
Figure 3.5. Children at Luisen-Kino in Berlin, circa 1910   180
Figure 3.6. A page from Lemke's *Die kinematographische Unterrichtsstunde* (The Cinematic Lesson Plan, 1911)   183
Figure 3.7. Hermann Häfker   185
Figure 4.1. Aesthetic experience as a series of interlocking dichotomies   198
Figure 4.2. A prewar audience enjoying a night at the Union-Theater in Berlin, 1913   219
Figure 4.3. Kammer-Lichtspiele Theater in Berlin, 1912   234

ACKNOWLEDGMENTS

This has been, to use an inappropriate medical metaphor, a long and difficult birth, but certainly not for lack of attendants or painkillers. After so many years of conceptions, reconceptions, and labor, however, it is hard to know who to thank or how far back to go, so I will simply begin to recite and offer my deepest apologies to anyone I inadvertently leave out. Various versions took shape in various locations, where I owe debts to the institutions and people who supported me and this project. The initial research was made possible by a stipend from the German Academic Exchange Service. Several people made my stay in Frankfurt more productive than it might have been. Heide Schlüpmann, Helmut Diederichs, and Martin Loiperdinger offered good advice and kind, if somewhat bewildered encouragement as I struggled to define my topic. Diederichs also provided the rare image of Hermann Häfker for chapter 3. Special thanks go to the late Eberhard Spiess for allowing me access to the periodicals and holdings of the Deutsches Institut für Filmkunde, and to Brigitte Capitain for her patience as I took advantage of this generosity. In Iowa City, I appreciate the attention David Depew, Kathleen Farrell, and Hanno Hardt gave to the first version of this project, while John Durham Peters and Dudley Andrew deserve special thanks for their guidance through the years. In Los Angeles, I am grateful to Linda Mehr and the late Robert Cushman of the Academy of Motion Picture Arts and Sciences' Margaret Herrick Library; I learned much from their example. Joe Adamson, Val Almendarez, Anne Coco, Steve Garland, Harry Garvin, Barbara Hall, Doug Johnson, Janet Lorenz, David Marsh, Howard Prouty, Lucia Schultz, Matt Severson, Warren Sherk, and all of my fellow librarians at the Herrick deserve thanks as well.

In Evanston, I owe thanks to the Department of Radio/Television/Film—especially my colleagues Bill Bleich, Michelle Citron, Laura Kipnis, Chuck Kleinhans, Larry Lichty, Hamid Naficy, Eric Patrick, Jeff Sconce, Jacob Smith, Jacqueline Stewart, Deb Tolchinsky, and the rest of the faculty—for having faith in me. Chairs Annette Barbier, Mimi White, Lynn Spigel, and David Tolchinsky were steadfast in their support. I thank Lynn Spigel, especially, for all she has done on my behalf (as far back as Los Angeles) as a mentor, model, and friend. Deans David Zarefsky and Barbara O'Keefe at the School of Communication offered resources in various forms, and I am grateful for their time, money, and patience. In Berlin, stays at the Max Planck Institute, thanks to Lorraine Daston, significantly sharpened my thinking about physics and observation, especially. In Weimar, Karl Sierek, Friedrich Balke, Daniel Eschkötter, and the graduate students at Bauhaus-Universität welcomed me and gave me space and time to work. In Innsbruck, Mario Klarer, Gudrun Grabher, Christian Quendler, Erwin Feyersinger, Robert Tinkler, Cornelia Klecker, Johannes Mahlknecht, Monika Datterl, and Maria Meth likewise gave time, space, and warm companionship freely, as well as trips to mountain cabins. In Doha, the entire staff and faculty of Northwestern University in Qatar made me feel welcome, especially program directors Mary Dedinsky and Sandra Richards, and colleagues Greg Bergida, David Carr, Susan Dun, Elizabeth Hoffman and Bob Vance, Joe Khalil, Muqeem Khan, John Laprise, Jocelyn Mitchell, Sue Pak, Christina Paschyn and Alex Demianczuk, Barry Sexton, Bianca Simon, Anne and Adam Sobel, Allwyn Tellis, Tim Wilkerson, and Ann Woodworth. Deans Jim Schwoch, Jeremy Cohen, and Everette Dennis displayed an inordinate amount of trust and confidence in my abilities; Dean Dennis, especially, offered whatever it took—and it took a lot—to get it done, and I am deeply indebted to him.

Along the way, a number of people deserve commendation for having read or commented on various parts of this project in various forms. Chapter 1 benefited greatly from the insights of Nancy Anderson, Charlotte Bigg,

Thomas Haakenson, Andreas Mayer, and others at the Max Planck Institute; Hannah Landecker kindly shared her research and insights on cell biology; Martin Carrier and Alfred Nordmann sharpened my thinking about atomistic physics; Richard Kremer and Ken Alder offered face-saving corrections from the historian of science's point of view; and Dan Morgan, Oliver Gaycken, and Frank Kessler were kind enough to read the chapter at various points and help me clarify the argument. Chapter 2 profited from the attention of Ken Alder and the Science in Human Culture Group at Northwestern, who prompted me to rethink it, while Nancy Anderson and Mike Dietrich provided time and space at Dartmouth College to rewrite it. Lisa Cartwright, Oliver Gaycken, Andreas Killen, Kirsten Ostherr, and Henning Schmidgen were at various points inspirational and instrumental in shaping this chapter; Lisa Cartwright was especially helpful at a key point in the process. Chapter 3 owes its life to Steve Wurtzler, Jennifer Barker, and John Belton, and I owe Frank Kessler and Sabine Lenk for keeping it alive. A much earlier, different version of chapter 4 was lucky to have the scrutiny of Ben Singer, David Bordwell, David Levin, and Marc Silberman. For its current form, I must thank Robin Curtis, Gertrud Koch, Dan Morgan, Inga Pollmann, and especially Kaveh Askari and Tony Kaes for all of their insight and encouragement. There are still others who provided valuable assistance along the way, including research assistants David Gurney, Dan Bashara, and Rebecca Barthel. John Carnwath kindly and expertly corrected my translations, rescuing me from many infelicities. Stefanie Harris, Jörg Schweinitz, and various anonymous readers offered important insights that prompted revisions and changes in argument. For their stalwart professional support and friendship over the years, I must offer my heartfelt thanks to Richard Abel, Rick Altman, Matthew Bernstein, Jane Gaines, Dilip Gaonkar, the late Miriam Hansen, Tom Levin, Charlie Musser, Jan Olssen, Patrice Petro, Lauren Rabinovitz, Eric Rentschler, Mark Sandberg, Vivian Sobchack, and Virginia Wexman.

All of the people named so far I count as my friends, but some friends deserve special mention for their unselfish and nonjudgmental acceptance of me and my book. Tom Gunning has been a friend and mentor for a very long time; more than anyone, he has shaped the contours of this ongoing investigation, usually without even knowing it. Tony Kaes has been a loyal fan and inspiration since I was a student. Both Tony and Tom have offered insightful, transformative commentary on several versions. Oliver Gaycken, Vinzenz Hediger, and Kirsten Ostherr are my fellow travelers on this interdisciplinary journey; I don't often take a step without consulting them.

Oliver read every word I have given him and always came back for more. Greg Waller and Brenda Weber have always offered valuable moral support and close friendship at just the right times. Ken Alder, Joe Carli, Lisa Cuklanz, Tracy Davis, Doug Johnson, Charlie Keil, and Will Schmenner have all been steady, life-long friends on whom I have leaned especially heavily at times. All the graduate students who have attended my seminars deserve note for their role in shaping my thinking over the years, but I thank especially Dan Bashara, Catherine Clepper, Beth Corzo-Duchardt, Alla Gadassik, Leslie Ann Lewis, Jason Roberts, Jocelyn Szczepaniak-Gillece, Kati Sweaney, and Meredith Ward. For their continuing friendship, I thank Richard Abel, Charles Acland and Haidee Wasson, Dana Benelli, Joanne Bernardi, Bill Bleich, Jeremy Cohen and Catherine Jordan, Kelley Conway and Matthew Sweet, Mark Garrett Cooper and Heidi Rae Cooley, Don Crafton and Susan Ohmer, Nick Davis, Leslie Midkiff DeBauche, Nico de Klerk, Carol Donelan and Shannon Spahr, Nataša Ďurovičová, Dirk and Myrna Eitzen, Jen Fay, André Gaudreault, Philippe Gauthier, Frank Gray, Alison Griffiths and William Boddy, Barbara Hall and Val Almendarez, Sara Hall and Monty George, Stefanie Harris, Micaela Hester, Chris Horak, Laura Horak and Gunnar Iverson, Rembert Hueser, Jenn Horne and Jonathan Kahana, Christopher Hurless and Rachel Henriquez, Zara Kadkani and Axel Schmitt, Jim Lastra, Tom Levin, Melody Marcus, Caitlin McGrath, Christie Milliken, Priska Morrissey, Tania Munz, Bill Palik, Anna Parkinson, Jennifer Peterson, Sarah Projansky, Christian and Grace Quendler, Isabelle Raynauld, Mark Sandberg, Ben Singer, Blane Skowhede, Jake Smith and Freda Love Smith, Stefan Soldovieri, Matthew Solomon, Shelley Stamp, Bing Stickney, Claudia Swan, Steve Tremble, Alison Trope, Mark Williams and Mary Desjardins, Tami Williams, Michael and Julia Wilson, Pam Robertson Wojcik, Robb Wood and Hanaa Issa, Steve Wurtzler, Harvey Young, and Josh Yumibe.

I should stress that this book would not have been published except for the efforts of John Belton and Jennifer Crewe, whose stubborn determination to wrest the manuscript from me finally overmatched my stubborn refusal to give it up. I also thank my editorial team at Columbia University Press: Ben Kolstad, Roy Thomas, Jennifer Jerome, Anne McCoy, and Kathryn Schell.

Of course, my family also deserves a large share of credit, not the least for their good-natured and bemused acceptance of a project that seemed never to end. Cindy and Randy Lee have been unconditional in their love and acceptance; Brandon and Noelle Lee, Trevor, Josh, and Michael Lee all are great relatives to have. Grant and Donna Boyles deserve mention for their love, care, and hospitality. I know my grandmother, Louise, and my aunt, Dian, would have been proud. I thank the Pike and Kelly families for welcoming me into their close-knit web of love and kindness: Ken and Elnora Kelly, Clayton and Carol Pike, Kerry and Kelli Graf (and Kaitlyn and Kayla), and Kory Pike; Kevin and Meredith Pike, especially, are not just relatives, but friends, which is a rare thing to say. But I owe most to Kirsten Pike, who has been my anchor, sail, and compass since I met her; this book and I would not be the same without her years of love and encouragement. If I had another book in me, I would dedicate it to her. But I must dedicate this work to my parents, Ray and Pat Curtis, whose inexhaustible patience, support, and love made everything possible.

There have been publications of parts of this project along the way, but what lies in the pages ahead is usually significantly different from what came before. Even so, we should note that parts of chapter 1 originally appeared as "Die kinematographische Methode. Das 'Bewegte Bild' und die Brownsche Bewegung," *montage/AV: Zeitschrift für Theorie & Geschichte audiovisueller Kommunikation* 14, no. 2 (2005): 23–43; and "Science Lessons," *Film History* 25, nos. 1–2 (2013): 45–54. Parts of chapter 2 originally appeared as "Between Observation and Spectatorship: Medicine, Movies, and Mass Culture in Imperial Germany," in *Film 1900: Technology, Perception, Culture*, edited by Annemone Ligensa and Klaus Kreimeier (New Barnet, U.K.: Libbey, 2009), 87–98; and "Dissecting the Medical Training Film," in *Beyond the Screen: Institutions, Networks and Publics of Early Cinema*, edited by Marta Braun, Charlie Keil, Rob King, Paul Moore, and Louis Pelletier (New Barnet, U.K.: Libbey, 2012), 161–167. Parts of chapter 3 originally appeared as "The Taste of a Nation: Training the Senses and Sensibility of Cinema Audiences in Imperial Germany," *Film History* 6, no. 4 (Winter 1994): 445–469. Parts of chapter 4 originally appeared as "Einfühlung und die frühe deutsche Filmtheorie," in *Einfühlung. Zur Geschichte und Gegenwart eines ästhetischen Konzepts*, edited by Robin Curtis and Gertrud Koch (Paderborn: Fink, 2009), 61–84.

THE SHAPE OF SPECTATORSHIP

INTRODUCTION

Probably no contemporary invention has generated quite as much discussion in the daily press and in daily conversation as the cinema. Everywhere new theaters shoot up overnight like mushrooms. Our cities at night can no longer be imagined without the beaming portals of the movie houses. But it's not just the simple folk pushing themselves through these "narrow gates of grace." The educated class, as well as science and the schools, the state, the city, and rural communities have all grasped the cultural significance of cinema and have taken a step closer to the establishment and utilization of their own film theaters. Who would look upon this burning question with indifference?

ADOLF SELLMANN (1912)[1]

Whatever cinema is, it has always been many things to many people. Even in 1912, it was clear to a reformer such as Adolf Sellmann that a variety of interest groups and interested parties, from scientists to educators to town councils, were using motion picture technology. Each group recognized cinema's "cultural significance" and power or acknowledged its inevitability, but not every group agreed on how motion pictures should be used, either in the public sphere or, especially, within the boundaries of the group or discipline. Everybody had their reasons for using motion pictures, and those reasons often diverged. This book attempts to understand how various disciplines or communities used motion pictures. What did cinema mean to these groups? What were the criteria for the acceptance of motion pictures as a tool within a given discipline? What problems presented themselves such that motion pictures were considered a solution? This book explores these questions to discover the criteria for the legitimacy of a new media technology within the disciplines of science (specifically, human motion studies, physics, and biology), medicine, education, and aesthetics in Germany before World War I.

These disciplines correspond roughly to the familiar historical trajectory of early cinema, from its roots in scientific research to its early bids for acceptance as an art form. Taken together, they also represent the heterogeneity of early cinema, not only in terms of the many types of films available during this period but also with respect to the varied venues, audiences, and uses of the medium. Additionally, they typify relatively well-defined communities with strong, native traditions, where, outside of the entertainment industry, the liveliest discussion of motion pictures took place during cinema's early period. This listing obviously leaves out the entertainment industry, but questions of appropriation and legitimacy are less interesting in this area (at least to me), where the criterion for acceptance of film within that industry was clear, even tautological, in that the medium only had to prove its commercial viability and little else. So with its focus on the way groups used film for purposes other than entertainment, this study is aligned primarily with recent work in nontheatrical uses of film.[2]

However, even within the framework of "useful cinema" and the good work that has been done to define that area of film history, questions of appropriation and legitimacy are not often explicitly asked. While we know much about the use of motion pictures in the classroom in the 1920s, for example, we still know comparatively little about the state of pedagogical theory and practice at that time and why some groups within the discipline saw motion pictures as a partial solution to a variety of problems, and why others did not. We know little of their disciplinary agenda. Perhaps inevitably, we approach these questions as film historians, not as historians of education. Yet to understand fully any given appropriation, we must fully understand the agenda that shapes it. There is an intimate and complex relationship between any technology and the agenda that makes use of it. The technology is not merely applied to a problem; the problem presents itself in part because of the technology. What any scientist investigates, for example, is partly due to what the available technology makes available for investigation. Historians of science are very good at demonstrating the dialectical relationship between tools, theories, and representations, which shows us that we cannot take "use" for granted; the criteria for use of any given tool within a given discipline are not obvious. As different groups used media technologies for their different purposes, the nature of those appropriations changed, and in a significant way, so did the medium. "What cinema is" for one group was not necessarily the same as for another. Indeed, the nature of the appropriation often depended on what the agents *thought* film was. So there was more to "use" than simply taking a camera, recording an event, and projecting it; the representational problems faced by any given discipline shaped its appropriation of (media) technology. Understanding those representational problems demands an intimate knowledge of the historical contexts and camps of that discipline.

So this project is not just about the encounter between other disciplines and film but the encounter between other disciplines and *film studies*. Specifically, each chapter stages a meeting between methods or approaches common to film studies and those of the history of science, the philosophy of medicine, the history of education, or the history of aesthetics. What does the result of such an assignation look like? What can we take away from such an encounter? Or, to put it another way, what can we reasonably expect from interdisciplinary research? Max Weber has his hand up: "With every piece of work that strays into neighboring territory . . . we must resign ourselves to the realization that the best we can hope for is to provide the expert with useful *questions* of the sort that he may not easily discover for himself from his own vantage point inside his discipline."[3] "Useful questions," however, are rarely presented as such; they are instead approaches or agendas that seem foreign at first, yet bear on our own. To formulate them as questions, we need to know the discipline well enough to recognize the pattern common to the approaches. So interdisciplinary research should be more than cherry-picking a few juicy quotations from an exciting discovery in another field; it must entail some significant level of immersion. What useful questions does the history of science, for example, provide film and media studies, and vice versa? As hinted above, one broad question could be: What is the relationship between technology and a disciplinary agenda? This ambitious question might be answered only after an accumulation of case studies, but it is useful for film and media studies, because it leads us to speculate about the tangible relationship between a representational technology and a community's conception of the object of study. Another question might be something like: What is the relationship between a technology and other elements of the experimental system? This question forces us away from our habitual focus on film and toward an understanding of film and media technologies as part of a larger experimental arrangement or as part of a technological group along the lines of what Germans call a *Medienverbund*, or "media ensemble."[4] With the help of the history of science, we can see film in these contexts as an important but nevertheless interdependent part of an experimental system or a larger institutional project.[5] Each disciplinary encounter in the following chapters results, I hope, in a different set of similarly useful questions.

What can film historians bring to the table? We can bring different kinds of answers. Our training in close analysis heightens our sensitivity to formal relationships that rely on analogies and homologies as well as causal or empirical connections. This would be useful, because the acceptance or legitimacy of any given technology for a discipline hinges not simply on the tool's function for a particular task but also on how that technology fits into a larger disciplinary system. A wrench is useful because it fits the bolt; certain parts of the bolt and wrench have similar shapes. In fact, the formal features of each determine their use, or vice versa; there is no reason to use the wrench on the bolt, except for this formal relationship. Likewise, film as a tool must have fit the object, task, and system of which it was a part. This fit was variable, because any given technology as complex as motion pictures is not a single thing but multiply adaptable to various agendas, so the relationship between them might be successful or not. The relationship was shaped discursively, too: the convergence between a technology and a discipline depended on the successful appropriation of the technology but also on the successful preparation of the discipline or agenda for that technology. If wrench and bolt are designed together from the start, the mutual shaping of film and discipline happened over time and was just as discursive as practical. That is, as mentioned earlier, the fit between a research task and film often depends on what the agents *think film is*, as well as the analogical relationship between a technology and a system (such as how the film frame isolates an object in a way similar to how experiments attempt to isolate objects and variables). Film and media studies is especially good at teasing out film-theoretical assumptions and finding formal connections. So a useful question could be: To what extent can the *formal* features of a technology and its representational products explain the relationship between that technology and the research agenda? This is a useful question for historians of science, because the significance of formal features of their objects of study rarely takes precedence.

More than a book about Germany, science, aesthetics, or even cinema, then, this is a book about historiography. How should we approach the relationship between a (media) technology and the group or discipline that uses it? This book insists that the formal features of the technology matter. It argues that the disciplinary legitimacy of a technology or the successful appropriation of a technology by an expert group depends on *a correspondence between the logic of the discipline and the formal features of the technology*. In this case, filmic form refers to two manifestations of the image: the projected image and the frames on the celluloid filmstrip. The logic of the discipline refers to its problem-solving protocols, its investigatory methods, and its ideological (in terms of its discipline) assumptions. Medical logic, for example, refers to a method of arriving at a diagnosis, etiology, prognosis, and treatment of a disease. It is a way of thinking that provides for a (more or less) consistent and balanced understanding of what is peculiar to the individual case and what is generalizable from it. As we shall see in chapter 2, this logic relied on series of cases for training and context. We will also see that the ambidexterity of film, its ability to be useful in both its still and moving forms, corresponded in unique ways to this medical logic and observational practice. The goal of that chapter, then, is in part to sketch the correlation between a "medical way of thinking," as philosopher of science Ludwik Fleck put it, and the specific formal features of film.[6] Likewise, in some scientific disciplines, the ability of film to frame and isolate the object under study and film's temporal malleability (its ability to expand or compress time) were analogous to features of scientific experiments, such as the ability to isolate variables and extend observational duration. In others, film's temporal discontinuity, indicated by the gap between film frames, matched theoretical ideals about the physical world. For education and aesthetics, the logic of the discipline was less about solving problems than about describing goals or mental operations to attain those goals. In education, the consonance between the detail and duration of the moving image and the richness of the natural world created a homology between the (perceived) realism of the image and the goal of visual instruction, which was to teach students to recognize objects and generalize from them. Aesthetics describes the terms of artistic production and, especially pertinent to this project, the conditions of aesthetic experience; it describes the set of presumed cognitive and emotional operations that accompany aesthetic experience, along with their ideological significance. With regard to film, then, the formal features of the object and the accompanying experience were compared with the prevailing understanding of aesthetic experience. In this case, the pace of the moving image worked against ideals of free will and agency embedded in common conceptions of the aesthetic experience, while the ability of the moving image to encourage emotional projection corresponded to different theories of aesthetic experience popular at the time.

Frame, gap, detail, duration, and pace: each discipline or group, then, saw something useful (or counterproductive, as the case may be) about filmic form but emphasized different features for different goals. In the early period, as motion picture technology emerged as a real possibility for various applications, individual researchers, teachers, reformers, or cultural pundits took it upon themselves to justify motion pictures as a tool or good object to their colleagues. They endeavored to demonstrate that motion pictures could indeed conform to the logic and practices of the discipline. This is why the early period of film's cultural dissemination is so productive for this kind of project: the reasons for appropriation—what the champions thought film was—are often clearly stated, before the use of motion pictures becomes naturalized and obvious, requiring no justification at all (which is the case today when scientists, for example, use moving images as a matter of course without discussion). This is also why this project is less concerned about individual films than the contours of the discussion about the use of film in general; close analysis of filmic style can tell us very little about the justifications mounted for film or the correlation between disciplinary logic, practice, and film form. Such rationalizations can tell us much, however, about the state of the discipline at the time. Whether the use of film addressed common representational problems within a discipline (as in the case of cell biology at the turn of the century) or the discipline faced larger philosophical problems that film seemed to exemplify (as in the case of fin de siècle aesthetics), the engagement with motion pictures tells us quite a lot about the priorities and changes that faced any given discipline. Again, this is especially the case during film's early period, when groups latched onto this new technology because of a pressing need or because of its prevalence. In other words, the timing of these appropriations is not coincidental: each group needed motion pictures in some way—even as a scapegoat, which was often the case. Within any given discipline, these maneuvers were far from unanimously approved; dissent was more common in the cultural than scientific spheres, but the debates sharpen our understanding of what was at stake for each group.

What was at stake exactly? As a primarily visual technology, motion pictures presented an aid or a challenge to *expert modes of viewing*, which were the most common manifestation of disciplinary logic and practice. Around what practices, ideals, and "ways of thinking" do disciplines coalesce? There are many answers, ranging from laboratory procedures to nationalist rhetoric. But a very common way to train new members of a discipline is to teach them what to look for and how to look for it. Ludwik Fleck insisted that "one has first to learn to look in order to be able to see that which forms the basis of the given discipline," but I would further insist that "learning to look" *is* what often forms the basis of a discipline.[7] Whether it is medical observation or aesthetic contemplation, disciplines have ways of looking that are assumed; if you have to ask what it is or how to do it, you obviously are not part of that group. It is a badge of honor or a barrier to entry. Scientific and medical disciplines, for example, invested much time and energy into training recruits to their modes of viewing, to what Fleck called "the directed readiness to certain perceptions."[8] British surgeon Sir James Paget emphasized in 1887 that "becoming scientific in our profession [requires] the training of the mind in the power and habit of accurately observing facts.... The main thing for progress and for self-improvement is accurate observation."[9] Likewise, Swiss pedagogue Johann Heinrich Pestalozzi organized his entire, influential educational program around close examination; he believed that direct perception of the world was *the foundation of all knowledge.*"[10] And perhaps it goes without saying that aesthetic experience, as expounded by German philosophers since Immanuel Kant, relied heavily on a mode of viewing that was leisurely, probing, and attentive—in a word, contemplative. Above all, these disciplinary modes of looking are more than individual, as Fleck notes: "We look with our own eyes, but we see with the eyes of the collective body, we see the forms whose sense and range of permissible transpositions is created by the collective body."[11] So expert modes of vision and disciplines orbit each other closely and inextricably, held together by ideological bonds.

This is not to say there is only one mode of viewing for each discipline. Indeed, there are two obvious examples of different modes of expert viewing that obtain regardless of discipline: the holistic, all-at-once, instant appraisal; and the roaming, penetrating, leisurely, detail-oriented contemplation of the object. These two modes can be found in a range of disciplines from scientific observation to medical observation to aesthetics. In art history, for example, Robert Vischer distinguished between *Sehen* and *Schauen*, two modes of viewing artworks, which can be translated as "seeing" (by which he meant the synthetic, intuitive, instant appraisal) and "scanning" (by which he meant analytic, detail-oriented contemplation). These correspond roughly to our "glance" and "gaze," as long as they are not confused with their current alignment in media studies with distraction and contemplation.[12] For Vischer and experts across disciplines, glance and gaze were complementary modes of viewing, both requiring expertise. In medicine, physicians in turn-of-the-century Germany who were troubled by the increasingly scientific approach to healing—represented in their minds by the analytic, focused gaze—advocated a more holistic view that took in the entire patient at glance rather than a penetrating gaze that examined only the localized disease in detail.[13] Even if in this case these modes of viewing were taken up as flags symbolizing different professional stances, in practice medical training encompassed both modes. An expert both synthesizes and analyzes.

In fact, as Michel Foucault has noted about medical observation, these modes of expert viewing were more than merely visual—they were also a set of logical operations.[14] Specifically, I find that expert viewing is largely a process of *correlation*. What happens during expert observation? The expert sees the parts and the whole and finds patterns between them and correlates those patterns to expert knowledge of the disciplinary context. Scientific observation, for example, connotes an analytic gaze, but the most important aspect of scientific observation is the context the scientist brings to it; the researcher assimilates observed data into an existing framework of knowledge. What the scientific observer already knows frames what he or she observes, and he or she incorporates or juxtaposes new data with old and thereby generates new insights. Observation therefore implies *the production of knowledge through correlative insight*. This is true even with art, in which the lynchpin of post-Kantian conceptions of aesthetic experience is the "free play of associations." As the expert gazes upon the artwork, the parts work together in pleasing ways to prompt in the viewer a correlation of textual and contextual patterns. As these associations come together, interlock, separate, and recombine with others, the viewer supposedly experiences a pleasant, even moving state of being between artwork, world, and his or her own subjectivity—all of which has been known as aesthetic experience, which depends fundamentally on a particular kind of viewing that is leisurely, active, attentive, and *correlative*.

Especially with regard to aesthetic contemplation, expert viewing was not only a disciplinary practice or a logical operation, it was also an exhortation. That is, expert viewing involved a measure of training and hence status, so it also often implied a way of looking at the world or an ideological stance. Immanuel Kant's and Friedrich Schiller's philosophies, for example, gave aesthetic experience a prominent ethical and moral position, in that it was both a mode of engaging with art and a way of bridging an impasse between competing forms of knowledge (Kant) or between conflicting human impulses (Schiller). Schiller's system, especially, suggested that aesthetic experience could lead the way to a better social organization, if only we applied this mode of engaging with art to our way of engaging with the world. Arthur Schopenhauer was even more insistent: aesthetic experience, for him, was one of the few ways by which we could escape the dreary consequences of our everyday impulses, and therefore in his system it became the fundamental model for interacting with the world in general. In the twentieth century, the discussion about the relative merits of contemplation and distraction was, as we know, highly politically charged precisely because of their implications for how one should conduct oneself; recall, for example, Georg Lukács's dismissal of the contemplative life as unconscionable in the face of changes that called for political action and initiative. Even scientific observation became a global ideal as "objectivity" came to mean more than refraining from theoretical speculation. I am not suggesting that these systems work or that they are universal. I am arguing that the operations and stances required by these expert modes of viewing also functioned widely as rhetorical instruments, as presumptions about how one should engage with the world.

How did motion pictures fit into these traditions of expert viewing? That is exactly the question the present project explores. If my larger argument claims that the legitimacy or success of a media technology depended on a correspondence between its formal features and the logic of the discipline, then the more specific argument holds that a discipline's expert viewing ideals and practices exemplified its logic and that *a discipline's successful appropriation of film hinged on its ability to accommodate the new technology to its mode of expert viewing*. To put it another way, the acceptance of film as an appropriate tool or good object within a discipline depended on an alignment between the formal features of the filmic image and the practices and ideals of that discipline's modes of expert viewing. The acceptance of film as a tool rested in part on the advocate's ability to adapt—physically and/or rhetorically—motion pictures to established modes of viewing. If proponents could prove, either through example or argument, that the filmic image could accommodate or even amplify established methods of observation, then the community accepted the justification and elaborated on it in subsequent literature.

For example, when teachers needed to justify the educational use of motion pictures, they aligned its formal features to already established conceptions of expert viewing and visual instruction; in Germany, these conceptions were encapsulated by the term *Anschauungsunterricht*, which translates roughly as "visual means of instruction." Their justifications were persuasive to the extent that certain features of film could match or accommodate the methods of this means of instruction. When cultural pundits argued against the idea that film could be art, their arguments were less often about its mechanical reproduction of nature and more about the relationship between the projected moving image and the possibility of aesthetic contemplation. In each case, the question was, can film be used in a manner familiar to our way of viewing and our way of viewing the world? Motion pictures also famously offered a view *different* from what individual viewers or disciplines were accustomed to. My argument about the correspondence between film and disciplinary logic accommodates this difference as well, because historically the tug-of-war between the familiar and the novel stretched across the literature in a given field; advocates of film technology came to it because it was new and different, yet understood it in terms of what had already come before. Indeed, any given experimental system or discipline moves forward in the same way: methods, technologies, and inscriptions generate new insights that are folded into the existing system, which is thereby subtly changed. In other words, the patterns of justification and dissemination of this new media technology, like that of knowledge itself, were incremental and correlative.

This implies that expert vision is not the only criterion by which disciplines judged motion picture technology. A survey of the literature in each of these fields reveals that the advantages of film for their projects were often expressed in two ways: in terms of inscription (what the camera can record) and reception (how the image can be viewed). Chapter 1, unlike the rest of the chapters, will focus on inscription. It will examine how researchers adapted motion pictures as an inscription technology to their objects of study, their theories, and their disciplinary needs. The following chapters will explore how experts adapted motion pictures to their different modes of viewing: medical observation, visual methods of instruction, and aesthetic contemplation.

Needless to say, expert viewing requires training that derives from and reinforces a community. Whether that community is a scientific discipline or an educated class of citizens or both, training to partake of the privileges of that group often came in the form of *visual* training as well as education in the relevant literature or canon. Sometimes that training was explicit, as in *Anschauungsunterricht*, which was specifically designed to teach children how to observe; this was not considered disciplinary training so much as one criterion for good citizenship and cultivation. Sometimes it was implicit, as in aesthetic contemplation (although the German tradition of *Bildbetrachtung*, or "image viewing," explicitly instilled principles of looking, art appreciation, and aesthetic contemplation). In any case, experts and their apprentices came together to form a community (disciplinary, class based, national) through shared values and practices, especially practices of viewing. These bonds were shaped not only by the training and values themselves but also by comparison with their opposite: the untrained, "valueless" spectator.

Indeed, the values of a group were often not explicitly stated except through reference to what was unacceptable. At stake was the proper method of processing visual information to achieve the desired end, yet the idea of *proper* viewing was articulated equally often, if not more often, via denunciations of *improper* viewing. The most obvious example, described in chapter 2, would be the medical experts' description of film audiences as passive, agog, and addicted, whereas their own practices of film viewing emphasized activity, detachment, and control. Yet the latter model was rarely explicitly stated. Instead, we must surmise that "proper" viewing model from descriptions of their practice as well as their explicit denunciations of "improper" viewing.

It is hardly surprising that a community of privileged white men in imperial Germany would condescend to audiences of a perceived lower-class amusement such as the movies. Yet perhaps we miss something if we leave it at that. Our understanding of the discursive construction of film spectatorship too often ignores the obvious *duality* of these constructions. If "proper" modes of viewing relied on "improper" modes as a useful foil, then the opposite is also true: our understanding of spectatorship is incomplete without an understanding of expert observation. Observation and spectatorship depended on each other (perhaps still) in a figure/ground relationship wherein the dominance of any given term was often difficult to determine. Spectatorship and observation functioned as each other's "negative space," the outline of one shaping the other by default. This book contends that the "shape of spectatorship" can be truly determined only by simultaneously examining the "shape of observation." Observation and spectatorship function akin to stillness and movement or analysis and synthesis: they are oscillating, dynamic dichotomies that are not opposites so much as necessary halves of a hermeneutic process. In this case the hermeneutic is one of self-identity; as experts worked to define their position with regard to disciplinary viewing, "spectatorship" was something of a by-product of the work process. More precisely, the work of (explicitly or implicitly) defining spectatorship or observation resulted in a coproduction of the opposite term in a strikingly consistent logic of the supplement. So yes, we know how privileged white men characterized spectators, and we have known that since audiences were aligned with crowds in discussions of early theater.[15] But if we want to know *why* any given characterization of spectators took its particular form, then we should be aware of the modes of viewing preferred *by the experts writing the description.*

If theories of film spectatorship in the 1970s and 1980s were preoccupied with the ideological dominance of Hollywood, then the historiographic innovations of film studies during the 1980s and 1990s demonstrated that the Hollywood model was, of course, not always dominant in production and reception; the paradigm of early cinema was not simply Hollywood in nuce but a positive model in its own right: these viewers saw film differently, and early film production and exhibition was heterogeneous. This heterogeneity also implied multiple types of textual address and hence multiple kinds of spectator in terms of gender, race, ethnicity, and class. So if the transitional era codified and standardized filmic address, then it also in a way standardized the spectator; the individual spectator of 1970s film theory was, then, an abstraction generated by filmmaking practice of the Hollywood era, just as the spectator of 1990s film history is a product of textual analysis of early film, or the latest theory of film spectatorship might be a derivation from journalistic accounts of audience experience or of the theorist's own experience. Theories of film spectatorship derive from either filmic textual address or written textual address, and these derivations are separate from historical, empirical viewers and audiences. While contemporary written discussions of viewers and audiences may seem closer to the historical viewer, the written account also produces an abstract notion of spectatorship. My point is that we must take into account the role and interest of the expert in the abstraction. We need to recognize our own expertise and role in producing spectators, but we also need to acknowledge the *disciplinary* context of historical experts as they wrote about observation and spectatorship. If histories of early film helped to establish the multiplicity of film spectatorship through the categories of gender, race, ethnicity, and class, I would like to add one more category, expert/lay, which seems a powerful determinate of the character of written discussions of audiences and of disciplinary modes of viewing. Expert/lay appears closely aligned with high/low class differences, but I would argue that it is more specific, because educational or disciplinary training provides a traceable, historiographically clear path to the fount of class-based distinctions.

Yet as Walter Benjamin argued, the line between expert observer and lay spectator was often fuzzy, especially as film viewers adopted "an attitude of expert appraisal."[16] Benjamin described a mode of reception in which the reactions of the individual were simultaneously amplified and regulated by the reactions of the mass, thereby giving the spectator the pleasure of viewing but also the detachment of the expert. Similarly, chapter 2 describes an expert mode of reception that was also simultaneously distanced from and thrilled by the moving image. Hence the dichotomy between expert and lay is heuristic; most often it was very fluid in practice, but the descriptions of spectatorship often functioned to chart a perceived difference that was helpful for self-identity or class identity. The significance of such descriptions has implications for the "modernity thesis" debate, which pivots on the validity of generalizing Benjamin's claim—that our "mode of perception" changes over historical periods[17]—to questions of film history and style. That is, can we claim that any film style of a particular era expressed the "mode of perception" of the time?[18] This debate has often been preoccupied with details of film style during the early period, or with periodization, or with the question of whether cognitive structures really change, but too often the debate has overlooked the obvious: experts of the time described the cinematic experience as precisely an expression of the state of urban life and as a reflection of the modern mode of perception. Writing about film was a primary means by which modern problems (of representation, observation, education, aesthetic standards) were articulated. In other words, Benjamin's claim did not come out of a blue sky: writers in the 1910s and 1920s described film in exactly these terms, so trying to *explain* those historical descriptions by constructing a homology between filmic style and modern urban experience—without overgeneralizing to include all films at all times during this period—is not only valid, but logical.

So descriptions of the cinematic experience in Germany before World War I often functioned as a means to articulate modern problems of representation, observation, education, or aesthetic standards. To describe the experience of attending a film screening was to work through questions associated with modern urban upheaval and transformation.[19] It was so in many places during the early period, when cinema was not really "cinema" but a heterogeneous mix of screen practices and ideals—one of many "cultural series," in André Gaudreault's term.[20] In Germany, however, the lines between expert and lay were perhaps more starkly drawn (hence more visible) than elsewhere. At the same time, the deeply ambivalent character of discourse during this time of transition, as Germany struggled to balance rapid modernization with traditional values, makes it an especially interesting case study for the mutual construction of expert observation and lay spectatorship. Historians of Wilhelmine Germany have long recognized the dichotomous yet strikingly fluid relationship between left and right political positions, progressive and reactionary attitudes toward the role of the state, male and female roles in public service, or middle- and working-class positions on the social hierarchy.[21] The same ambivalence is evident in the reactions to new media forms such as motion pictures, which makes it an intriguing and challenging case from which we can perhaps infer broader principles about the mechanism of cultural legitimacy.²² Of course, arguments for and against the use of motion pictures in Germany at this time are similar, perhaps identical, to arguments for and against in other countries. Yet every agent marshals evidence and rhetorical weight from his or her native tradition in support of his or her argument; the Germans were no different. So this book straddles the line between a specialist study on early cinema or German culture and a broader investigation of the nature of expert appropriations of new technologies. The mixture of these two goals differentiates it sufficiently, I hope, from other works on the discourse on early cinema in Germany or elsewhere.

This project therefore takes up the dual challenge of describing *how* experts cleared a cultural space for motion picture technology while explaining *why* that work (rhetorical or practical) took the form it did. Both questions require not just primary research into specific appropriations but an understanding of the disciplinary expectations and needs faced by the experts. So each chapter will explore the *history* of that discipline (such as biology) by also using the *method* of a corresponding discipline (such as the history of science). Each chapter, then, is intended to be a hybrid between the methods of film studies and those of another discipline. Explaining why the work of appropriation took the form that it did also requires an understanding of the historical and cultural context of the experts, in this case imperial Germany. So each chapter also attempts to provide a social context, but because arguments for and against carried the weight of past traditions and values, a detailed philosophical context is sometimes required as well. For example, to understand fully the visceral reaction to film's perceived accelerated pace, we will need to examine that reaction's relationship to Schiller's temporal understanding of aesthetic experience, which was accepted by experts and the middle class implicitly. Over time, philosophical ideas sometimes spread and harden into ideological dogma.

I should emphasize that this project focuses on the question of cultural legitimacy or the acceptance of a new media technology by a given discipline at a historical moment. It therefore concentrates on the *correspondence between* film form and discipline, rather than the *impact* of form *on* that discipline. Each chapter charts the relationship between the material or form of cinema (which may include the celluloid image, the projected image, the apparatus, or the venue) and the logic of a discipline by demonstrating how experts conformed that material to their disciplinary agenda. Charting this relationship synchronically is, I believe, preliminary to surveying the diachronic impact of film on any given discipline's understanding of its object of study. That is, any determination of the influence of moving images on a discipline's agenda or conception of its object would require, in my opinion, an examination of the changes in the discipline over time. An accumulation of case studies on a particular topic in which motion pictures were vital would be necessary to determine the change in research questions and the extent to which moving images shaped those questions. The conclusion hints at what such a diachronic or longitudinal study might look like, but the chapters ahead focus instead on the criteria for acceptance (or refusal) of film within each expert group, rather than on any transformation in that group's thinking as a result of using film. I believe we need first to clarify the pertinent elements of film form and disciplinary logic and their potential points of contact before we can argue for the impact of film form on that logic, if any.

Chapter 1 examines the use of chronophotography and motion picture technology for scientific research in the fields of human motion studies, physics, and biology. After a survey of early scientific filmmaking, the chapter leverages Henri Bergson's critique of modern science's "cinematographical thinking" to underscore the philosophical affinity between film and scientific method. Given that "science" is just as varied as "film," however, the chapter finds that different disciplines used motion picture technology for different reasons and in ways that emphasized film's different formal features. In these case studies, the filmic material consisted of the chronophotographic image, the projected image, and the technological apparatus, all of which were adjusted to conform to various elements and processes of scientific inquiry. Accordingly, the chapter examines how film became evidence in relation to the discipline's object of study, its theories, and its representational options. This chapter does not focus on observation but emphasizes experiment instead.

Chapter 2 continues this investigation of film for research purposes but extends the argument to the field of medicine. The chapter provides an overview of early medical filmmaking that emphasizes not the different uses by different specialties but instead the common functions governing the use of film technology, such as the use of film to explore, document, or educate. Any given film functioned multiply, depending on who was using it for what purpose at the time; this rubric allows us to glimpse research films in tandem with their other uses. The chapter then explores the relationship between the formal features of film technology and expert medical observation. The use of motion pictures as both still and moving images—the way researchers described their viewing of and extraction of data from these images—subtly expressed the researchers' mode of observation. In this case, the chapter explores the similarity between a "medical way of thinking"—its disciplinary logic—and how investigators took advantage of the unique features of film technology, especially the celluloid image and the projected image. It also underlines the ideological investment in that mode of viewing, which was a sign of training and disciplinary allegiance. That investment explains diagnoses of film spectators that accompanied the general medicalization of society during the Wilhelmine period. That physicians dismissed movies in their mass cultural incarnation but lauded their use in the laboratory is best explained through the dual, mutually dependent character of observation and spectatorship.

Chapter 2 functions as a pivot between the specialized, rarified use of motion pictures in the laboratory (which was more common than we might think) and the emergence of film in the public sphere. If physicians diagnosed the cultural ills of a nation, reformers of all sorts took to the streets to carry out the treatment. Chapter 3 surveys the film reform movement of the early period and its relationship to larger reform movements, especially those that emphasized a renewal of cultural priorities through education. In this case, then, the filmic material was the cinema venue itself as reformers and educators attempted to mold exhibitor and spectator habits to certain educational standards or aspirations. The art education movement was especially pertinent for efforts to "ennoble" cinema and its audiences through an aesthetic education. The history of German pedagogy, moreover, reveals a deep investment increasing over the nineteenth century in a mode of visual instruction known as *Anschauungsunterricht*, which spread quickly across Europe and the United States, where it was known in English as "the object lesson." Film reformers such as Hermann Lemke worked hard to fashion film screenings for educational use that would conform to this method of visual instruction and observational training. Reformers such as Hermann Häfker, on the other hand, created film presentations that attempted to conform to models of aesthetic contemplation, which were not incompatible with these educational and ideological goals.

Häfker's efforts to create a film screening that would make any middle-class aesthete proud speak to the perceived importance of aesthetic values for unifying German culture and ameliorating its social ills. Chapter 4 argues that the explosion of writings about cinema around 1912, which Anton Kaes called the *Kino-Debatte*, was not only about the danger cinema presented to literary values and territory, as has long been argued, but also about the relationship between the cinematic experience and aesthetic reception in general. How could film be considered an aesthetic object? What aspects of film viewing fit or not with established conceptions of aesthetic contemplation? It is something of a critical commonplace that the advent of mass reception, of which the filmic experience was emblematic, prompted a change in the standards of "traditional aesthetics." This chapter examines the validity of that claim by demonstrating what was precisely at stake in aesthetic contemplation as it was usually understood, and what exactly about the cinematic experience presented a challenge. So this chapter divides the ideological components of aesthetic contemplation into four main areas—issues of agency, identity, time, and space—to show where the objections to and applause for film borrowed from "traditional aesthetics" and where they diverged. Furthermore, the chapter shows that professional aesthetics of the day—as encapsulated in discussions of *Einfühlung*, or emotional projection during the aesthetic experience—was wrestling with some of the same issues that challenged traditional aesthetics, such as emotional projection, synesthesia, and movement, even while descriptions of the cinematic experience also touched upon these issues. The *Kino-Debatte*, then, anticipated the theoretical proclamations of the 1920s and 1930s that declared contemplation inadequate for modern living. Chapter 4 explicates the precise nature of the change in preference from contemplation to distraction and suggests that the *Kino-Debatte* was an important turning point for this transition. Chapter 4 therefore focuses on how film's projected image could be discursively managed to fit (or not) into the logic of aesthetic contemplation. Finally, the conclusion emphasizes the importance of disciplinary context and expert/lay distinctions for histories of nontheatrical film and suggests a historiographic approach that highlights how experts "handled" motion picture images, technologies, and audiences—and the impressions those efforts left on their own institutions and on the history of cinema.

Each chapter functions more or less on its own, but together I hope they will add to the conversation about nontheatrical film, spectatorship, and interdisciplinary work in film and media studies. Indeed, each chapter demonstrates that the appropriation of motion pictures was a series of constant adjustments between writing and reading, or between inscription and reception. Looking at film in this way, we can begin to explore how cinema fit into a cultural space, the work that went into making it fit, and what that work tells us about the people and institutions who did the work. In other words, how experts managed to call film their own while using it to manage the scientific and social problems they faced is the subject of this book.

1

SCIENCE'S CINEMATIC METHOD

MOTION PICTURES AND SCIENTIFIC RESEARCH

Time would flee, I subdue it.
—CHARLES CROS (1877)[1]

In the early 1890s, after Étienne-Jules Marey's successes with chronophotography ensured the continued viability of his "graphic method" of recording motion,[2] two German physiologists, Wilhelm Braune and Otto Fischer, set out to improve upon Marey's techniques for the study of human locomotion. No mere dilettantes, Braune and Fischer were already well known in their field for studies of the human center of gravity and other investigations into the fundamentals of human motion. From 1889 to 1904 they published a series of studies of human locomotion, culminating in their landmark work, *Der Gang des Menschen* (The Human Gait).[3] In their Leipzig laboratory, they dressed their experimental subject, a military recruit, in a black jersey and attached to him an elaborate mechanical scaffolding of incandescent tubes designed to illuminate his stride with short, intensely bright bursts of light (fig. 1.1). They then photographed the subject as he walked, as comfortably as could be expected, across the darkened room. As he walked, the tubes fired, producing a strobe effect that was recorded through the camera's open shutter. The resulting image—a series of white lines across a black background—became the basis for hundreds of calculations concerning the specific mechanics of human motion (fig. 1.4). Interested in the economy of the laboring body—its most efficient conservation and expenditure of energy—they hoped their scientific analysis of human movement would lead eventually to a more efficient and productive society (or, at least, a more efficient soldier). Likewise, their improvements over Marey's system of photographic measurement led to a much more analyzable and therefore "productive" chronophotographic image. After Braune and Fischer, photographs became even more efficient and productive tools of scientific research.

In Marburg in 1907, a young physicist named Max Seddig presented his dissertation on "Measurement of the Temperature Dependency of Brownian Motion."[4] In an attempt to clarify the arguments for and against the atomic–kinetic model of heat, while also trying to provide experimental confirmation of Albert Einstein's theories of Brownian motion, Seddig fashioned a device that could record chronophotographic images of microscopic particles affected by molecular activity. Seddig's microscope–cinematograph combination supplied an objective record of Brownian motion from which he could calculate the velocity of these particles. Yet it was not the objective record that Seddig emphasized and praised but the machine's ability to measure time intervals precisely. Alternatively, Heidelberg biologist Hermann Braus presented in 1911 the results of his application of a similar microscope–cinematograph combination to record and explore the growth of a tissue culture of a frog's heart.[5] Using time-lapse cinematography, Braus sought to demonstrate that the culture actually grew rather than merely survived outside the organism. Unlike Seddig, he was not so concerned with measurement but with the temporal record of the event. With his motion picture record Braus was able to challenge competing claims about the growth of nerve cells.

These three cases—from human motion studies, physics, and biology, respectively—represent a fair sampling of the scientific use of film for research purposes in Germany before 1914. From these case studies we can draw some preliminary conclusions about the practical and philosophical connections between film and science. We might also be tempted, of course, to make general theoretical claims about the nature of cinema's relationship to modern science and temporality.[6] But if we are to understand film's role in modern scientific inquiry, we must temper expansive claims with more historically localized analyses of how film technology was actually *used*.[7] What "film" meant to any given scientist depended very little on a theoretical conception of cinema in general; instead it depended primarily on how some specific incarnations of motion picture technology could be applied to the specific and historically contingent problems the scientist faced in his or her discipline. Hence any study of scientific research films must incorporate methods common to both the history of science and film history by investigating the manner in which research questions and media technology mutually influenced each other. What questions does a given discipline privilege at a given time? How do those questions shape—indeed, how are they shaped by—experimental design and available instruments? The appropriation of motion picture technology as a scientific research tool, its specific use within the laboratory, reveals the researcher's assumptions about that which the camera is designed to capture. And these assumptions vary as widely as the different uses of film. But rather than making broad claims about what "science" or "cinema" is, thereby concealing this variety under top-down theoretical categories, we should let individual experiments reveal their assumptions and make our generalizations from those, if necessary. Design and deployment are themselves implicit theoretical statements.

If the cinematograph were an especially flexible tool that could be adapted to a number of different needs, these needs existed in the first place because of changes in the sciences themselves at the turn of the century. Biologists interested in cell development, for example, searched for new modes of visualization that would allow researchers a clear view of movement in time to resolve some heated disputes about the nature of cell growth; the techniques of tissue culture, on one hand, and those of motion pictures, on the other, offered two kinds of solutions, as we shall see. Physicists, too, looked to cinematography as a tool for investigating previously invisible phenomena, such as the effects of molecular movement, a topic of debates about the behavior of atomic phenomena. Seddig's use of chronophotography and motion pictures reflects a common application of this technology in scientific research; while scientists admired cinematography's ability to capture fleeting phenomena, they prized most highly film's ability to decompose the event into discrete, regular units, which permitted measurement of its temporal and spatial components. Indeed, this particular use of motion picture technology, especially in the physical sciences, betrays an assumption about the event or phenomenon under study as itself discrete, divisible, determined by classical laws, regular, and—just like the filmstrip—reversible. Seddig's use of motion pictures therefore demonstrates at once the value of film for scientific research and his (and Einstein's) commitment to a particular understanding of the relationships between movement, time, and space—an understanding that French philosopher Henri Bergson was criticizing at precisely this moment in *Creative Evolution*, his 1907 landmark study of the philosophy of biology. In short, as scientific disciplines reconceptualized the nature of matter, time, space, and the organism, they seized upon tools that could visualize these phenomena in accordance with these new concepts.

But film has never been just a convenient device, patiently waiting on the shelf as the scientist thinks up a new use for it as a solution to a new problem. Its availability and its existence also generated questions. Investigators' interest in temporal phenomena was in part spurred by the arrival of a machine that could make the phenomena amenable for analysis. Furthermore, those scientific research programs that featured film technology were not only shaped by that technology, but their scientific method itself had certain features in common with the filmic apparatus. Scientific experiment shares with motion pictures an impulse to record immediately and directly, a willingness to manipulate time, and an inclination to isolate and quantify phenomena.[8] There are good practical reasons for using motion picture technology, of course. But in the late nineteenth century there developed also a *philosophical* affinity between science and film that went beyond mere convenience. Cinema and science have come to share a certain way of thinking, so the history of the application of motion pictures in science can offer us a valuable opportunity to explore the relationship between science and modernity. Bergson was the first to suggest that science, through its parsing of continuous movement into discontinuous moments, proceeds in a way analogous to cinematography. If we are to understand fully the implications of the appropriation of motion picture technology by the scientific community, we must maintain a balance between the theoretical and the practical by considering this philosophical affinity alongside the way investigators put film to work in the laboratory. Bergson's conception of science as inherently "cinematic" offers us a logical point of departure for such considerations. True enough, Bergson was not as popular in Germany as he was in France and the United States. But because I am interested in his thoughts on cinema and science in general—and not in his thoughts (if any) on particular applications in Germany or elsewhere—his historical impact on German culture is actually not relevant to my approach. Bergson's theoretical framework can help us answer the question "What did cinema and science see in each other at this particular moment?" The individual case studies can reveal *how* film was used; reading Bergson alongside these cases can help to reveal *why* film was used. This chapter, then, will survey the use of motion pictures in the three case studies mentioned above.

Let me first reiterate the role of this chapter in the book's overall goals. As I mentioned in the introduction, the larger argument (the "general theory," if I may) concerns the correspondence and mutual accommodation between the logic of a discipline—its problem-solving patterns, its investigatory methods, its ideological assumptions—and film's formal characteristics. This implies not just taking advantage of a medium-specific formal feature, such as temporal malleability (for example, time-lapse cinematography), to solve a representational problem, it also implies a functional or productive homology between this formal feature (for example, the linear regularity of time-lapse recording as a statistical sample of a larger unit of time) and a way of solving problems (the primacy of mathematics, for example) or of viewing the world (as naturally divisible into equal, regular units, for example). This match—between the formal features of the representational technology and the investigatory presumptions—matters, because it provides the researcher, community, or discipline with the reassuring sense that the tool will fit the task to which it is assigned. However, we must note that the match is ideally never perfect; otherwise there would be no new information. Experimental systems are designed to generate new questions, so there should be a dislocation or displacement between the more or less ideological assurance of this tool's "worldview," so to speak, and the strangeness of the view it provides.[9] Time-lapse cinematography can, for example, confirm an understanding of nature as regular and divisible, but it also offers a surprising, even thrilling new image that prompts new questions. The larger argument of this book is that the acceptance of any new (media) technology depends in part on this correspondence between some set of its formal features and the logic of the discipline. The specific argument (the "special theory") is that expert vision expresses this disciplinary logic especially well and that film's legitimacy within disciplines depended on its accommodation to expert modes of viewing.

But expert viewing is not the only way that disciplinary logic is expressed; the experimental system itself is also a manifestation of the discipline's problem-solving patterns and theoretical presumptions. As Gaston Bachelard and others have argued, instruments are "theories materialized": the design and deployment of experimental technology carries a preconception or preunderstanding of the phenomenon it is designed to isolate.[10] If we extend this system to include chronophotography or motion picture technology, we can see how the work of accommodating their formal features already made a statement about the relationship between the system and the object of study. Indeed, the work of creating a legible image—that is, understandable and acceptable to the discipline—reveals this relationship between system and object quite clearly. Likewise, the theory guiding the experimental observations is an expression of disciplinary logic. If instruments are theories materialized, then, as Hans-Jörg Rheinberger has noted, theories are also "machines idealized."[11] Einstein's theory of Brownian motion, for example, made several "shortcuts"—including the presumption of molecular two-dimensionality and velocity without direction, as we shall see—to manage the object mathematically and accommodate the phenomenon to both experimental confirmation and a particular disciplinary understanding of nature. Einstein's mathematics, in other words, simultaneously became an "instrument" to guide observation *and* a theoretical rendering of the experimental situation. To offer one more example, disciplinary logic can be expressed through the experimental system's representational options. Film is only one part of an experimental system, of course, and only one in a range of representational tools that includes writings, sketches, graphs, and photographs. Selecting film as part of this media ensemble already implied a certain set of questions or needs. Hermann Braus, for example, found that using film was an especially powerful means of engendering belief among colleagues, while at the same time expressing, through its formal features (such as the duration of the projected image and its temporally forward motion) theoretical presumptions about cell growth. In this case, film and the new technology of tissue culture—also a representational tool—combined to express a new direction in the discipline's visual needs and strategies.

So the correspondence and accommodation between experimental system or discipline and technology can happen in several ways, depending on the researcher's or discipline's goals and needs, which are locally and historically determined. This chapter will explore the "general theory" rather than the "special theory" of media technology's disciplinary legitimacy. Specifically, this chapter will focus on three ways in which experimental systems incorporated chronophotographic or motion picture technology, especially on the correspondence and mutual accommodation between a given set of formal characteristics (of photography and/or film) and (1) an *object of study*, (2) a *theory*, and (3) the *representational options* of an experimental system and discipline. Or, to put it another way, each of these three adaptations required a certain kind and amount of work to make the chronophotographic or filmic image into evidence. A close analysis of Braune and Fischer's method will show how they created evidence in relation to their object of study (in this case, the human body). The discussion of

Seddig's experiment stresses the creation of evidence in relation to a theory (Einstein's theory of Brownian motion). And the section on Braus emphasizes the creation of evidence in relation to a set of representational options within a discipline (here, cell biology). Each example deals to some extent with all three adaptations, of course, because they are intertwined, but the emphasis changes. In general, I will demonstrate that while the scientific community readily accepted chronophotography and film as valid instruments, investigators still had to perform considerable labor to adapt these devices to their tasks and to transform the resulting images into acceptable scientific evidence. Motion pictures may have allowed scientists to manage time and movement, but researchers first needed to manage film's temporal rush and excessive detail. The nature of this work was shaped by its historical context. The subject of this chapter is therefore the way that investigators were continuously obliged to adapt as they juggled chronophotographic or motion picture technology, the needs of the individual experiment, the theories shaping the experiment, and the discipline's priorities shaping the representational choices.

While observation is not an explicit focus of this chapter, it is inescapable. Braune and Fischer decomposed movement to train expert vision to phenomena that it might not see or have been able to see. Einstein's theory of Brownian motion focused experimenters' attention, showing them what to look for. For Braus, film was an observational tool that forced researchers to abandon previous theories and modes of analysis (which emphasized discontinuity) for an approach that emphasized synthesis and continuity. These case studies demonstrate that, as Ian Hacking has noted, experiment and observation are only heuristically separable.[12] Nevertheless, this chapter emphasizes other ways, besides observation, that disciplines accommodated the formal features of film and chronophotography. Expert vision is the explicit focus of the following chapters.

The chapter has five sections. It starts with a general overview of scientific research films (as opposed to popular science films) before World War I, in which I outline various prominent applications as well as some hypotheses about how motion pictures fit with the rhetorical goals of scientific enterprise. (A majority of scientific applications of film during the early period were devoted to various medical fields, which I will explore separately in chapter 2.) The next section continues to explore why motion picture technology was intriguing to researchers; it focuses on Bergson's discussion in *Creative Evolution* of the relationship between cinema and science. In the third section, I place Braune and Fischer's work on human locomotion in the context of both social modernity and the science of work. This section contains a detailed explication of Braune and Fischer's method to show exactly how the merging of apparatus, image, and object actually occurs. The fourth section relates Seddig's work to the general problems of atomistic physics at the turn of the century. In the fifth section, I sketch Braus's cinematic contribution within the context of fin de siècle changes in the discipline of cell biology. The concluding paragraphs of this chapter compare the different cases directly to emphasize the mutual dependence between motion picture technology and emerging scientific agendas.

EARLY SCIENTIFIC FILMMAKING: AN OVERVIEW

Scientists from a wide variety of disciplines—including but not limited to botany, military engineering, meteorology, neurology, psychology, and medicine, as well as the three already mentioned—were among the earliest users of motion picture technology.[13] Most histories of research films start with Jules Janssen, Eadweard Muybridge, and Étienne-Jules Marey, who were important transitional figures between scientific photography and motion pictures.[14] The initial adoption of moving images was relatively smooth because so much work had already gone into creating scientifically valid photographic images during the middle decades of the nineteenth century. The slow process by which still photography had been standardized—setting norms for emulsions, exposure times, preparation techniques, image interpretation, and so on—meant that the enthusiasm for photography often collided with rapidly evolving disciplinary requirements for scientific documentation. As Jennifer Tucker and others have shown, photography was not immediately and unconditionally accepted as a completely objective and scientific record.[15] Photography's evidentiary status depended on its ability to meet several criteria of production and reception. Photographs had to withstand cross-examination by experts from any discipline to which they wished to testify; they were as subject to expert judgment as drawings or other illustrations.[16] In Germany, for example, microphotographs were not generally accepted as proper evidence by the scientific community until respected bacteriologist Robert Koch's innovations and rhetorical efforts made "reading" photographs of bacteria a truly collective activity among an international group of scientists and physicians.[17] Yet by the 1890s, the protocols for generating acceptable scientific photographs had been established in most disciplines, so the innovation of motion pictures was greeted enthusiastically, their way smoothed by Marey's renowned chronophotographs and graphic inscription methods.[18]

Marey is indeed the most important figure in any history of early scientific film, because his efforts and resources shaped the application of motion pictures to scientific experiment. Marey was intrigued by Muybridge's serial photography when he first encountered it in 1881 but ultimately disappointed in its scientific value: Muybridge's 24-camera, trip-wire method of recording was prone to mechanical inaccuracy and incapable of managing the exact time intervals required for careful research. So at the Collège de France in the 1880s and 1890s Marey explored photographic and chronophotographic methods for visualizations of movement that accounted for distances and times more precisely. The Institut Marey was founded in 1901 to carry on his work; researchers such as Lucien Bull, Pierre Noguès, and Joachim-Léon Carvallo continued to explore the relationship between experiment and visualization, especially in the areas of slow and high-speed cinematography and X-ray cinematography. Marey's assistant at the Collège de France, Charles Émile François-Franck, continued his work on microcinematography in particular; collaborating with Lucienne Chevroton, Fred Vlès, and others, François-Franck published widely on the chronophotographic and cinematographic recording of microscopic and macroscopic movement. These two sites were also magnets for individual researchers searching for novel ways to make visible their objects of study; French biologist Antoine Pizon and Swiss biologist Julius Ries both worked with the team at the Institut Marey to capture visually the process of cell division, for example, while François-Franck helped French physicist Victor Henri with his research into Brownian motion (I will have more to say about all of these examples later in the chapter).

Whether scientific cinema grew out of team efforts focusing on new visualization techniques or out of the work of individual researchers focused on experimental problems that motion pictures might solve, both models had one thing in common: the need for resources. Needless to say, the early motion picture apparatus was expensive, often cumbersome, and usually difficult to adapt. Its use in scientific circles, therefore, was generally restricted to established researchers who had the necessary financial and technical resources at their disposal. Indeed, the distribution of resources largely dictated the spread of motion picture technology in the laboratory. Thanks to Marey's considerable political and scientific skill, France could boast at least two sites with the necessary budget and expertise to pursue such a program. Germany also had a university research infrastructure in place that allowed individual researchers to explore the use of motion picture technology, but no single site dominated.[19] Aside from the many medical applications, which we will examine closely in the next chapter, we can count botanist Wilhelm Pfeffer's time-lapse chronophotography of plant growth at Leipzig University and Carl Cranz's high-speed studies of ballistics at the military academy in Berlin-Charlottenburg as notable intersections of science and film at the university level.[20]

Research film also received a boost from manufacturers who recognized that lending their equipment and funds provided not only a measure of legitimacy and good press but entertainment for movie-going audiences as well. Pathé Frérès, for example, funded the filmmaking of microcinematographer Jean Comandon and then distributed his films to theaters around the world.[21] Occasionally a German manufacturer would lend a hand to researchers; film pioneer Oskar Messter, for example, had pretensions in this arena, and companies such as Ernemann were sometimes acknowledged as technical patrons.[22] But even researchers who purchased the basic apparatus from a manufacturer such as Lumière or Ernemann were still obliged to make modifications to the equipment in their own laboratory setting. Carl Zeiss's optical laboratory in Jena, for example, seems to have been especially suited to this kind of work. Generally speaking, despite the involvement of some film manufacturers, these films were usually made by specialists to be shown to other specialists at scientific conferences or in the lecture hall.

This is not to say, however, that scientific films were limited exclusively to an elite audience. Scientific titles were often part of the program at the local movie theater. Companies such as Pathé, Gaumont, and Éclair in France or Urban in England, especially, produced their own series of scientific films for general audiences.[23] German audiences would have been familiar with this genre from the titles imported by these companies. (Foreign manufacturers generally dominated the German exhibition market until the 1910s.) In addition, the German periodical *Film und Lichtbild* (a German *Popular Science* for film enthusiasts) acted as something of a clearinghouse and ad hoc distribution company by publishing lists of scientific films and offering discounts to its readers.[24] There were also dedicated screenings occasionally,[25] as well as theaters devoted to the genre, such as the Fata Morgana theater in Dresden, which opened in August 1912 and showed only scientific, industrial, and nonfiction films.[26]

But what of the research films of a Seddig or a Braus? Did they ever find their way to the public? It is very difficult to say without a complete survey and correlation of all films made in the laboratory with those screened in public or semipublic venues—a task that is likely impossible to complete. There are good reasons for both possible answers, yes and no. On one hand, these films were the result of considerable expense and effort, so investigators might have been reluctant to give them up for duplication.[27] (Comandon and others like him were exceptions, because they made legitimate research films for hire.) Also, some scientists might have hesitated to cross the line between serious research and its popularization. Motion pictures already had acquired the yellow tinge of mass culture, so some investigators were probably reluctant to have their work completely jaundiced by a matinee showing at the local nickelodeon. Not that the scientific application of motion pictures encountered serious objection; while film histories often repeat legends of academic hostility to researchers using motion pictures in their experiments (usually limited to French medical films and the Doyen controversy, described in chapter 2), evidence of such protests is actually rare in comparison with the general enthusiasm displayed for the new technology. On the other hand, research films were the topic of a considerable number of screenings and discussions, and teachers, reformers, and even scientists encouraged these ventures into the popular realm as important efforts at public outreach.[28] Furthermore, there were already a number of screenings of these films in the semipublic realm of the university lecture, so it would not have been too much of a leap to take the next step.[29] And as manufacturers such as Messter and Ernemann lent their equipment and expertise to researchers, they may have asked for copies of the films, which might have been made available for rental in the manufacturer's catalog.[30] It seems that the public screening of any given laboratory film depended on the resources and predilections of the individual researcher or the manufacturer; I have not yet found a consistent conduit in Germany between the scientific laboratory and the movie theater.

By and large, then, these films served primarily as a form of scientific evidence. Despite its mass culture connotations, its high cost, the high level of technical expertise required, and the often futile results, motion picture technology offered several tempting benefits to the researcher. Like still photography, the motion picture camera provided a mechanical, automatic, hence "objective" record, thereby adding substantial evidentiary weight to scientific claims.[31] The photographic image, like other graphic inscription devices (such as the electrocardiograph), seemed to provide researchers with an unmediated and permanent record of any given phenomenon, one that could be stored and disseminated with ease. And because it could be projected and reproduced, the photographic image proved useful for demonstrations as well as experiments; indeed, a motion picture projector was just as likely to be found in a lecture hall as in a laboratory.[32]

Unlike still photography, however, the motion picture had the capacity to record events as they occurred over time. This singular feature offered several advantages. The camera itself could act as a mechanical, indefatigable prosthesis of the scientist's eye, a tireless observer of events that was able to catch the slightest change without the interruption of a blink. Furthermore, motion pictures offered the scientist the option of manipulating time by recording (or projecting) the film at different speeds. Slow-moving events could be sped up with time-lapse techniques; fast-moving phenomena could be slowed down through high-speed cinematography. As a result, temporal events invisible to our ordinary perception became "visible"; film became a kind of temporal microscope or telescope, bringing nature's aloof wonders closer to our level. Finally, as implied above, the motion picture camera could also act as a precise measuring tool. By controlling the rate at which the film passed through the camera as the phenomenon traveled a set distance, the scientist could then calculate the speed of the recorded movement. This ability was by far the most intriguing aspect of cinema's scientific potential, and researchers spent considerable energy perfecting it.

Motion picture technology, then, was an especially flexible tool that could be adapted to a number of different tasks. In this respect, however, it is no different than any number of technical innovations adapted for scientific use, from perspective drawing to the computer. Successful adaptation depends on a variety of circumstances, but as sociologist of science Bruno Latour has argued, the most salient predictor of which technologies the scientific community will adopt is the extent to which the adoption will aid the community in rhetorical struggles. Latour maintains that technologies that serve as inscription devices, or "writing and imaging procedures," function rhetorically in debates between authors and groups as tools that help "in the mustering, the presentation, the increase, the effective alignment, or ensuring the fidelity of new allies."[33] Struggles between scientists are the same as those between generals and politicians, Latour argues; those with the most allies win. Therefore, the essential characteristics of any inscription device—the qualities that will ensure its success in the scientific arena—have less to do with its ability to provide accurate inscription (visualization, writing) per se than whether those properties can be put to use in rhetorical struggles. Latour explains:

> In other words, it is not *perception* which is at stake in this problem of visualization and cognition. New inscriptions, and new ways of perceiving them, are the results of something deeper. If you wish to go out of *your* way and come back heavily equipped so as to force others to go out of *their* ways, the main problem to solve is that of *mobilization*. You have to go and to come back *with* the "things" if your moves are not to be wasted. But the "things" have to be able to withstand the return trip without withering away. Further requirements: the "things" you gathered and displaced have to be presentable all at once to those you want to convince and who did not go there. In sum, you have to invent objects which have the properties of being *mobile* but also *immutable, presentable, readable,* and *combinable* with one another.[34]

Motion picture technology had all of the qualities of this sort of immutable mobile, a good indicator of its success as a scientific instrument. The instrument itself was mobile, but more importantly, so were the films. "Immutability," in Latour's sense, refers to the permanency of both the inscription process and the object or condition represented. Simply, the inscription must be relatively permanent, as films were, while providing a translation without any seeming corruption of the thing represented. The photographic image's necessary physical connection (via light and chemical processes) to the object represented served to guarantee that the object was relatively uncorrupted by the recording process. The films were meant to be projected, so they were of course presentable, but because they were also photographs, they could be presented in a wide variety of ways, as illustrations in journal articles or as lecture slides, for example. In this way, the films could also be combined with other technologies, such as print technology, but the apparatus itself could be combined with others as well, such as the microscope. The "readability" or legibility of the technology is the most contentious aspect of any innovation, because the interpretation of new forms of inscription always requires negotiation within a discipline. Experts and innovators haggle over the meaning of signs until standards of production and protocols of interpretation emerge.[35] Generally speaking, however, motion pictures, like photography before them, were considered very legible for scientific purposes.

BERGSON, CINEMA, AND SCIENCE

Without a doubt, motion pictures presented scientists with new analytic techniques that could manipulate time and space. However, at the same time, science's very mode of analysis was newly subject to debate. At least since the romantics, some German thinkers had come to associate science's empirical, experimental, and materialist method with a cold, mechanistic, and "disenchanted" view of the world. As Anne Harrington has argued, the rebellion against mechanistic, reductionist science often consisted of calls for "wholeness" in various forms, from Hans Driesch's biological vitalism to Gestalt psychology to Richard Wagner's *Gesamtkunstwerk*.[36] This reaction was not limited to Germany, of course; perhaps the most prominent figurehead of this rebellion was Henri Bergson, whose philosophy—or its popularization—spread like wildfire through the parlors and lecture halls of Europe and the United States in the years before World War I. If film was still a relatively marginal research technology at this time, Bergson's *Creative Evolution* brought it to the center of the debates about scientific method.[37]

Creative Evolution, which appeared in 1907 and quickly became Bergson's most famous work, was concerned with evolutionary biology in the same manner that his earlier *Matter and Memory* (1896) was concerned with psychology and his later *Duration and Simultaneity* (1922) dealt with physics. In *Creative Evolution*, he took the evolution of life as a fact, but expressed dissatisfaction with scientific explanations of it. According to Bergson, the mechanistic approach to biology perpetuates a common analytic mistake: it divides up organisms and living processes in order to understand their parts, but does not, because of this division, fully understand their total, living reality.[38] Consequently, the analytic, mechanistic, reductionist approach cannot satisfactorily explain change or the creation of new forms or new solutions. Bergson was intent to bind the organism's "living reality" to the flow of time. That is, Bergson made the flow of time, which we all experience as incessant and forward moving, the model of life itself. The central concept here is his notion of *durée*, or "duration." Bergson insisted that we make a category mistake when we divide the continuity of lived experience into the discontinuity of a series of separate points. Bergson reversed the traditional or commonsensical view that change is a succession of states, a view that makes those states logically anterior to change. Bergson's view, instead, was that change is primary and that to analyze change by breaking it into a succession of states misrepresents the true reality of the world, which exists in constant flux, "becoming," or *durée*. Organisms exist in time, in *durée*, and cannot be separated from it without losing an understanding of their lived reality.

This same logic applies to the examination of motion. Bergson saw the universe in a constant state of flux; change and movement are the only constants, the only true reality. Matter, form, or solidity are only stable views of this essential instability. Unfortunately, our common, ordinary, analytic perception cannot grasp this flux; it can only extract determinate moments, which we then mistake for an accurate picture of reality. Even our body is not solid but "is changing form at every moment; or rather, there is no form, since form is immobile and the reality is movement. What is real is the continual *change of* form: *form is only a snapshot view of a transition*. . . . [Therefore,] our perception manages to solidify into discontinuous images the fluid continuity of the real" (328, emphasis in original). It is not so much that this mode of perception is wrong, it is just that it is incomplete and should not be mistaken for a true understanding of the world. Bergson acknowledged that human beings cannot help but think this way, dividing the flow of the world and time into discrete intervals, but he insisted that this habit of thought should be overcome: "to invert the habitual direction of the work of thought."[39] That is, given that analysis or this way of viewing the world is more or less habitual and necessary, Bergson declared that to recognize *durée*, "The mind must do violence to itself, has to reverse the direction of the operation by which it habitually thinks, has perpetually to revise, or rather to recast, all its categories."[40] Bergson's philosophy, then, was an attempt to save us from our logical errors of thought so that we could see the world in a different way. Specifically, he argued against two category mistakes: mistaking discontinuity for continuity and space for time.

Bergson used the example of motion pictures, which consist of a series of still images projected to imitate real movement, to illustrate the relationship between the continuity of *durée* and the discontinuity of our ordinary, analytic perception. The mechanism of cinema was analogous to, even expressed, our ordinary perceptual process. This process, Bergson elaborated,

> consists in extracting from all the movements peculiar to all the figures an impersonal movement abstract and simple, *movement in general*, so to speak: we put this into the apparatus, and we reconstitute the individuality of each particular movement by combining this nameless movement with the personal attitudes. Such is the contrivance of the cinematograph. And such is also that of our knowledge. . . . Whether we would think becoming, or

express it, or even perceive it, we hardly do anything else than set a kind of cinematograph inside us. We may therefore sum up what we have been saying in the conclusion that the *mechanism of our ordinary knowledge is of a cinematographical kind.* (332, emphasis in original)

What does it mean to "set a kind of cinematograph inside us"? It means that when we try to think movement or change, we cannot help but to conceive it first as a series of individual states of being or "snapshots" of form. The accumulation of these "snapshots" provides us with a conception of movement, a conception as illusory as the movement of the cinematic image at twenty-four frames per second. Thinking of movement or change as a series of states—rather than as a thing in itself, as Bergson urged us to do—forces us to extrapolate movement from the series. With that extrapolation, movement becomes universal, ideal, or "movement in general." We attempt to reconstitute the particularity of the movement by an effort of synthesis, which is what the accumulation of snapshots amounts to. Out of habit, our ordinary way of thinking is cinematic.

In the same way, modern science "proceeds according to the cinematographical method," in that "it is the essence of science to handle *signs*, which it substitutes for the objects themselves" (357). These signs "denote a fixed aspect of the reality under an arrested form. To think movement, a constantly renewed effort of the mind is necessary. Signs are made to dispense us with this effort by substituting, for the moving continuity of things, an artificial reconstruction which is its equivalent in practice and has the advantage of being easily handled" (ibid.). According to Bergson, science is not *really* able to understand movement per se, movement as continual flux. Yes, it can understand it as the difference between the changes of two stable states, but it cannot grasp the essential dynamism of change. Instead, the scientific approach creates abstractions—the geometric understanding of movement as a line or the mathematical understanding of movement as an equation—to help us grasp that which is in constant motion. For Bergson, the normal scientific approach to movement proceeds by leaps from moment to moment, from arrangement to rearrangement; science may increase the number of moments it isolates "but it always isolates moments. . . . It does not bear on the interval, but only on extremities" (357–358). To the extent that science followed this tendency to divide movement into stable units and to treat the interval as an enabling ellipsis and not as precisely the problem, modern science for Bergson was essentially cinematic.

Bergson characterized the history of science in these terms as well. Aristotelian science was also cinematic, but it differed from modern science in the way it broke up time. For the ancients, the time of the movement of a falling body, for example, had certain determinate periods, whose natural articulations, like puberty, presented moments when there occurred the natural release of a new form. For modern science, according to Bergson, time "has no natural articulations. We can, we ought to, divide it as we please. All moments count. None of them has the right to set itself up as a moment that represents or dominates the others. And, consequently, we know a change only when we are able to determine what it is about at any one of its moments" (360). Time, then, has become democratic in the modern age; there are no more privileged moments. This democratization marks the difference between the *qualitative* description of the ancients and the *quantitative* measurement of the moderns (361). Modern science works only with a view to measure, and it selects as its objects only those phenomena that can be measured. "It retains only the events or systems of events that can be thus isolated without being made to undergo too profound a deformation, because only these lend themselves to the application of its method. Our physics dates from the day when it was known how to isolate such systems" (371–372). In the same way, motion picture technology divides temporal events evenly, into a neat twenty-four frames per second, for example. The precise and democratic division of time common to the cinematic apparatus mimics modern science's insistence that no moment be privileged over another.

In this way, Bergson outlined the deep affinity between cinema and science, one that at least partially explains their immediate mutual attraction. But if Bergson condemned our "ordinary perception" and therefore cinema (and scientific method) as incomplete—thereby joining the rebellion against mechanistic science—he nevertheless recognized that cinema had potential beyond its analytic character.[41] In an interview from 1914, Bergson regarded motion pictures somewhat more sympathetically:

> Several years ago, I went to the cinema. I saw it at its origins. Obviously, this invention, a complement to instant photography, can suggest new ideas to a philosopher. It could be an aid to the synthesis of memory, or even of thought. If the circumference [of a circle] is composed of a series of points, memory is, like cinema, a series of images. Immobile, it is in a neutral state; in movement, it is life itself.[42]

Bergson recognized that cinema had the capacity to provide both analysis and synthesis, that it had an inherently ambiguous character precisely because it is both still and moving. Indeed, Bergson's invocation of the cinematographic apparatus was never an outright condemnation of the cinema but instead a way of describing a tendency in our habitual, everyday way of thinking that science had codified into experimental method.[43] If the point of his philosophical project was to "reverse" this way of thinking toward a truer, less alienating view of the world—toward an embrace of "intuition"—then there was also the possibility that cinema could somehow participate in that reversal. In other words, if modernity—"the alienating, blinding experience of the age of large-scale industrialism" to which, Walter Benjamin claims, Bergson's work responded[44]—exacerbated this alienating perception, cinema's analytic character aligned it with that alienation, while its synthetic nature associated it with authentic experience. Cinema was both a social irritant and amelioration. In other words, the Bergsonian tension between continuity and discontinuity, between unity and fragmentation, was an expression of "the alienating, blinding experience of the age of large-scale industrialism," but it was also a tension expressed in the very form of film itself. The aspect of cinema that made it a significant tool for scientific analysis existed alongside the aspect of cinema that made it useful for philosophical thinking.

While Bergson's philosophy struck a chord throughout Europe and the United States, its resonance was muted in Germany, where the reception of his work was not quite as overwhelming.[45] Nevertheless, many German thinkers struck notes in much the same key. Wilhelm Dilthey, for example, argued that positivist philosophy and science alone could not grasp the whole of life's rich variety: "The basic conception of my philosophy is that up to now no one has put whole, full, and unmutilated experience at the basis of philosophizing, that is to say, the whole and full reality."[46] According to Dilthey, this "whole and full reality" could not be fully understood using the "mutilating" knives of reductionist analysis and causal explanation.[47] So he proposed his famous division between the physical sciences, which rely on causal explanation, and the human sciences, which derive their insights from empathetic or hermeneutic understanding: "The basis of the human studies is not conceptualization but total awareness of a mental state and its reconstruction based on empathy."[48] This recalls Bergson's conclusion about the incompleteness of the scientific enterprise:

> It seems then that, parallel to this physics, a second kind of knowledge ought to have grown up, which could have retained what physics allowed to

escape. On the flux itself of duration, science neither would nor could lay hold, bound as it was to the cinematographical method. This second kind of knowledge would have set the cinematographical method aside. It would have called upon the mind to renounce its most cherished habits. It is within becoming that it would have transported us by an effort of sympathy. (372)

In other words, both philosophers were interested in opposing the machine of scientific method with the wholeness offered by a certain intuitive, empathetic understanding. Cinema, via Bergson, stood squarely in the middle of these debates about scientific method in fin de siècle Germany. Its inherently ambiguous character was reflected in the different ways scientists used it in their experiments and in the different ends to which it was employed. It is true that Bergson in *Creative Evolution* emphasized only the analytic potential of motion picture frames—that is, the use of celluloid as a "mutilating" dissection knife trained on living movement. Yet scientists in their actual use of film also effectively employed the *projected* film as an end in itself, revealing what Bergson would later recognize (and Gilles Deleuze would still later describe) as cinema's and science's capacities for synthesis within a persuasive image of movement itself. In the rest of this chapter, I will outline some of these different means and ends by examining a series of case studies of the appropriation of cinematic technology in the science of work, in physics, and in cell biology.

THE SCIENCE OF WORK AND THE WORK OF SCIENCE

To illustrate his point that "with immobility set beside immobility, even endlessly, we could never make movement," Bergson suggested this thought experiment:

> Suppose we wish to portray on a screen a living picture, such as the marching past of a regiment. There is one way in which it might first occur to us to do it. That would be to cut out jointed figures representing the soldiers, to give each of them the movement of marching, a movement varying from individual to individual although common to the human species, and to throw the whole on the screen. We should need to spend on this little game an enormous amount of work, and even then we should obtain but a very poor result: how could it, at its best, reproduce the suppleness and variety of life? (331)

This "little game" bears a remarkable similarity to the serious experiments of Braune and Fischer, who used the marches of a military recruit to extract principles of motion "common to the human species" from his "individual movement." Their results, after "an enormous amount of work," look very much like "jointed figures," although Braune and Fischer would protest that their representations were not meant to convey "the suppleness and variety of life" but rather the universal laws that subtend it. Interested primarily in isolating moments in motion and reducing them to universal principles, they were unapologetic contributors to the mechanistic, positivist, analytic trend in the sciences. In this section, I will use their work as an example of the "cinematographical method" in science that Bergson interrogates. But I also want to demonstrate how their agenda fits into larger social concerns about modernity. So I will first survey the science of work and its promise to offer a way to manage social problems and class relations. The military heritage of this discipline leads to a consideration of the "docile body" in the science of work and scientific photography or cinematography. Then I will turn to the actual process by which scientific images are rendered, by reading their technique as exemplary of the transformation of the chronophotographic image into acceptable evidence.

Before the image could be accepted as evidence, the subject itself had to be made manageable. At each stage of Braune and Fischer's experiment, they had to make adjustments to the subject, the apparatus, and the means of analysis. The recruit's body, for example, had to be continually manipulated and adjusted to create a legible image. Indeed, just as adapting the subject's body to the apparatus, and vice versa, played a large role in creating and legitimating a scientific image, so adapting labor to the machine age was the primary goal of the science of work. Likewise, Braune and Fischer's intricate efforts to prepare the military recruit for his mission correspond nicely with the mathematical contortions required to translate the image into acceptable scientific data. Understanding the mechanics of the human body required its submission and disassembly on both sides of the equation, before and after the chronophotographic inscription. As we shall see, disassembly, or analysis, was the first step to rebuilding the body and society. Braune and Fischer used the apparatus (both the camera and its accompanying scholarly translations) to break down human movement into its constituent parts. But the painstaking work of analyzing each movement would be rewarded only if they could find a way to eliminate needless effort and bring all the elements back together again so that they worked more efficiently, all in the name of conserving energy and forestalling fatigue.

As Anson Rabinbach has persuasively demonstrated, the twin concepts of "energy" and "fatigue" were enormously powerful tropes for scientists and reformers of fin de siècle Europe.[49] New scientific models of energy consumption and conservation, along with the ubiquitous technologies and techniques of mass production that accompanied the second industrial revolution, led many scientists to think of the human body as a motor, a machine governed by the same laws of physics and chemistry as its man-made counterparts. The centuries-old battle between vitalism and mechanism heated up once again as this new mechanistic theory of the human motor encountered the critical vitalism, or *Lebensphilosophie*, of Bergson, Driesch, and others.[50] As the demands of adapting human labor to new industrial techniques grew, a new science—the science of work—adopted these mechanistic principles. The science of work studied the "human motor" in order to define its laws and track down the origin of fatigue.

Yet the trope of fatigue was more than a scientific mania of the age; it expressed a profound concern over decline and social disintegration. As the structural changes wrought by the Industrial Revolution rippled through society, there arose a tendency "to locate the body as the site where social deformations and dislocations can be most easily observed" (21). Metaphors of health and sickness were used to express national anxiety. Fatigue became more than a physical ailment, it became a *moral* problem, a sign of weakness and absence of will. According to Rabinbach, "In fatigue the physical horizon of the body's forces was identified with the moral horizon of the species; the moral infirmity of the population was directly proportional to the debilitating effects of fatigue. . . . [Furthermore,] fatigue represented the membrane between morally sacrosanct labor and the violent, irrational impulses that constantly threatened to disrupt social order" (43). The tensions between human labor, capital, and the machine age demanded some sort of solution, and European scientists studying the nature of work believed that greater productivity was the key to social harmony. Investigations into the nature of fatigue were hopeful steps toward resolving class conflict scientifically, that is to say, "neutrally."

The science of work, which emerged in the nineteenth century on the periphery of scientific study, compensated for this marginal status by becoming a self-consciously international phenomenon. The relatively small group of researchers interested in this topic around the 1880s had very little support, institutional or otherwise. Ernest Solvay, a Belgian chemist, who conceived of society as "an enormous industrial enterprise dedicated to increasing overall productivity while encouraging social justice," endowed the Institut de Sociologie in Brussels as part of his plan to study the nature of energy and fatigue.[51] Angelo Mosso of Turin, an Italian physiologist and educational reformer, invented the first efficient and accurate measure of fatigue, the ergograph, in 1884.[52] But it was Hermann von Helmholtz of Germany and Étienne-Jules Marey of France, in particular, who gave direction and means to the fledgling science.

Helmholtz's contributions to science are legion, but he is perhaps best known for his version of the first law of thermodynamics, or the law of conservation of energy, which holds that energy is neither created nor destroyed but simply transferred.[53] His mathematical formulation of this fundamental law of physics provided science with its most substantive version yet, and it soon became more than a truism. This law's companion axiom, the second law of thermodynamics, which was formulated by other scientists, including Rudolf Clausius and William Thomson (Lord Kelvin), also gained credence beyond the world of physics. The second law explained the process of *entropy*, which holds that although energy cannot be destroyed, it tends over time to be degraded from useful forms to uselessness. The best example of entropy is the phenomenon of heat transference: when a hot body is placed next to a cold one, the potentially useful energy of the former will transfer to the latter until both have equal, and less useful, temperatures. The relevance of these laws of conservation and entropy was soon extended from the study of molecules to that of the human body and even to society as a whole. Fatigue became the corporal analog of the second law of thermodynamics, and degeneration became its social equivalent (45–48). Scientists, reformers, and opinion makers of all sorts soon began to depict the nature and problems of society in terms of energy, attention, will, and utility, on one hand, and fatigue, degeneration, entropy, and uselessness, on the other.[54] The laws of thermodynamics therefore provided the metaphors that motivated social plans for managing the conflicts and excesses of industrial society.

On the other hand, Marey provided the means by which scientists could study the laws of energy and entropy as they were played out through the human body. Marey, a physiologist who made lasting contributions to the fields of cardiology, physiological instrumentation, and aviation, as well as to the craft of photography and the science of work, provided not only the means for a minute analysis of the movements of the human body but the basis upon which these studies could be counted as legible and legitimate areas of inquiry.[55] His transformation of ephemeral phenomena, such as movement, into scientifically acceptable (that is, legible by disciplinary standards) visual evidence through his "graphic method" prompted a flood of motion studies that overran journals in the 1890s. In this respect, Marey is an exemplary figure in late nineteenth-century European positivism. His graphic method signaled the ascendancy of the process-oriented approach in physiology that has dominated the field for the past 150 years. His focus on the disassembly and reformation (in the broadest possible sense) of the human body is representative of the general goal of positivistic, inductive, physical sciences: analyzing individual instances of natural phenomena and provisionally concluding from them an ideal model or set of laws. As Rabinbach demonstrates, Marey represents the forging of a crucial link between cultural and social modernity, between late nineteenth-century disruptions in the perception of time and space and the efforts to manage the contemporary social crises (84–88). His kymograph and chronophotographs codified, even embodied, the historical confluence of these forces, giving the science of work direction, means, and legitimacy.

Designed to calculate "the mechanical work expended in different movements,"[56] Marey's chronophotographs were also a link between cultural and social modernity, providing the basis for both a new leisure activity and a science of labor. Marey intended his "ergonomics"—or science of efficient movement—to lead to greater productivity. The concept of "training" was very important for Marey's ergonomics and the science of work in general. Marey maintained that all animal locomotion is characterized by the transformation of abrupt and disjointed movements into consistent motion. (This is also an accurate description of the technological principle of cinema.) The central feature of all work—whether of humans or machines—is the transformation of irregular, inconsistent, and jarring shocks into regular and uniform activity. The body's own elasticity permits the suppression of shock into regular effort. Muscles, for example, act to turn abrupt movements into dynamic work. Marey believed that animal and human motors are naturally efficient yet capable of improvement. His chronophotography attempted to demonstrate the potential for greater economy to be attained from "training"—essentially a program of scientific and bodily discipline. Through careful study of the movement of the body in various stages of a work process—whether forging iron, for example, or pole-vaulting—scientists could spot and correct inefficient movements, thereby showing the worker how he or she might expend the least amount of force and consequently accomplish the task with the least amount of fatigue.[57]

Braune and Fischer's studies of human motion extended this tradition.[58] To "investigate the influence of a relatively heavy load on gait," they asked their experimental subject to carry "an army regulation knapsack, three full cartridge pouches and an 88 rifle in the 'shoulder-arms' position" in a number of different experimental settings.[59] These studies, ostensibly dealing "only with the experimental determination of the process of movement, without considering the cause," were designed to recreate the process of labored movement in the hopes that these activities could be improved ergonomically.[60] In this they shared the approach and goal of other practitioners of the physiological branch of the German science of work, or *Arbeitswissenschaft*, such as Nathan Zuntz and Wilhelm Schumberg, whose *Studien zu einer Physiologie des Marsches* "surveyed all aspects of military drill" in the hopes of pinpointing the causes and consequences of fatigue and march-related illnesses.[61] The military orientation was a common motif in the science of work: Braune and Fischer's experiments were supported by the German High Ministry; Marey stressed the potential benefits of motion study for military training in his numerous appeals for support from the French government;[62] Wilhelm Weichardt tested his infamous "fatigue vaccine" on the Austro-Hungarian army;[63] and Mosso tested his ideas about the relationship of mental and physical fatigue on Italian soldiers.[64]

It is not too surprising that European theories of efficiency, elasticity, and fatigue often took an explicitly military orientation, considering that a military motto might be "efficiency through training." As a matter of fact, the science of work was the industrial application of forms of discipline first deployed in the military. The temporal decomposition and reconstitution of the human body through chronophotographic or cinematic means also fit well into such agendas. *Moving Picture World* once noted, "The United States Army has had [motion] pictures taken of a soldier going through the manual of arms. Thumb books with these pictures are made up and furnished to the recruit, who by looking carefully through them can easily trace every minute movement that goes to make up the completed action."[65] In this example the recruit was expected to incorporate the lessons of the cinematic image in much the same way that he was expected to embody military ideology. The point is not so much that cinema and the science of work cooperated with the military but that the military application of motion pictures and of these scientific principles point to a coincidence of techniques that provide further insight into the relationship between science and cinema. Michel Foucault cast some light onto the larger intellectual history connecting the science of work and the state:

> The great book of Man-the-Machine was written simultaneously on two registers: the anatomico-metaphysical register, of which Descartes wrote the first pages and which the physicians and philosophers continued, and the technico-political register, which was constituted by a whole set of regulations and by empirical and calculated methods relating to the army, the school

and the hospital, for controlling and correcting the operations of the body. These two registers are quite distinct, since it was a question, on one hand, of submission and use, and on the other, of functioning and explanation: there was a useful body and an intelligible body.[66]

But these registers, Foucault continues, overlap in the notion of " 'docility,' which joins the analyzable body to the manipulable body. A body is docile that may be subjected, used, transformed and improved." The scientist's efforts to survey the human body and the colonel's attempts to modify it both required that the body submit to a regimen of exercises. This may have included a series of measurements, tests, recordings, or it might have meant calisthenics in the morning and full-load drills in the afternoon. Either way, whether under the rubric of science or the state, docility, according to Foucault, "implies an uninterrupted, constant coercion, supervising the processes of the activity rather than its result and it is exercised according to a codification that partitions as closely as possible time, space, movement."[67]

This last sentence is certainly an apt description of Braune and Fischer's experimental procedure, especially in that their chronophotographs "partition as closely as possible time, space, movement." Foucault called these methods "disciplines," and we would include science among them, because its work is essentially that of domestication. We may, after Ian Hacking, divide the work of the scientist into two types: representing and intervening.[68] Hacking equated this division, generally speaking, with the split between theory and experiment, even while acknowledging that the two are inseparable. Indeed, experiment is tightly bound with the process of representing, as sociologists of science have shown. But whether through representation or experimentation, any phenomenon to be studied must be "tamed" before it can become scientific data. In the attempt to analyze (or, more accurately, render analyzable) natural phenomena, the work of the scientist involves any number of phases, such as selecting, partitioning, measuring, or representing. It is impossible to present a phenomenon in its "natural" state; it must be rendered into material images, such as graphs, photographs, tables, charts, and diagrams—representations that function as Bergson's "snapshot" of duration. Sociologist of science Michael Lynch, following Foucault, calls the product of these scientific procedures a "docile object":

> It is an object that "behaves" in accordance with a programme of normalization. This does not mean that it fails to resist, or that its recalcitrance does not serve to adumbrate its objective news for science. It is to say that, when an object becomes observable, measurable and quantifiable, it has already

become *civilized*; the disciplinary organization of civilization extends its subjection to the object in the very way it makes it knowable. The docile object provides the material template that variously supports or frustrates the operations performed on it. Its properties become observable-reportable in reference to the practices for revealing them. If the object was not compliant to such a programme, its attributed properties would be incompletely or "unscientifically" observable.[69]

This same civilizing process applies equally to the human body in motion studies, the paramecium under the biologist's microscope, or the astronomer's optical pulsars.[70] I would argue that it also applies to the apparatus created or transformed to view or inscribe these phenomena. Motion picture technology also underwent a certain domestication as its image was transformed into acceptable scientific evidence. For the cinematic image to function legitimately as scientific evidence, it must undergo a transformation, a *rendering* that is common to all scientific practice and "docile objects." The army's flip books, for example, were used to train the soldier's body, to corral its forces into a productive and useful activity. The books were part of the system of discipline. Yet the very creation of the flip books was also one way of making the cinematic image analyzable by controlling its size, speed, and impact. In the same way, Braune and Fischer took specific, formal steps to domesticate both the human body and their chronophotographic apparatus. The next section will describe these steps in some detail and discuss how they fit into the broader process of representing natural phenomena in science.

BRAUNE AND FISCHER'S *THE HUMAN GAIT*

The Human Gait appeared as a series of papers published between 1895 and 1904 in the *Proceedings of the Royal Saxon Society for Sciences* (Abhandlungen der königlich sächsischen Gesellschaft der Wissenschaften). Although Braune died immediately after the experiments described in their first chapter (perhaps testifying to the exhausting nature of the work), Fischer honored his teacher by making him the first author, and both names have been associated with the studies ever since. Chapter 1 appeared in 1895 and chapters 2 through 6 appeared in 1899, 1900, 1901, 1903, and 1904.[71] Sponsored by the German High Ministry of War, the essays were the first quantitative analysis of human locomotion, and their precision set a standard that is still frequently cited today. That an English translation appeared in 1987—and is still considered an important text in the modern field of human motion studies—testifies to their continued relevance in the science of human movement. In addition to formulating important axioms for this field, Braune and Fischer developed new techniques and instruments for analyzing images. In fact, they are recognized as the originators of analytic, close-range photogrammetry, the science of measurement from photographs.[72]

Previous investigators of human motion, such as the Weber brothers, Marey, Muybridge, Carlet, and Hermann Vierordt, were primarily concerned with qualitative, two-dimensional motion studies. That is, their largely descriptive studies focused on motion through only two axes—the horizontal and vertical lift of the human leg, for example. Braune and Fischer, on the other hand, realized that animal locomotion takes place along a z-axis as well, that is, *sideways*. They strove to create, through experimentation and exact measurement, an ideal, three-dimensional model of human movement, something that had not been done up to that point. Several researchers had tried to represent the intricacies of human movement graphically, but their methods lacked the scientific rigor to which Braune and Fischer aspired. Braune and Fischer noted, for example, that the achievements of Muybridge, Anschütz, and Londe[73] "are very important for artists, particularly those who depict people and animals in motion," but "the use of photography as a scientific research tool and the improvement of cameras to this end are due, above all, to Marey."[74]

The primary difference between studies "important for artists" and those that could be counted as "scientific" was measurement. Muybridge and Londe had in common the instinct for "automatic writing," that is, some method of creating automatic, mechanically inscribed signs of movement. But such a sign alone would be useless if it could not be held up to ever stricter scientific protocols. Marey's single-camera setup represented considerable progress over Muybridge's series of cameras, simply because, for Muybridge's series method to be successful, "the distances between axes of the cameras, standing side by side, had to correspond to the phases of movement and the different cameras had to be optically similar" (6). This distance depended upon the velocity of the moving body, which could not be known beforehand and is different for each body and each type of movement among the various body parts. Therefore, precise comparison between the pictures was nearly impossible. Yet Marey's single-camera system had problems as well. "If all the points of a human or animal body moved in one plane during walking or running, these series of pictures would represent not only a one-sided projection but also a true picture of the whole process of movement. However, movement in space has to be taken into account, whereby the centers of all joints describe double curves. Thus, the projection achieved by Marey's method is insufficient to describe completely the movement in space" (8). In other words, while Marey's pictures provided information about the horizontal and vertical motions of the entire body during walking, they provided no insight into its sideways oscillations. Braune and Fischer's contribution, then, was a four-camera system that produced "two-sided" chronophotography, because "two simultaneous photographic exposures of the same movement are sufficient to determine the movement in any direction" (10).

Two-sided chronophotography had its own peculiar difficulties, however. It "requires the cameras to be opened and shut at short intervals at precisely the same time. This requirement can only be achieved using a highly complicated mechanism. Therefore, to interrupt the exposure, we relied not on shutting the camera but on *altering the photographic object itself*, so that it was possible to dispense with a particular mechanism for shutting the camera" (10, emphasis added). From the beginning, then, Braune and Fischer confronted the dilemma of simultaneously regulating subject and apparatus. This was not an either/or situation; the subject was not completely subdued in favor of an implacable apparatus and method. Rather, both were altered until their properties *merged* in the representation. That is, features of the body that could be graphically enhanced (such as a limb's straight line) were coordinated with characteristics of the camera (such as the two-dimensional film plane) to create a usable image. This stage of their experiment, then, was crucial for adapting the body to the chronophotographic image and vice versa.

Braune and Fischer's alterations included, first of all, highlighting the recruit's body with a series of strategically placed Geissler tubes—long, thin, straight tubes filled with rarefied nitrogen that, when exposed to an electric current, become incandescent. The recruit wore a black jersey, similar to the one used by Marey, which "provided a dark background for the tubes and permitted better attachment of the tubes to the body" (12) (fig. 1.1). When fired by a regulated electrical circuit, the flashes of light provided an ideally manageable strobe effect. The black jersey offered not only a dark background for the tubes, it effectively erased all extraneous details—meaning most of the body itself—from the picture and presented only the most graphic qualities of the process under examination. Significantly, Braune and Fischer also took into account the points that were not illuminated (fig. 1.2).

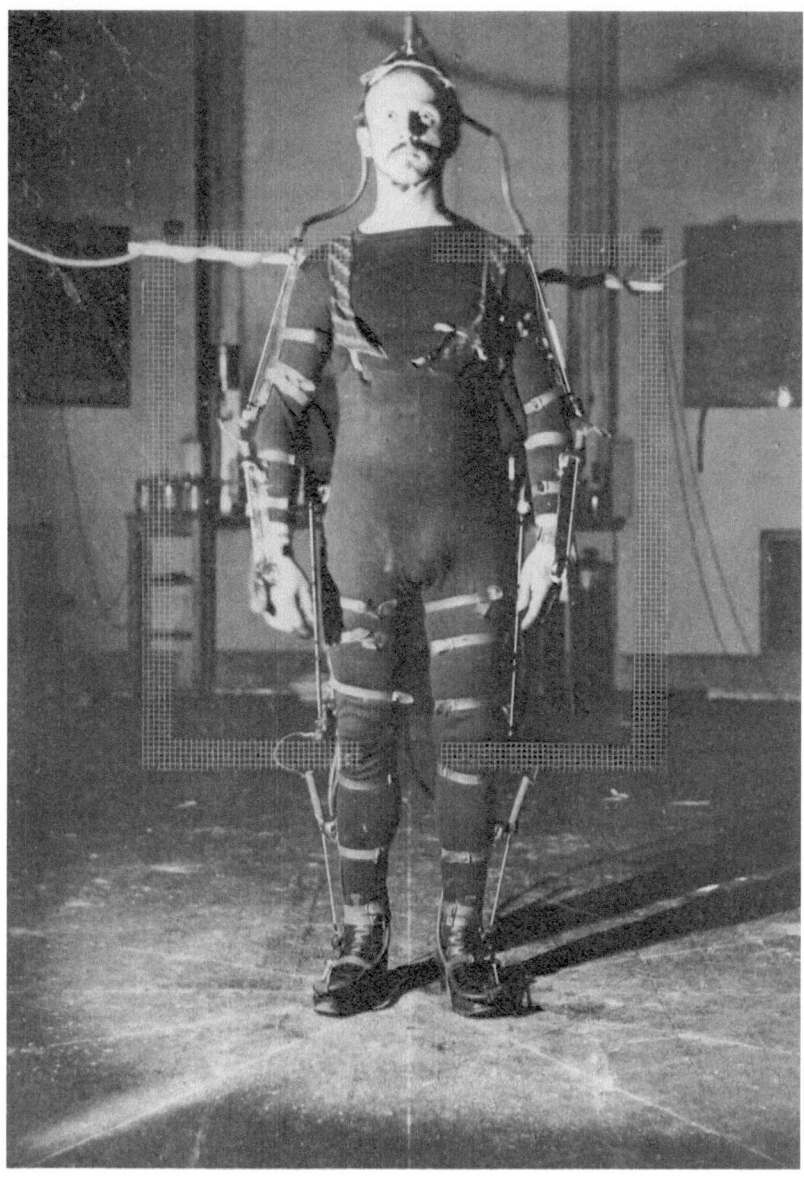

FIGURE 1.1. Braune and Fischer's military recruit in the experimental suit

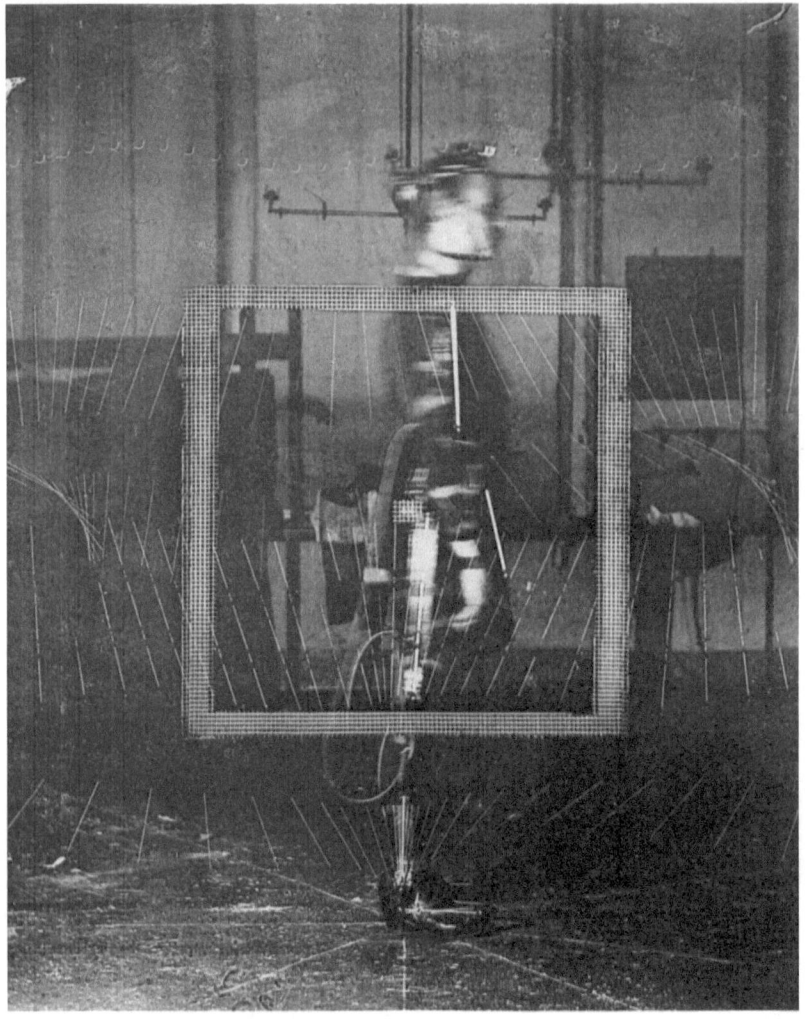

FIGURE 1.2. The subject at rest with the grid superimposed

In some places the Geissler tubes were surrounded by narrow rings of black Japan varnish; these places thereby appeared as short interruptions in the line of light in the pictures. They were located near the ends of the tubes and at the same level as the center of the joint. They therefore marked the corresponding positions of the joints as isolated points of light in the photographs. Similar black rings were also located at the places corresponding to the center of gravity of each segment of the body; the different centers of gravity appeared, then, as black dots on the white lines in the photograph (14).

Generally speaking, Braune and Fischer were "marking" the object in preparation for its representation, a process that is common in scientific rendering practices. In recounting how a "natural" space is rendered into a geometricized workplace, Lynch identified a number of themes or processes. Exploring how representations—graphs, tables, diagrams, and photographs—come to embody the "natural object," he asks "how science initially determines what is natural on the basis of what its graphic qualities disclose."[75] "Marking" is a first step toward identifying and cultivating those graphic qualities. Lynch finds that marking occurs in two phases: "labeling" and "upgrading visibility." In the first phase, the visibility of the object is initially consolidated by first-order techniques, for example, dyeing a cell so that its constituent parts show up under microscopic study. In the second, scientists mark instances that stand out as clear examples, which is a process of selective perception while the project is underway. Braune and Fischer's Geissler tubes served this dual purpose, then, by both enabling the visibility of the object and making visible only what already had been identified as significant.

The second step common to this rendering process, so closely related to the first as to be only theoretically distinct, is "the constitution of graphic space." Here a "'mathematical' space comes to dwell within the 'natural' terrain."[76] Scientists mark the space of the experiment in such a way that it can be measured or formally structured. In the happy coincidence of anatomy and classification, or the way a specimen is appropriated so that its visible properties are brought together with the graphic qualities of the representational medium, natural objects are prepared for "mathematicization." In other words, the surfaces of the natural object are brought in line with ideal, geometric properties that can be mathematically useful. The placement of the Geissler tubes illustrates nicely how the protomathematical properties of the human body merge with the graphic qualities of the chronophotographic image to become an intelligible, analyzable, measurable scientific representation.

Braune and Fischer also constituted the graphic space by careful camera placement and superimposition of a coordinate grid. They placed two cameras across from each other and perpendicular to the x-axis (the recruit's path) and two other cameras at 30-degree angles to the x-axis (fig. 1.3). The two simultaneous exposures (of each side of the body) could therefore

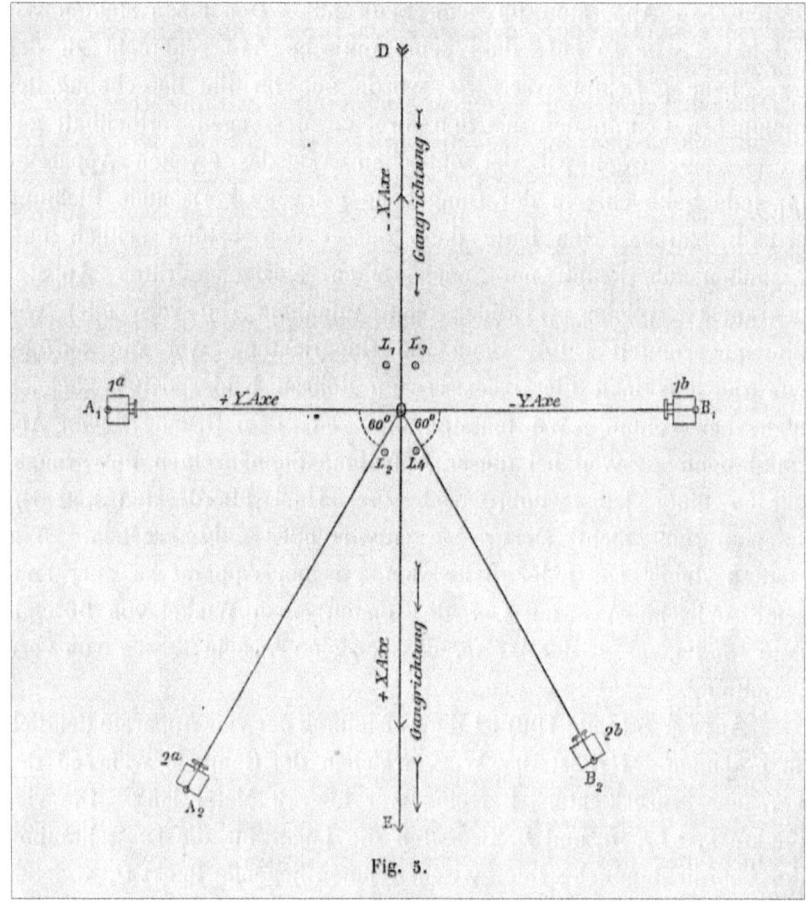

FIGURE 1.3. Braune and Fischer's camera placement

determine the position of any given point in three-dimensional space. To ensure that the points on the different exposures were registered exactly in relation to one another, Braune and Fischer created a grid that could be superimposed upon the image, so that the points of light could be matched to an established set of coordinates (see fig. 1.2).

> To enable us to draw the trajectories in a system of tridimensional coordinates, after photographing the phases of movement we photographed on the same plate a network of 1-cm squares printed on a glass plate covered

with Japan varnish. In order to improve accuracy we built a large wooden frame on which a 1-m square table of co-ordinates could rotate about a vertical axis. This frame rested on four small screws for which four metal recesses were prepared in the floor. The frame and the table of co-ordinates could thus be brought into exactly the same position at any time. (16)

So the wooden frame held a one-meter-square glass grid that could rotate parallel to the plane of each of the four cameras. After the initial exposures of the recruit were finished, Braune and Fischer would cover the cameras, screw the frame into its predetermined spot on the x-axis, and superimpose the grid onto each of the four photographic plates, thereby creating a "graphic space."

It was also necessary that the different phases of movement be regulated *temporally*. To coordinate the firing of the strobe effect, Braune and Fischer connected the primary circuit of the induction coil to a large tuning fork (15). They determined that the fork vibrated at a frequency of twenty-six vibrations per second, which meant that between any two phases of movement approximately 0.04 seconds elapsed (16). The frequency of the fork did not matter so much as its regularity.

Connecting the subject to an electrical source also regulated him in a more indirect way. Braune and Fischer sewed long strips of gutta-percha (a substance resembling rubber but containing more resin) into the jersey where the tubes were to be placed to protect the recruit from electric shock. "Perhaps we were overconcerned with regard to the insulation and could have saved ourselves a great deal of work since it usually took us between 6 and 8 hours to dress the experimental subject. However, we thought that the subject would walk naturally if he knew that the electric current, of which so many people are afraid, would not come into contact with his body" (14). Indeed, it would be surprising if the recruit actually did walk naturally. The eleven Geissler tubes were connected in series and powered by a large induction coil in the laboratory. Wires from the coil hung from the ceiling, were draped over a light wooden rod fixed across the shoulders of the recruit (see fig. 1.1), and connected to the circuit on the body. The recruit could therefore walk freely for about ten meters, "the length of the room necessary for the experiment" (14). Camera placement played a crucial role in the success of the experiment, so it was essential that the subject not waver from a specified path, which the constraints of the wires and the threat of electrical shock certainly ensured.

FIGURE 1.4. The resulting chronophotograph

Clearly, then, by the time the exposures were made, the phenomenon of human movement had already been transformed from an unreadable, "natural" object into a regulated, mathematicized process. The photographic image's unruly detail had also been restrained in a number of ways: the strobe effect reduced the image to only its most graphic, mathematical components; the camera placement and the coordinate grid regulated the space of the experiment; the tuning fork and Geissler tubes marked the temporality of the images. The resulting chronophotographic image, however, was far from self-explanatory (fig. 1.4). No image is meaningful without another set of interpretive routines.

If they were to create a three-dimensional model from these images, Braune and Fischer needed to measure the distances between the points of light and darkness. Figure 1.5 can only hint at some of the spectacular trigonometric operations involved in determining three-dimensional coordinates from two photographs, operations that I will not gloss in any detail. Generally, however, they used triangulation, the basic principle of photogrammetry, to find the position of any point in space from the bearings of two fixed points a known distance apart. This required, of course, that they take measurements directly from the photographic plates. To accomplish this task, they created an instrument designed especially for this purpose (fig. 1.6).

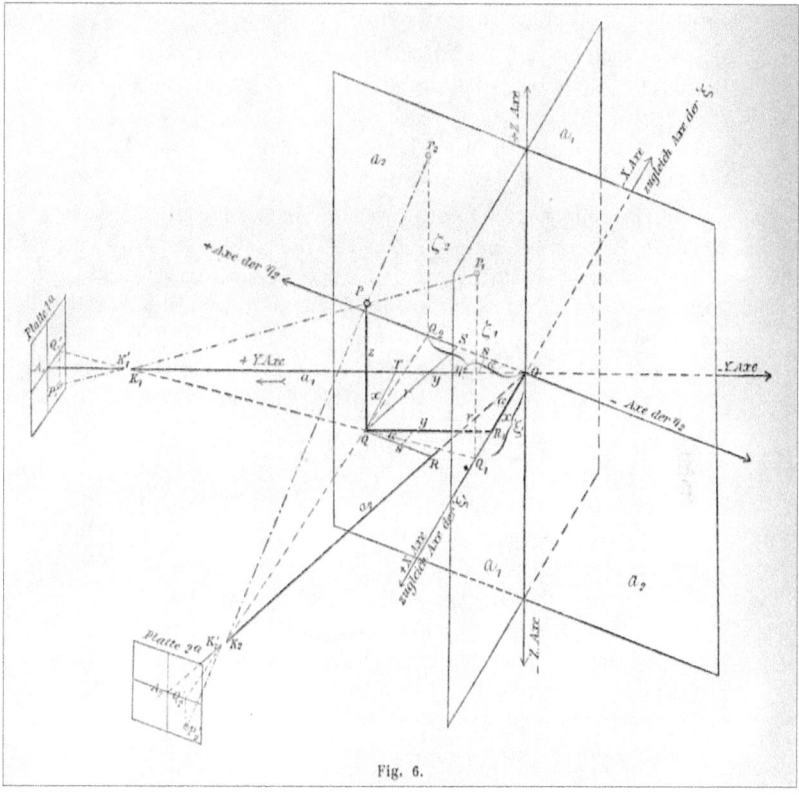

FIGURE 1.5. Determination of the coordinates of a point *P* from the projections of *P* on two planes as seen from the two cameras

The photographic plate was fixed upon a mobile ring and viewed through a microscope, which could slide along a track and bring any point on the plate into view. The microscope gave a view of both the image on the photographic plate and a ruler placed alongside the plate, thereby allowing easy measurement of the points on the plate (fig. 1.7).

These measurements for all four views were then collected and, "to avoid an accumulation of data" (!) (43), tabulated for only nine points on the human body: the shoulder, elbow, wrist, hip, knee, ankle, center of gravity of foot, tip of foot, and the point on the head (fig. 1.8). This reduction of the recorded points to a workable number was the first step in transforming the wealth of data into an ideal, that is, a theoretical model. It should also be noted that the gaps in the tables represent coordinates that for some

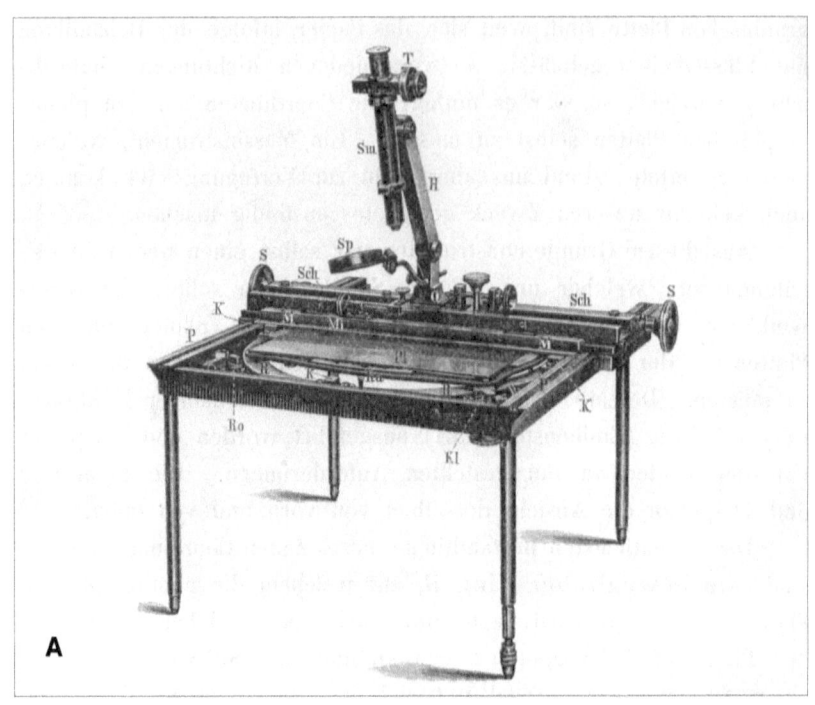

FIGURE 1.6. Side (A) and top (B) views of the instrument used to measure coordinates

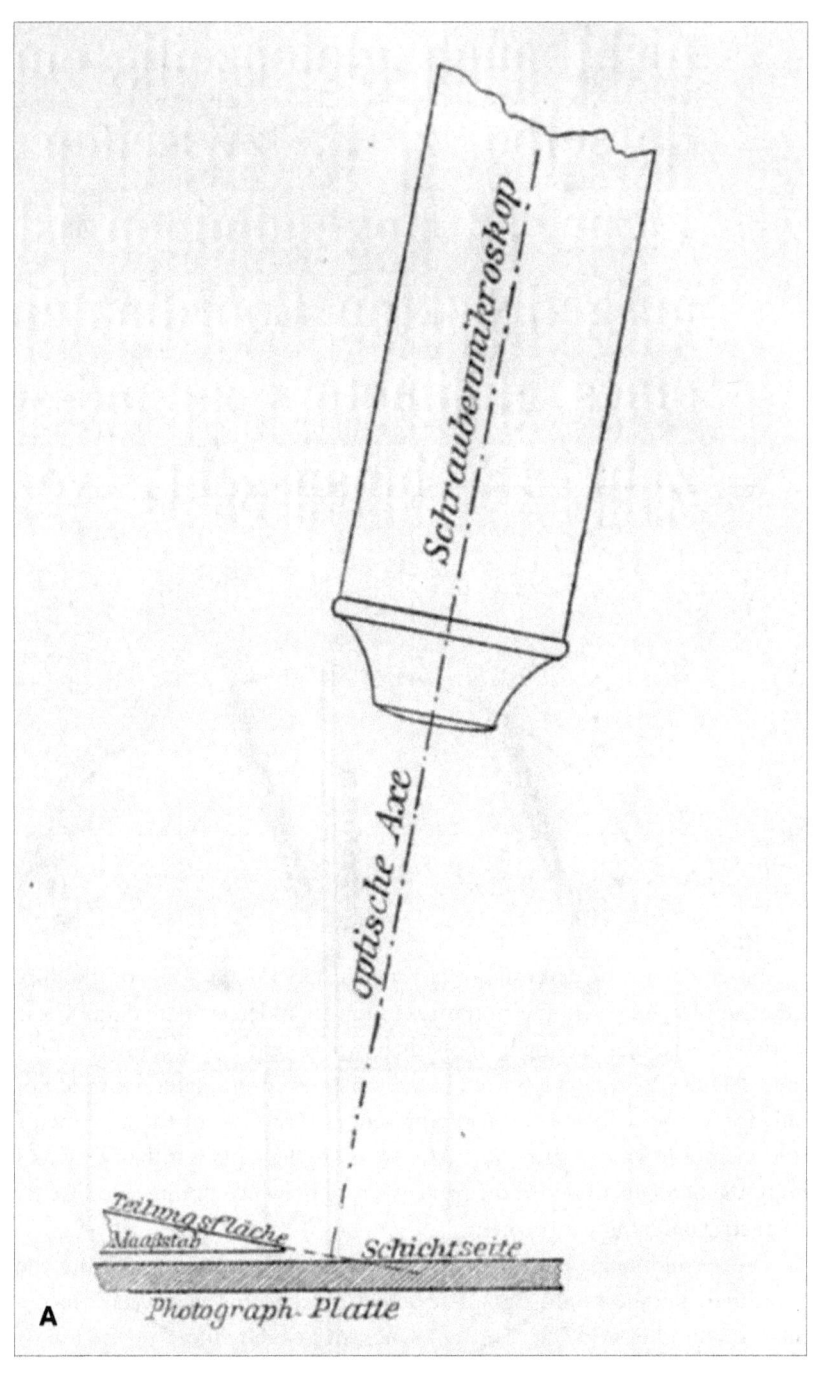

FIGURE 1.7. Measurement of a coordinate: The instrument trained on the photograph (A) and the view through the instrument (B)

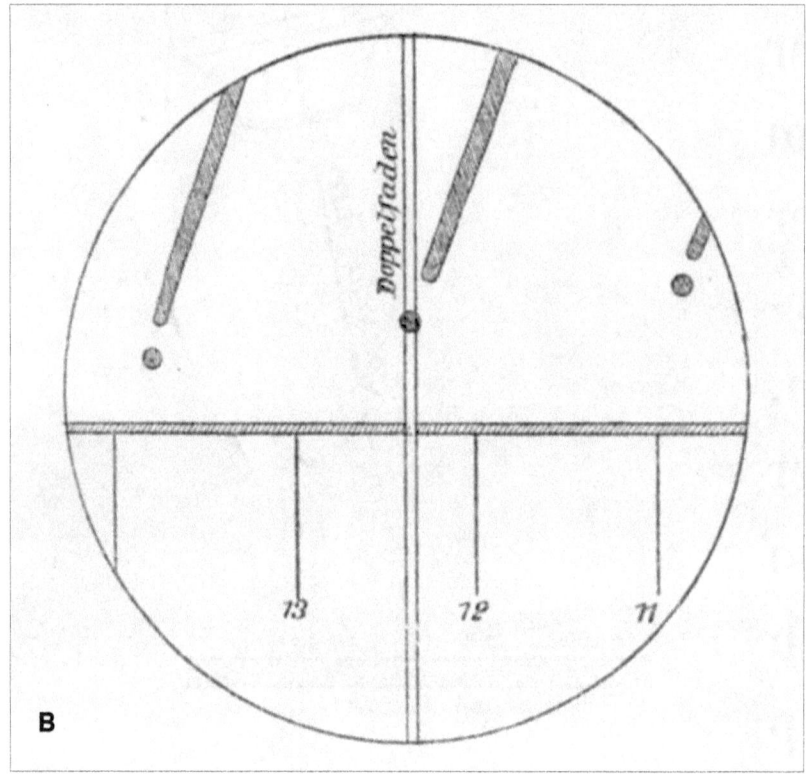

FIGURE 1.7. (Continued)

reason could not be determined (a forearm blocking a point on the hip, for example); likewise, question marks indicate indeterminate data. These statistics, although varying by individual experiments, were consistent over time. "Thus, the results reported in the tables of co-ordinates are valid not only for an individual. They also represent general laws of the movements of the limbs in human gait" (80). At once more than and less than a human body, the numbers provided the principal tool by which Braune and Fischer reconstructed human movement.

The second step in "polishing" the rough data involved plotting the coordinate numbers onto a graph, essentially recreating the recruit's movement on another grid (fig. 1.9). Idiosyncrasies were reduced by the use of straight lines and points. Comparing one of the photographs (see fig. 1.2) with the graph, we can see clearly how theoretical presuppositions informed

Tabelle 1.   I. Versuch, rechts (Tafel 1ᵃ und 2ᵃ).

Nr.	Schultergelenk rechts				Ellbogengelenk rechts				Handgelenk rechts				Nr.
	ξ_1'	η_2'	ζ_1''	ζ_2''	ξ_1'	η_2'	ζ_1''	ζ_2''	ξ_1'	η_2'	ζ_1''	ζ_2''	
1	−59,437	+32,529	+28,626	+22,872	−56,797	+35,340	+10,342	+ 8,100	−43,555	+31,854	− 6,695	−5,394	1
2	−55,366	+31,127	+29,735	+23,950	−53,831	+34,065	+11,068	+ 8,760	−44,704	+32,872	− 7,892	−6,316	2
3	−51,271	+29,727	+30,731	+	−51,852	+33,186	+11,824	+ 9,409	−46,125	+34,139	− 7,938	−6,356	3
4	−47,211	+28,347	+31,370	+25,601	−50,211	+32,697	+12,615	+10,050	−47,333	+35,290	− 7,517	−6,060	4
5	−43,109	+27,003	+31,604	+25,961	−48,255	+32,176	+13,242	+10,590	−47,573	+35,960	− 7,014	−5,440	5
6	−39,078	+25,634	+31,394	+25,981	−45,744	+31,165	+13,512	+10,833	−46,714	+36,010	− 6,721	−5,460	6
7	−34,966	+24,237	+30,762	+25,664	−42,562	+30,552	+13,235	+10,610	−45,975	+36,950	− 6,683	−5,530	7
8	−30,496	+22,750	+29,753	+24,982	−38,456	+29,291	+12,373	+10,000	−42,358	+35,767	− 7,104	−5,730	8
9	−26,110	+21,272	+28,763	+24,360	−33,941	+27,806	+11,304	+ 9,210	−38,502	+34,963	− 7,776	−6,340	9
10	−21,821	+19,926	+28,143	+24,000	−29,156	+26,348	+10,454	+ 8,650	−33,318	+33,482	− 8,573	−7,100	10
11	−16,889	+18,423	+27,870	+23,960	−23,512	+24,555	+ 9,893	+ 8,210	−26,076	+31,050	− 9,829	−8,100	11
12	−11,887	+16,812	+27,970	+24,220	−17,331	+22,617	+ 9,614	+ 8,090	−17,626	+27,970	−10,413		12
13	− 7,014	+15,111	+28,421	+24,920	−10,611	+20,419	+ 9,695	+ 8,230	− 7,767	+24,180	−10,545	−8,990	13
14	− 2,825	+13,528	+29,291	+25,900	− 4,302	+18,250	+10,340	+ 8,960	+ 2,224	+19,940	− 9,511	−8,300	14
15	+ 1,671	+11,739	+30,678	+27,430	+ 2,047	+16,060	+11,802	+10,200	+12,934	+14,943	− 6,683	−5,912	15
16	+ 5,603	+10,446	+31,721	+28,230	+ 7,189	+14,369	+13,247	+11,560	+21,470	+10,625	− 3,417	−3,040	16
17	+10,146	+ 8,283	+32,350	+29,450	+13,251	+12,165	+14,084	+12,680	+30,669	+ 5,660	+ 0,615	+9,750	17
18	+14,185	+ 6,550	+32,260	+29,490	+18,833	+10,680	+14,577	+13,300	+38,134	+ 1,140	+ 4,050	+1,030	18
19	+18,451	+ 4,570	+31,678	+29,320	+24,519	+ 8,500	+14,682	+13,600	+44,986	− 2,935	+ 6,852	+6,790	19
20	+22,882	+ 2,490	+30,880	+28,880	+30,521	+ 6,150	+14,376	+13,330	+50,939	− 6,850	+ 8,432	+8,300	20
21	+27,469	+ 0,250	+29,979	+28,330	+35,742	+ 3,570	+13,662	+12,580	+55,269	− 9,540	+ 8,662	+8,650	21
22	+31,957	− 2,100	+29,154	+27,560	+40,366	+ 0,950	+12,730	+11,620	+60,016	−11,876	+ 7,588	+8,030	22
23	+36,362	− 4,420	+28,653	+27,560	+44,281	− 1,300	+11,871	+11,200	+64,701	−13,700	+ 5,392	+5,660	23
24	+40,804	− 6,820	+28,409	+27,750	+47,786	− 3,540	+11,168	+10,800	+67,614	−15,270	+ 2,521	+2,270	24
25	+45,205	− 9,270	+28,614	+28,240	+50,812	− 5,470	+10,794	+10,600	+69,399	−15,620	− 1,123	−1,430	25
26	+49,064	−11,800	+28,997	+28,855	+54,230	− 7,842	+10,739	+10,700	+70,102	−15,100	− 4,851	−4,750	26
27	+53,915	−13,800	+29,550	+29,500	+57,124	−10,060	+10,961	+10,840	+69,701	−14,000	− 7,395	−7,370	27
28	+57,843	−15,870	+30,441	+30,830	+60,015	−12,090	+11,453	+11,464	+68,752	−12,560	− 8,146	−8,050	28
29	+61,360	−17,640	+31,197	+31,830	+61,615	−13,300	+12,031	+12,100	+67,824	−11,300	− 7,962	−8,020	29
30	+65,119	−19,520	+31,691	+32,460	+63,177	−14,270	+12,633	+12,740	+67,153	−10,470	− 7,532	−7,160	30
31	+68,888	−21,500	+31,812	+32,960	+65,047	−15,230	+13,068	+13,165	+67,484	−10,000	− 7,153	−6,770	31

Nr.	Hüftgelenk rechts				Kniegelenk rechts				I. Fussgelenk rechts				Nr.
	ξ_1'	η_2'	ζ_1''	ζ_2''	ξ_1'	η_2'	ζ_1''	ζ_2''	ξ_1'	η_2'	ζ_1''	ζ_2''	
1	−58,332		−5,793		−58,330	+31,672	−33,384	−26,885	−79,137	+35,487	−46,956	−36,855	1
2	−53,410	+29,490	−4,680	−3,876	−50,024	+28,637	−32,029	−26,200	−69,702	+32,342	−47,499	−37,943	2
3		+27,990			−42,324	+25,787	−30,467	−25,243	−59,243	+28,689	−18,912	−39,861	3
4	−43,987	+26,285	−2,947	−2,157	−35,445	+23,072	−29,164	−24,500	−48,220	+24,713	−50,766	−42,211	4
5	−39,643	+24,805	−2,875	−2,393	−29,249	+20,487	−28,387	−24,142	−36,577	+20,306	−52,430	−44,558	5
6	−35,723	+23,370	−3,471	−2,915	−24,130	+18,329	−28,431	−24,416	−24,963	+15,854	−53,401	−46,391	6
7	−31,979	+22,020	−4,744	−3,980	−19,955	+16,327	−29,424	−25,590	−14,163	+11,734	−53,328	−47,260	7
8	−27,801	+20,630	−6,289	−5,331	−15,677	+14,130	−30,800	−27,000	− 4,681	+ 8,162	−52,447	−47,351	8
9	−23,197	+19,083	−7,101	−6,072	−10,611	+12,100	−31,310	−27,770	+ 1,020	+ 5,617	−52,595	−48,037	9
10	−18,461	+17,360	−7,219	−6,240	− 4,792	+10,165	−30,954	−27,770	+ 3,695	+ 4,008	−54,005	−49,697	10
11	−12,891	+15,560	−7,071	−6,168	− 0,121	+ 8,504	−31,324	−28,210	+ 6,567	+ 2,771	−54,577	−50,521	11
12	− 8,077	+14,116	−6,345	−5,530	+ 4,702	+ 7,615	−30,703	−27,890	+ 7,652	+ 2,336	−54,968	−50,984	12
13	− 3,927	+12,827	−5,792	−5,095	+ 8,291	+ 6,310	−30,530	−27,880	+ 8,111	+ 2,049	−55,060	−51,269	13
14		+11,576		−1,413	+ 9,472	+ 5,870	−30,708	−28,110	+ 8,202	+ 1,985	−55,019	−51,127	14
15	+ 3,751	+ 9,756	−3,972	−3,600	+10,943	+ 5,410	−30,834	−28,210	+ 8,213	+ 1,920	−54,945	−51,050	15
16	+ 6,663		−3,622		+10,474	+ 4,770	−30,815	−28,290	+ 8,242	+ 1,850	−54,863	−51,000	16
17	+10,257	+ 7,139	−3,532	−3,229	+11,134	+ 4,100	−30,890	−28,470	+ 8,276	+ 1,780	−54,809	−50,970	17
18	+13,749	+ 5,500	−3,631	−3,374	+11,927	+ 3,500	−30,907	−28,620	+ 8,297	+ 1,700	−54,804	−50,980	18
19	+17,670	+ 3,700	−4,069	−3,840	+13,214	+ 2,820	−31,025	−28,860	+ 8,394	+ 1,623	−54,798	−50,990	19
20	+21,543	+ 1,930	−4,797	−4,523	+15,106	+ 1,720	−31,243	−29,270	+ 8,581	+ 1,511	−54,733	−50,997	20
21	+25,418	+ 0,140	−5,342	−5,110	+17,216	+ 0,748	−31,355	−29,170	+ 8,978	+ 1,303	−51,409	−50,764	21
22	+29,570	− 1,992	−5,608	−5,420	+20,273	− 0,500	−31,350	−29,650	+ 9,743	+ 1,071	−53,763	−50,202	22
23	+34,173	− 4,461	−5,808	−5,640	+24,118	− 2,130	−31,574	−30,120	+11,050	+ 0,635	−52,734	−49,366	23
24	+39,118	− 7,100	−5,873	−5,750	+29,717	− 4,120	−31,900	−30,840	+13,492	− 0,444	−51,210	−48,223	24
25	+44,080	−10,010	−5,912	−5,840	+36,888	− 6,840	−32,733	−31,920	+17,377	− 2,153	−49,519	−46,843	25
26	+49,451	−13,100	−6,133		+46,004	−11,120	−33,721	−33,482	+23,822	− 4,990	−47,364	−45,557	26
27	−54,342		−5,674		+54,664	−15,890	−33,457	−33,834	+31,850	− 8,869	−47,967	−46,086	27
28	+59,008	−18,470	−4,635	−4,750	+62,802	−20,410	−32,188	−33,154	+41,515	−13,895	−47,826	−47,852	28
29		−20,680		−3,750	+69,805	−24,170	−30,799	−32,180	+50,892	−19,126	−49,270	−50,366	29
30	+68,082	−23,030	−3,076	−3,120	+76,433	−28,250	−29,588	−31,400	+61,630	−25,091	−51,183	−53,584	30
31	+72,028	−25,600	−3,080	−3,170	+82,137	−31,610	−28,964	−31,107	−72,616	−31,234	−52,901	−56,771	31

FIGURE 1.8. A table of the coordinates derived from experiment 1

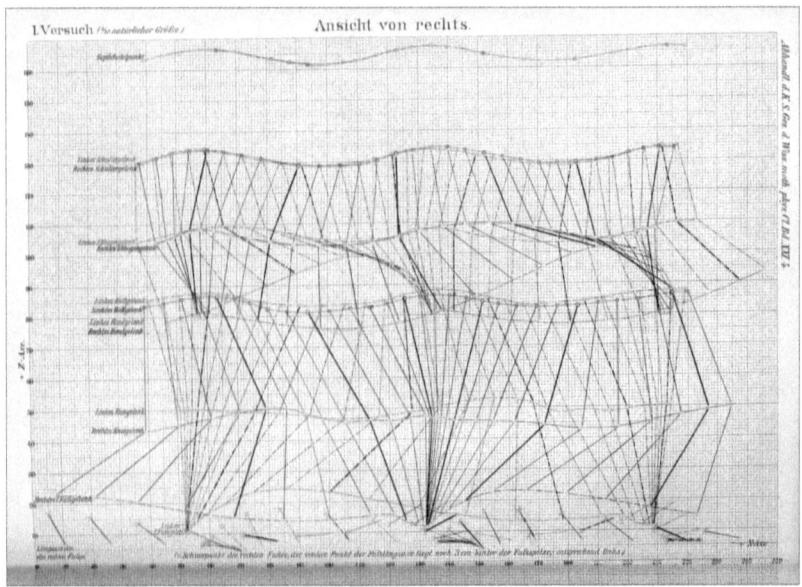

FIGURE 1.9. The graph of the coordinates (view from the right side)

the creation of the image. Yet Braune and Fischer maintained, "It must be stressed that the lines drawn in the diagram do not represent the Geissler tubes used for the experiment but the long axes situated inside the limbs" (81). Their coordinates also provided for a series of horizontal, or overhead, views of human movement. Figure 1.10 provides successive overhead views of the movements of the nine sections of the body. The views were separated, of course, to make them easier to read.

Edmund Husserl, writing on "The Origin of Geometry," outlined how material objects in the real world are identified with idealized geometric forms:

> First to be singled out from the thing-shapes are surfaces—more or less "smooth," more or less perfect surfaces; edges, more or less rough or fairly "even"; in other words, more or less pure lines, angles, more or less perfect points; then, again, among the lines, for example, straight lines are especially preferred, and among the surfaces the even surfaces; for example, for practical purposes boards limited by even surfaces, straight lines, and points are preferred, whereas totally or partially curved surfaces are undesirable for many kinds of practical interests. Thus the production of even surfaces and their perfection (polishing) always plays its role in praxis.[77]

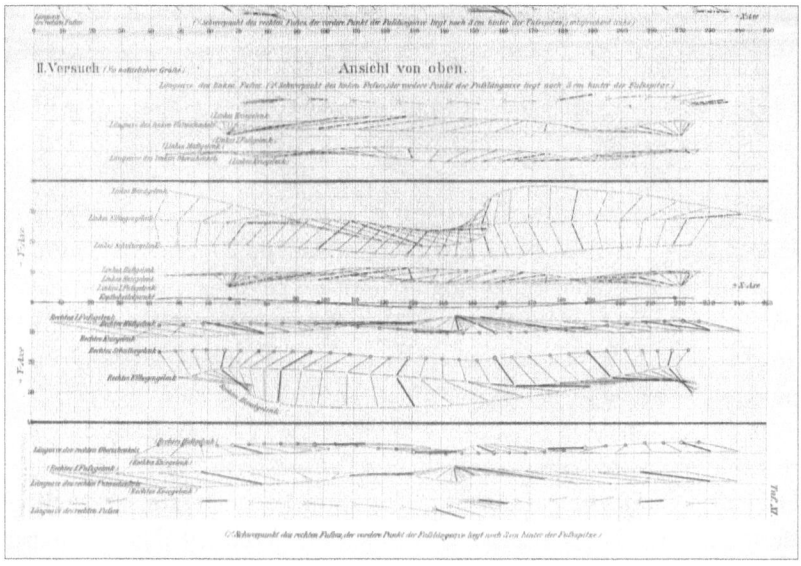

FIGURE 1.10. The graph of the coordinates (view from above of different body parts)

Lynch suggests that we read Husserl's account "as a description, not of a once-and-for-all historical movement from proto-science to science, but as an account of what scientists do every time they prepare a specimen for analysis in actual laboratory work."[78] We can see this mathematization, this "polishing," take place as Braune and Fischer smoothed out the rough edges of their data from table to graph. The body/image became progressively less recalcitrant as the process continued. Like a student who erases the rough pencil sketches and calculations only after finishing the final graph in pen, so Braune and Fischer's polishing erased the contingent, individual body for a generalized, ideal one. It became an *eidetic* image. *Eidos* is the Greek word for "idea" or "species," but Bergson elaborated on this definition:

> We might, and perhaps we ought to, translate *eidos* by "view" or rather by "moment." For *eidos* is the stable view taken of the instability of things: the *quality*, which is a moment of becoming; the *form*, which is a moment of evolution; the *essence*, which is the mean form above and below which the other forms are arranged as alterations of the mean; finally, the intention or *mental design* which presides over the action being accomplished, and which is nothing else, we said, than the *material design*, traced out and contemplated beforehand, of the action accomplished. To reduce things to Ideas

is therefore to resolve becoming into its principle moments, each of these being, moreover, by the hypothesis, screened from the laws of time and, as it were, plucked out of eternity. That is to say that we end in the philosophy of Ideas when we apply the cinematographical mechanism of the intellect to the analysis of the real.[79]

The culmination of Braune and Fischer's experiment, the final eidetic image, the reduction of real movement to an Idea—arrived at through an application of the cinematographical mechanism—was their three-dimensional model of human movement (fig. 1.11). With this model, they attempted to reconstruct movement by means of a series of discontinuous images. Their final operation, then, was essentially cinematic.

In this case, science's successful appropriation of cinema or chronophotography seems to have depended upon the ability to analyze the image and derive translatable information from it. The indeterminate detail of the image needed to be regulated, interpreted, and translated into data that could be mathematicized, tabulated, graphed, and modeled. Quantitative frame analysis, then, has been the traditional means of extracting this information or of disciplining the image. Anthony Michaelis, in his 1955 survey of the history of the scientific applications of motion pictures, declared that "only the quantitative use of cinematography, combined with frame analysis, has produced the maximum amount of research data of which the motion picture film is inherently capable."[80] In other words, if the image were to be a fully productive member of the scientific community, it needed to be, like the worker, broken down and reconstituted.

Braune and Fischer's efforts confirm Bergson's idea of the cinematic nature of science. In their attempts to chart the movement of the human body over time, Braune and Fischer were not concerned with duration so much as with isolation of particular moments. (They were also very interested in training the expert eye; their three-dimensional model in this respect functions as a guide to what to look for when studying human movement.) Film's success as a scientific instrument depended above all on its malleability and on the ability of scientists to manage the excessive detail within the image by reconfiguring the apparatus and image through various methods of frame analysis. Braune and Fischer's example is instructive because it provides a relatively clear view of the work involved in making a malleable and legitimate image. They also provide a good example because their simultaneous work on the image, the apparatus, and the human body replayed certain themes significant to cinema's relation to science and especially

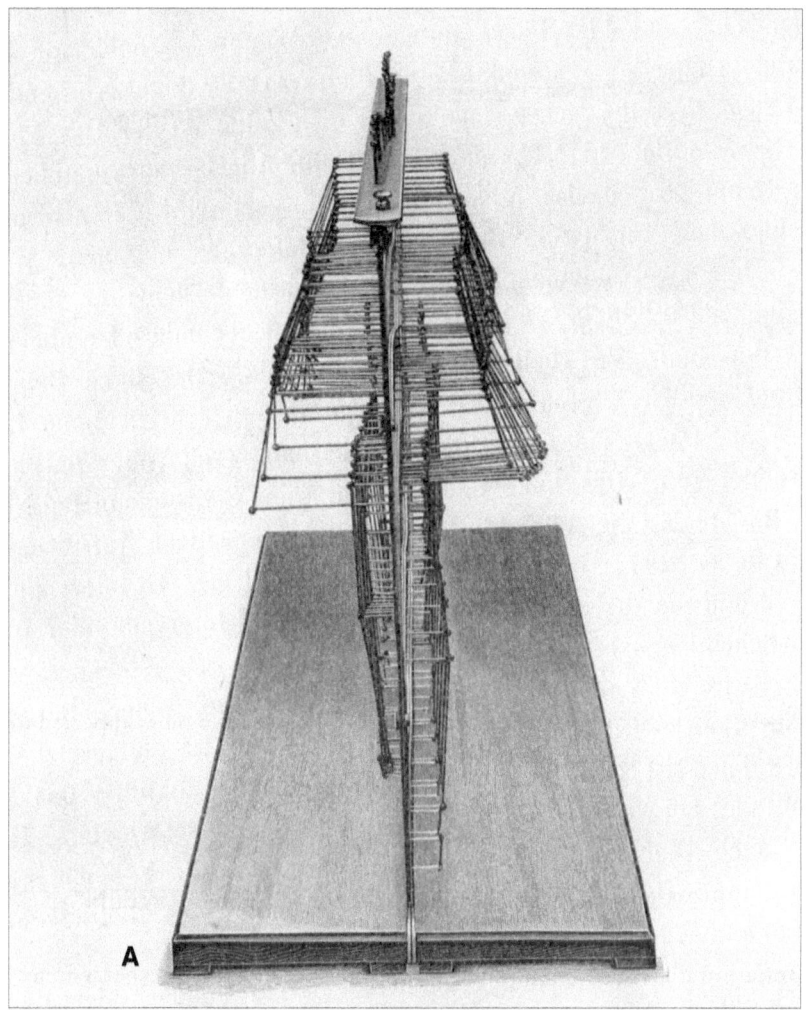

FIGURE 1.11. Left (A) and back (B) views of the tridimensional model representing the attitudes of the human body during walking

the science of work: the domestication of the object and apparatus, the isolation of physiological functions, and the disassembly and theoretical re-formation of movement and process. Braune and Fischer's example suggests that chronophotography's legitimacy as a scientific instrument rested not so much on its ability to record movement through time as its ability to stop that movement—that is, not so much its continuity as its *dis*continuity.

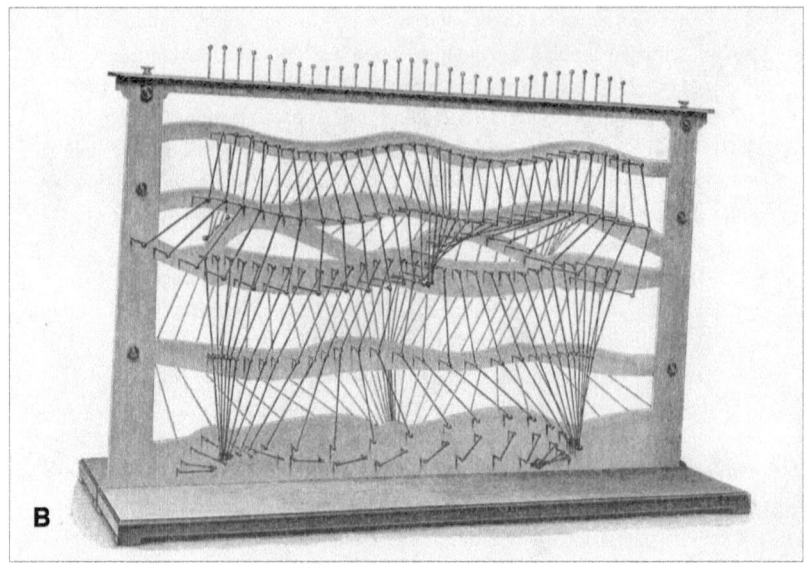

FIGURE 1.11. (Continued)

Above all, the ability to divide processes into ever smaller units appealed to the late nineteenth-century researcher.

BROWNIAN MOTION AND "THE SPACE BETWEEN"

Braune and Fischer's method depended upon finding the geometry inherent in their subject—in this case, the human body—and translating it into mathematical terms. Chronophotography was essential to this operation, because it could isolate the "determinate moments" (in Bergson's words) of the flux of movement so that the inherently geometric aspects of the body could be highlighted and studied. As Bergson argues,

> There is an order approximately mathematical immanent in matter, an objective order, which our science approaches in proportion to its progress. For if matter is a relaxation of the inextensive into the extensive and, thereby, of liberty into necessity, it does not indeed wholly coincide with pure homogenous space, yet is constituted by the movement which leads to space, and is

therefore on the way to geometry. It is true that laws of mathematical form will never apply to it completely. For that, it would have to be pure space and step out of duration.[81]

Here Bergson is building on his argument that matter, or more precisely form, is a "snapshot view" of reality in transition. Yes, everything is in continual flux, but this constant movement (hence contingency, which, for Bergson, implies liberty) sometimes, to our perception, "relaxes" into a more or less stable shape. So from the perspective of this moment—the "moment" during which the chair I am sitting on appears to be and acts solid and stable, which is a longer or shorter moment, depending on one's viewpoint—we can map and measure this form and make use of this momentary "interruption" in duration. We can therefore make use of the geometry inherent in form, or what Bergson calls the "mathematical immanent in matter."

Both René Descartes and Isaac Newton recognized this "mathematical immanent in matter." Descartes worked toward a mathematical physics when he argued that certain qualities (such as extension, shape, and motion) were more knowable, objective, or certain than other qualities (such as color, sound, heat, and cold).[82] He posited that we could know these "primary" qualities through measurement and ground our knowledge in the more certain world of mathematics. He thereby mapped geometry onto the physical world more thoroughly than his predecessors and opened the way for Newton's mechanics as mathematical formulations of motion. Classical physics was therefore founded upon matching the material world to mathematics. But, as Bergson noted, the pure form and space of geometry never completely match the matter of the object under study, simply because the object exists in time; in math, the elements of the equation do not age and change. Braune and Fischer's method indicates the lengths to which scientists must go to reconcile the *duration* of the object (however momentary) with the *space* of geometry and thereby harness the "mathematical immanent in matter." Hence Bergson's conclusion that "there is something artificial in the mathematical form of a physical law" (238).

Yet science succeeds. Bergson offered this explanation: "One hypothesis only, therefore, remains plausible, namely, that the mathematical order is nothing positive, that it is the form toward which a certain *interruption* tends of itself, and that materiality consists precisely in an interruption of this kind."[83] In other words, Bergson recognized a homology between science's method of arresting time and the momentary interruption of duration that is materiality. Or, to put it another way, science's tendency to *spatialize* time is analogous to the "relaxation" of duration into form and matter. In any case, according to Bergson, science's success in finding mathematical correspondences depends on a temporal interruption.

Max Seddig's attempt to confirm Einstein's theory of Brownian motion is an interesting case study, because his cinematic and chronophotographic method corresponded so closely to important features of the theory.[84] Einstein's theory, as I will explain later, emphasized a gap or interruption: an elision of the path of particles in favor of two independent but related points along that path. In other words, Einstein stressed the *displacement* of particles rather than their actual path. Likewise, Seddig's experimental method focused on the interval *between* filmic exposures, so that two separate exposures represented two independent but related points. Seddig's results captured more fully than most the specifics of Einstein's theory precisely because his method was able to accommodate this gap or displacement that previous researchers missed or ignored. Seddig harnessed the inherent discontinuity of the chronophotographic/cinematographic method to create an experimental system that corresponded to Einstein's important displacement equation. He experimentally created a "temporal interruption" akin to what Einstein had created theoretically, thereby instantiating the spatialization of time and form that Bergson found to be central to science's success. More than opening up new realms of the visible or even confirming important theories, this use of motion pictures in physics also illuminates certain features and limits of the scientific method as it had been traditionally understood. Indeed, examining Seddig's experiments alongside Bergson's critique allows us to see the deep compatibility between this application of motion pictures and the state of the physical sciences at the turn of the century. This section will proceed in the following manner: after a general discussion of the stakes involved in the discussion of atomistic physics, I will explicate Einstein's theory of Brownian motion as a watershed moment that collected and focused scientific observation. A survey of the experimental attempts to capture or confirm this phenomenon will lead us to Seddig's case and a close analysis of the connection between his experimental system and Einstein's theory. Finally, we will return to Bergson in order to ruminate on the homology between film form and the physics of Einstein and Seddig.

Einstein's theorization of the phenomenon of Brownian motion, for which he assumed the atomic–kinetic theory of matter, is generally considered to be "one of the fundamental pillars (or even the main one) supporting atomism in its victorious struggle against phenomenological physics in the early years of this century."[85] Throughout the nineteenth century, two approaches to scientific explanation had been in competition in classical physics: the atomic–kinetic theory of heat, on one hand, and on the other, a phenomenological thermodynamics, which assumed that all natural phenomena could be encompassed by the application of Newton's laws of mechanics, but which did not presume to have access to the ultimate constituents of matter. (This approach to physics is called "phenomenological" because it deals with the description and classification of phenomena while refusing to indulge in any claims about causation.) This approach allowed physicists to describe certain phenomena (such as heat transference) and derive laws (such as the first and second laws of thermodynamics) from observation without explaining *why* matter and energy acted in this way. These laws of thermodynamics therefore functioned much like other laws of classical physics, such as Newton's law of gravity, in that they presumed that phenomena would obey these laws no matter what their size or situation. These laws, in other words, were generally regarded as both scalable and absolute.

The rival approach to explaining thermal phenomena was the atomic–kinetic theory of matter, which started with specific assumptions about the constituents of matter, that is, "that it was discrete, molecular, ultimately atomic, and that heat was a 'concealed' form of motion associated with the molecules of a substance."[86] This theory had the advantage of actually trying to *explain* the nature of heat, but it had the disadvantage of being unobservable. While atomic theories of matter date back to the Greeks, many eminent nineteenth-century scientists, such as Ernst Mach and Wilhelm Ostwald, resisted building theories of physics on the supposed behavior of matter consisting of particles so small that they were invisible.[87] If advocates of the kinetic theory were to succeed, they would need to deduce the existence of these particles from their effects. Einstein and others theorized that these effects could indeed be deduced mathematically from the random fluctuations in an observable system, such as a container full of a certain kind of gas. These random fluctuations would not occur, by definition, if the absolute laws of thermodynamics held. In other words, the proof of the real existence of molecules was tied to proving that classical thermodynamics was true only in a statistical—not absolute—sense. And at the turn of the century, the atomic theory of matter and its statistical correlation were far from universally accepted. The theory and confirmation of Brownian motion was a crucial victory in this struggle.

Brownian motion, the irregular motion of microscopic particles suspended in fluid, had been known long before Scottish botanist Robert Brown described it in 1828, when he turned his microscope to cytoplasmic granules extracted from pollen. His achievement was "to show that the motion could not be attributed to any supposed vitality of the particles themselves, since all kinds of inorganic as well as organic substances behaved similarly."[88] This meant that the cause of the movement had to be external to the particles themselves. There were a variety of attempts to explain this phenomenon, but it was not until the close of the century that French physicist Louis-Georges Gouy suggested that Brownian motion constituted a clear demonstration of the existence of molecules in continuous, albeit random movement.[89] However, Gouy did not work out any mathematical theory that could lead to experimental confirmation. This was Einstein's contribution in 1905. Not that previous scientists hadn't tried. But the attempts to clarify the phenomenon experimentally were hindered by the lack of agreement among the observations. Researchers simply could not agree on the principal features of Brownian motion. Not only were the movements too quick and irregular to submit themselves to steady observation, but researchers could not concur on even such basic presumptions as whether the movements were dependent on the temperature of the fluid. This tangle of confusion indicates, to Roberto Maiocchi at least, "how difficult it is to make a meaningful and conclusive scientific 'observation' and, as a result, how any inductivist conception, which claims to start from an empirical base in order to then construct theories of some importance, is unsustainable."[90]

Photography and cinematography offered at least the possibility of creating an "objective" record that could be compared with previous and current descriptions of Brownian motion. But again, without sufficient technical means and a theory to guide them, this hope was fool's gold. Take, for example, the case of Austrian scientist Felix Exner, who attempted "to measure the size of the particles and their speed" via direct photographic exposure.[91] Unfortunately, the photographic plates at the time were not sensitive enough to register the small amount of light coming through the microscope lens. Undaunted, he observed the movements and traced them *manually* onto a blackened photographic emulsion, projected these traces onto a screen, measured the distances between the points, and divided by the observation time. "Of course, the values are not very exact," he admitted.[92] Indeed. Exner was interested in establishing the numerical relationship between the temperature of the fluid and the velocity of the suspended particles. His article was the first attempt to quantify features of Brownian motion that previously had been described only qualitatively. Even though his results were the most accurate studies to date, given his method, it is not entirely surprising that they were not decisive. But he did spotlight the need for systematic observation and measurements and the role photography could play in this process.

But previous observation had very little impact on Einstein's formulation of his theory. The jumble of conflicting observations was of limited use. Einstein was not interested primarily in generating a theoretical explanation of a puzzling phenomenon. Instead, he saw Brownian motion as an observable system that could conceivably resolve the debate between phenomenological thermodynamics and the atomic–kinetic model. Specifically, he derived equations governing motion based on the assumption of random molecular movement and then argued that their confirmation would demonstrate the existence of molecules and also show that the laws of thermodynamics did not apply *absolutely* to particles of molecular dimensions. In his 1905 paper on the topic, Einstein indicated both the inadequacy of previous observations and the stakes involved for future observations:

> In this paper it will be shown that according to the molecular-kinetic theory of heat, bodies of microscopically visible size suspended in a liquid will perform movements of such magnitude that they can be easily observed in a microscope, on account of the molecular motions of heat. It is possible that the movements to be discussed here are identical with the so-called "Brownian molecular motion"; however, the information available to me regarding the latter is so lacking in precision, that I can form no judgment in the matter.
>
> If the movement discussed here can actually be observed (together with the laws relating to it that one would expect to find), then classical thermodynamics can no longer be looked upon as applicable with precision to bodies even of dimensions distinguishable in a microscope: an exact determination of actual atomic dimensions is then possible. On the other hand, had the prediction of this movement proved to be incorrect, a weighty argument would be provided against the molecular-kinetic conception of heat.[93]

Einstein tackled the big problems; his equations would either prove or disprove the atomic theory of matter.

Because he did not invoke previous studies or conduct experiments of his own, Einstein's vision of Brownian motion was not empirically based; instead, "Brownian motion" was for him a system that he had mathematically pared down, even created, which could then be observed and measured.[94] Einstein did not so much *describe* Brownian motion as *manage* it mathematically, thereby providing researchers with the theoretical guidance they needed, giving them mathematical pointers so that they would know what to look for. As Maiocchi notes, "Only when Einstein had constructed, *independently* of the experimental accounts, a sufficiently articulated theory, did the experimenters know *what had to be observed* and only after this theoretical clarification did the observations turn out to be conclusive."[95] Einstein's theory, more than any technology, corralled subsequent observations into a stable of usable data.

Of course, before they could be so guided, the researchers first had to *understand* Einstein's work, and this was far from a foregone conclusion. As Mary Jo Nye remarks, Einstein's 1905 and 1906 papers included mathematical derivations that "were certainly beyond the ken of even the more precocious experimentalist."[96] The novelty of Einstein's equations lies in his emphasis on the *displacement* of the particles rather than their actual path or velocity. As noted earlier, Einstein was interested in proving the real existence of molecules and also in demonstrating that thermodynamics was true only statistically and not absolutely. These two goals were intimately related, in that any kinetic theory of molecules had to incorporate a thorough understanding of statistical mechanics. The observed properties of a gas, for example, depend on the *average behavior* of its molecules. But in any system that is sufficiently random—and the model of gas as a system of molecules assumes this randomness—there will be fluctuations from this average behavior. If such fluctuations were large enough, then they would put into doubt the stability of the measured properties and thereby raise questions about classical thermodynamics. In Brownian motion, Einstein had found a system of observable fluctuations that could prove the existence of molecules.[97]

These fluctuations were expressed in his equations as the mean square displacement $(\lambda_x)^2$ of a particle in any given direction in any given time interval. Einstein realized that particle velocities were so great that they would never be observed directly and therefore could not be measured accurately. So his solution was to factor the displacement of the particles.[98] Maiocchi explains:

> While previously the attempt had always been made to estimate the length of the trajectory actually traversed by a particle, Einstein's theory deals with the *displacement* effected in a given time, i.e., the intervening distance between the points of departure and arrival, *independently of the path followed*. This is a change of radical importance because it changes completely *the object of the observation*: it is no longer a matter of trying to measure the velocity of the Brownian movements (obtained by dividing the length actually traversed during the observation time by the time itself), but of a different quantity.[99]

This solved the problem that had bedeviled researchers in their previous attempts to describe the phenomenon, as Stephen Brush argues:

> Einstein showed that there is no possibility of observing this velocity. . . . Hence any attempt to measure the "instantaneous" velocity of particles in Brownian movement will give erratic and meaningless results. It is for just this reason that all the efforts of the experimentalists . . . had failed to lead to any definite conclusion about the average speeds of suspended particles. They were simply measuring the wrong thing until Einstein pointed out that only the ratio of mean square displacement to time could be expected to have any theoretical significance.[100]

Previous experiments, such as Exner's in 1900, had focused incorrectly on the *path* of the particles. Since the invention of the ultramicroscope in 1903, researchers had turned to Brownian motion with renewed interest, further spurred by Einstein's paper in 1905.[101] Some followed Exner's lead in measuring the actual paths, not catching Einstein's insight. French physicist Victor Henri's results, for example, published in 1908, are notable in that he too used cinematography to record particle movement and to trace particles' actual paths.[102] His results did not agree with Einstein's theories and momentarily cast doubt on Einstein's equations. But it soon became clear that experimental error negated Henri's results.[103] The winner in the Brownian motion sweepstakes was Henri's countryman Jean Perrin, who ran a series of experiments that were also published in 1908. Initially, Perrin paid little attention to Einstein's papers, but when he examined them more closely after 1908, he realized that his experiments confirmed Einstein's predictions. Further work by Perrin produced the most persuasive experimental confirmation of Einstein's theory; after that, virtually no one in the scientific community doubted the existence of atoms, and in 1926 Perrin won the Nobel Prize for his efforts.[104]

Which brings us to Seddig. Even though Seddig achieved results comparable with those of Perrin, they were, according to Nye, "neither as accurate nor as convincing."[105] While Seddig is therefore little more than a footnote in the traditional accounts of this moment in the history of science, he did discover an important homology between an element of his experimental method (cinematography and chronophotography) and an aspect of the new theory of physics (displacement). Like Exner, Seddig was interested in verifying the dependence of particle velocity on the temperature of the fluid. And like Exner, Henri, and others, Seddig first attempted to record the actual paths of the particles via photographic exposure through an ultramicroscope. He hoped that with long exposures of the particles, the traces left on the photographic plates would correspond to the paths:

> It seemed obvious to attempt to photograph the moving particles, which show up in the ultramicroscope as luminous points, on a stationary plate for a certain exposure time (around one second). The luminous, moving points should then sketch black lines on the plate, which should correspond to the horizontal path covered during this time. The lengths of the resulting curves obtained at *different* temperatures but during *identical* intervals should then stand in the reciprocal relation predicted by the theory.[106]

But again, as with Exner, this method failed due to the weak light from the ultramicroscope and the relatively insensitive photographic emulsions available at the time.

However, after this failure, Seddig employed a regular microscope, which provided stronger light and therefore better exposures, and an attached cinematograph. Again, like Henri, he tried to record the movements of the particles cinematographically. As he notes in his first presentation of his work, "A cinematic method with a precision camera gave some results that were also sufficiently precise."[107] While his modified cinematic apparatus seemed to achieve a greater degree of exactitude, he was eventually dissatisfied with the instability of the cinematic image in his efforts to trace the paths. So his final results were determined with the use of an ultramicroscope and a series of multiply exposed photographic plates (fig. 1.12). But while experimenting with cinematography, Seddig came upon the solution to a vexing problem. He had noticed that the strong light required to illuminate the particles also heated the suspending fluid, generating a temperature increase that could not be predicted and that therefore jeopardized his data. He minimized this problem by substituting an intermittent light source for a continuous one.[108] The lamp fired two successive flashes one-tenth of a second apart, while the camera was rigged with a system to measure exactly the time interval between the flashes (fig. 1.13). The light source moreover was projected through a series of cooling and polarizing filters and reflected off a mirror into the microscope, which was attached to the camera. The camera was attached to an electrical circuit so that with each exposure the circuit would open and close, creating an electrical spark that was graphically recorded on a rotating drum covered with blackened paper. The graphic record served as a control for the frame rate so that the interval between exposures could be measured precisely.

FIGURE 1.12. Seddig's photographic rendering of Brownian motion

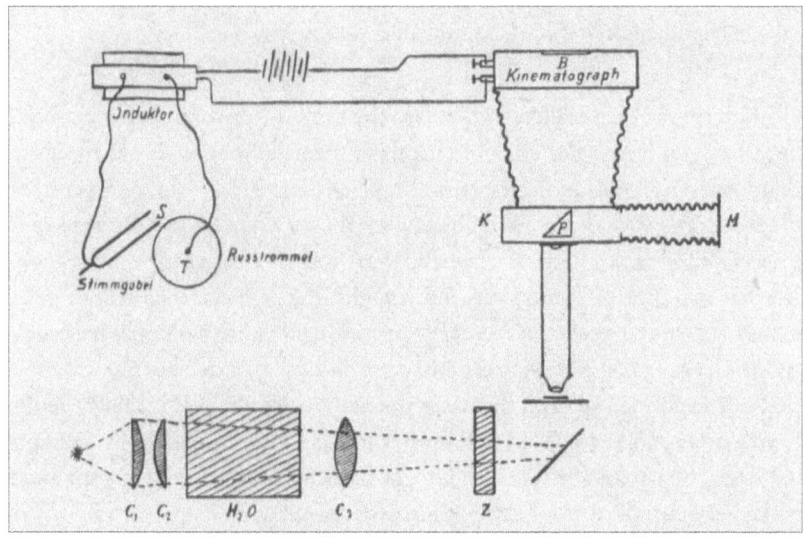

FIGURE 1.13. Seddig's cinematic apparatus for measuring Brownian motion

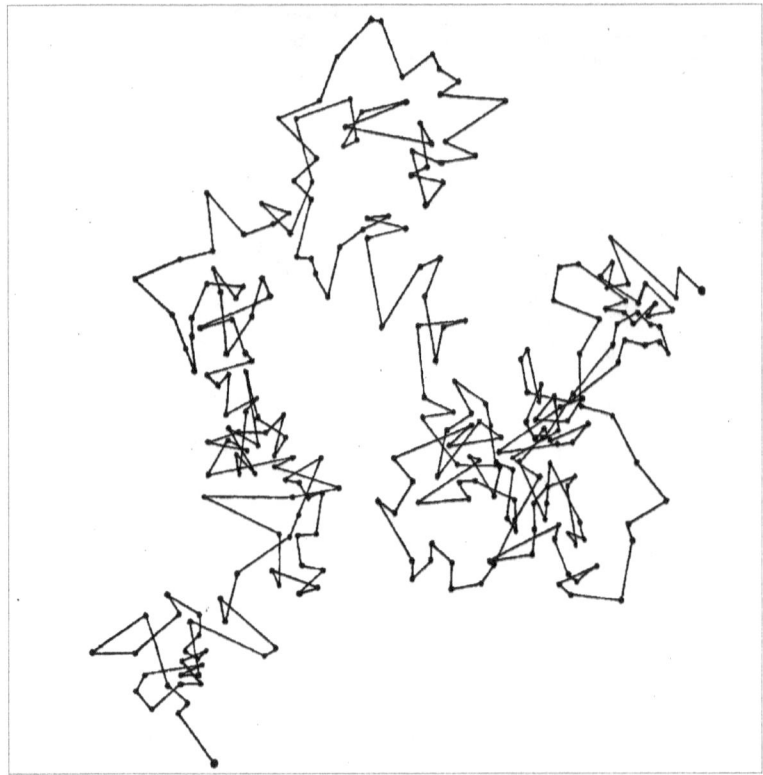

FIGURE 1.14. The irregular paths of Brownian motion

With this setup, Seddig could record the movement of any given particle, but not completely: the intermittent flash recorded only two points along that highly irregular path (for an example, see fig. 1.14)—say, point A_1 and point A_2. The path of the particle between these two points, which occurred *between* the frames, left no trace and was therefore experimentally elided. All that could be measured was the straight line between the two points—*which was exactly the empirical translation of Einstein's displacement equation*. In other words, Seddig had fashioned an experimental method that corresponded to (and therefore partially confirmed) Einstein's theory. Einstein praised Seddig's efforts, even if he was a bit befuddled by Seddig's method: "I have read Seddig's paper. He has done it very well. I cannot quite make head or tail of his descriptions of the results."[109]

Why was "displacement" so crucial to Einstein's theory? Because it worked around one of the major objections to the kinetic theory: the disparity between theoretical velocity and observational velocity at the molecular level. As Polish physicist Maryan Smoluchowski pointed out, experimental observation of particle velocity would never correspond to the theory, because theoretical velocity was simply not measurable by observational procedures: "What we see is only the mean position of the particle, driven 10–20 times a second, each time in a different direction, by that velocity. Its center will describe an unpredictable zig-zag path made up of straight lines *much shorter in length than the size of the particle*. Its displacement becomes visible only when the geometric sum of these lines is raised to an appreciable value."[110] Jean Perrin put it in another way:

> The apparent mean speed of a grain during a given time varies *in the wildest way* in magnitude and direction, and does not tend to a limit as the time taken for an observation decreases, as may easily be shown by noting, in the camera lucida, the positions occupied by a grain from minute to minute, and then every five seconds, or, better still, by photographing them every twentieth of a second, as has been done by Victor Henri, Comandon, and de Broglie when kinematographing the movement.[111]

That is, the particle, buffeted randomly by molecules, constantly moves ever so slightly in a variety of directions—these zigzags, smaller than the particle itself, cannot be observed and therefore measured or empirically confirmed. With the reduction of the equation to a question of displacement *without direction*, velocity (speed + direction) was no longer an issue.

Thus the gap. Einstein's theory elided the actual path of the particles over time, erasing it from the equation, thereby creating a theoretical *interruption* (à la Bergson) that was matched by Seddig's cinematic interruption. By focusing on the dark space *between* the frames—or, chronophotographically, between exposures—Seddig's method mimicked Einstein's theoretical attempt to erase the space between two points along a particle's unpredictable path. The gap between exposures was long enough to allow Seddig to measure the displacement of the particle from point A_1 to point A_2. If it had been too small a gap, Seddig would have traced (erroneously) the path of the particle, as Henri did. Seddig therefore found in his apparatus a specific use that serendipitously suited his needs.

Seddig's application of photography, chronophotography, and motion pictures to the problem of Brownian motion represents a particularly resourceful partnership between film and physics. We could leave it at that and still note with satisfaction the especially neat fit of technology and theory that this application evinces. Seddig's techniques matched certain elements of Einstein's equations and while lacking the precision and expertise of Perrin's experiments, still counted as partial confirmation. But this correspondence of technology and theory also testifies to Bergson's insight into the deeper connection between motion pictures and science. If science's success in finding mathematical correspondences takes advantage of what might be considered a "temporal interruption" in the constant flux of becoming—a "temporal interruption" we call "form"—by finding the mathematical, atemporal line of geometry in that form, then the scientific use of motion picture technology often matched this process by creating its own temporal interruption—the space between frames, for example—and finding mathematical correspondences to the resulting form. The temporal character of motion was therefore reduced to a series of static, discrete, two-dimensional images; time and motion were then "understood" by the measurement of differences between them. Motion pictures, in this application, reinforced the tendency to mistake the category of space for the category of time.[112]

Furthermore, the spatialization of time allows us to conceive of time as reversible. Our felt experience of time is that it is irreversible, but our experience of space is that we can pursue it in any direction. So when we represent the flow of time as displacement along a homogenous axis, "Nothing prevents us in this abstract representation," as French physicist Louis de Broglie remarked, "from supposing that we may reverse the course of time, contrary to the most certain property of real duration."[113] Likewise, a mathematical equation also implies this equivalence and reversibility. As Suzanne Guerlac explains, "In geometrical terms, if we depict a movement from left to right, we can reformulate it as passing from right to left. In algebraic terms, equations are commutable."[114] But when we conceive physical processes as reversible, we might fail to consider certain essential properties of real time, namely its directionality. Bergson, of course, argued for a dynamic ontology of irreversible time so that important properties of duration do not escape our understanding.[115]

By erasing the direction of the particle—whether the movement is from point A_1 to point A_2 or from point A_2 to point A_1 is irrelevant for the purposes of the theory—Einstein's equations implied a temporal reversibility that was also mimicked by Seddig's use of photographic and motion picture technology. In more than one way, the scientific use of film (especially in the physical sciences) was mathematical; in Seddig's hands, film functioned mathematically. How was this accomplished? Three aspects of this kind of application—the use of a flash, the use of the frame, and the form of the celluloid strip—combined to create a formal, almost geometric support for the mathematical rendering of the phenomenon. Seddig used his flash as an intermittent light source that would not raise fluid temperature as much as constant illumination. But the flash also immobilized the particle by reducing exposure time to an instant and the spatial fullness of the phenomenon to a two-dimensional position. The flash focused both time and space to the sharpness of a point.[116] Furthermore, the regularity of the flash created a series of equivalent points—$A_1, A_2, A_3 \ldots$—each of which corresponded to a single frame. The frame, in other words, determined the boundaries of the event, both spatially, in that the event took place within the field of the frame, and temporally, in that the frame represented a unit of time. So the flash and the frame reduced, limited, and focused the fullness of the event to a series of separable, measurable points—which is to say that the technology accommodated a mathematical understanding of the phenomenon.

While the photographic series is not strictly commutable—one cannot rearrange the order of the units and still obtain the same result—it is reversible, in that $[A_1, A_2, A_3 \ldots]$ is mathematically the same as $[\ldots A_3, A_2, A_1]$. In this respect, the celluloid strip itself functioned as a line, as a trajectory of completed motion. Bergson distrusted the representation of movement as a line, because such a rendering confuses movement as we experience it—that is, as something indivisible—with our imaginative recreation of that movement as a series of stages or points. More precisely, it confuses movement with the line traversed; it conflates what is happening right now with what just happened. The path is past, while movement exists in the moment. We may divide the path into points, but these points have no reality outside of the line drawn. Bergson argued that the difference is simple: "At a stage we *halt*, whereas at these points the moving body *passes*."[117] We therefore mistake immobility for mobility, discontinuity for continuity, and past for present. Representing motion as a line, especially one that predicts future motion, is an illusion, which rises from the conception that "motion, *once completed*, has deposited along its course an immobile trajectory on which one may count as many immobilities as one wishes. From this, one concludes that motion deposits at each instant a position with which it coincides."[118] In truth, however, no such deposit occurs, because movement itself does not halt; otherwise it would not be movement. But motion pictures reinforce this illusion, not so much through the projected image, which, as Gilles Deleuze pointed out,[119] *is* movement, but through the celluloid strip of individual frames. What is an exposed film frame except a "deposit" of a position coinciding with movement? And just like a line, the series of immobile instants on the celluloid strip is completely reversible. Bergson might therefore have argued that cinematic exposure itself is a category mistake.

NERVE FIBERS, TISSUE CULTURES, AND MOTION PICTURES

Toward the end of his dissertation, Seddig remarked on the close similarity of Brownian motion and cell movement, calling for further investigation into the relation between temperature fluctuation and "biological motion" that might lead to an understanding of the deeper "causes of movement itself."[120] Even if this particular research program never really caught on, in a way it points to the interesting history of cooperation between biologists and physicists during the early days of microcinematography. The precise date of the first moving pictures taken through a microscope has not been recorded, but Marey devotes a chapter to the technique in his 1894 book, *Le mouvement*.[121] Marey's experiments were continued by various disciples, most of them working at the Institut Marey in France. Lucien Bull, for example, is known best for his high-speed cinematography of fast-moving objects, which was an important technique in physics for research on surface tension phenomena. In 1903 he helped biology Professor Antoine Pizon microcinematically record the multiplication of a colony of *Botryllus*, or sea squirts.[122] Another disciple of Marey and a pioneer in the field, Charles François-Franck, adapted Marey's apparatus and made chronophotographic plates of a variety of biological phenomena from 1902 to 1908.[123] Lucienne Chevroton, who worked at the Collège de France alongside François-Franck, similarly added a motion picture camera to a microscope apparatus in 1909 to record microcinematographic investigations of a sea urchin egg.[124] François-Franck and Chevroton offered their apparatus to their colleague Victor Henri, a physicist at the Collège de France, who used it to record Brownian motion, as we saw in the previous section.[125] He, in turn, may have introduced it to Jean Comandon, who during the early years became the most successful microcinematographer of cell movement.[126] Lucien Bull also helped outfit Swiss biologist Julius Ries with a Lumière microscope–cinematograph combination for his research on the fertilization of sea urchin eggs.[127] As early as 1900, with the help of Charles Gaumont, French physicist Henri Bénard used cinematography to study dynamic, convective systems—specifically, the spontaneous formation of inorganic matter into hexagonal shapes in a heated liquid—and went so far to comment on the analogy between this type of motion and biological phenomena.[128] In Germany, physicist Henry Siedentopf, who ran Carl Zeiss's optical laboratory in Jena and collaborated with Richard Zsigmondy on the invention of the ultramicroscope in 1903, actively used motion picture techniques to investigate crystallography.[129] Siedentopf also helped Heidelberg biologist Hermann Braus with his cinematic investigations of tissue culture, thereby continuing this trend of cinematographic collaboration between biology and physics.

This cooperation between physicists and biologists interested in adapting motion picture technology is not surprising, given that investigators in both fields had set out to study motion as a way of answering difficult questions about causality. Brownian motion, as we have seen, functioned as an observable system that could provide clues to the constituents of matter itself. Biologists similarly searched for observable, temporally continuous systems in which the analysis of motion could resolve previously intractable disputes about the processes governing the organism. Braus, a morphologist and experimenter, used motion pictures for precisely this purpose.[130] His case is interesting because it illustrates the close formal relation between cinema and biology at the turn of the century; motion picture technology not only helped solve evidentiary problems in biology, but the introduction of cinematic means into experimental technique matched broader changes in the discipline itself. Furthermore, Braus's use of motion pictures demonstrates especially dramatically cinema's *rhetorical* power in science; Braus altered his viewpoint in a dispute because of what he witnessed in a filmic record. This section will place Braus's cinematic contribution in the context of certain broader debates and experimental trends in the discipline of biology.[131]

The dispute in question—between Braus (fig. 1.15) and American experimenter Ross Granville Harrison (fig. 1.16)—concerned the development of nerve fibers. Braus and Harrison were not the only figures in the debate, which had been brewing since the mid-nineteenth century. The problem was this: while it was clear that nerve fibers grew, it was not clear *how* they grew. Did a fiber grow from one cell or from a chain of cells? How was the connection between fibers and tissues established? Did the nerve grow from the center to the periphery, or was the connection already there in the embryo? The disputants offered at least three theories. Many believed that the formation of the nerve required the participation of many cells, not just one. This was known as the "multicellular" theory of nerve development.

FIGURE 1.15. Hermann Braus

Courtesy of the National Library of Medicine

FIGURE 1.16. Ross Granville Harrison

Courtesy of the National Library of Medicine

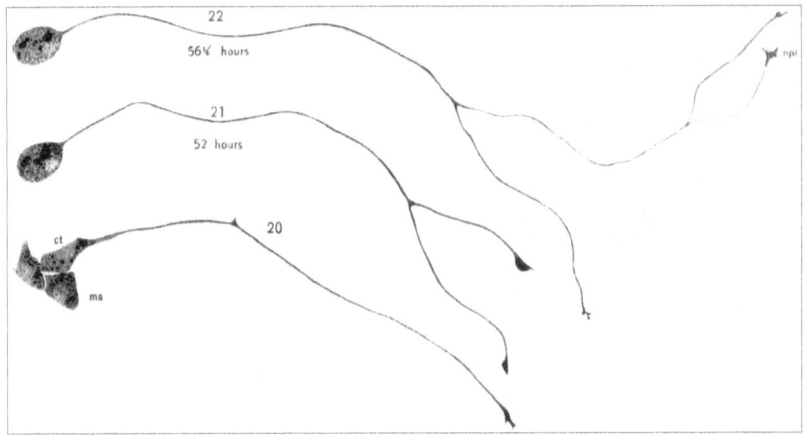

FIGURE 1.17. Harrison's sketches of the elongation of frog nerve fibers grown in culture

From Ross G. Harrison, "The Outgrowth of the Nerve Fiber as a Mode of Protoplasmic Movement," *Journal of Experimental Zoology* 9, no. 4 (December 1910): 787–846

Others, such as Braus, maintained that nerve fibers had already been present in embryonic form and that as the organism developed and cells grew and multiplied, the nerves stretched across them, making bridges of nerve fibers that differentiated as they became functional. This was the "protoplasmic bridge" theory. Still others, such as Harrison, argued that nerve fibers grew outward from a single nerve cell into the interstices between other cells. This became known as the "outgrowth" theory, and most biologists would come to agree, but not until around 1930, that this theory was the most accurate and persuasive depiction of nerve fiber development (fig. 1.17).[132]

There were a number of reasons why investigators in the early twentieth century could not agree on this topic, including prevailing concepts of the structure of the nervous system. But a large part of the problem derived from contemporary methods of observation. As Harrison complained in 1906, "Prior to the year 1904 all attempts to solve these problems were based on observations made upon successive stages of normal embryos. When one compares the careful analyses of their observations, as given by various authors, one cannot but be convinced of the futility of trying by this method to satisfy everyone that any particular view is correct."[133] That is, researchers attempted to observe the growth of nerve fibers by slicing, staining, and mounting successive stages of an embryo, which were then viewed through a microscope. This technique had long been dominant in biology and developmental embryology.[134] Crucial questions about the origin, direction, and character of growth had to be inferred from what was effectively a series of still images.

Because investigators basically had to speculate about any movement taking place between images, disputes over interpretation were common. Among the most famously frustrating was the debate between Santiago Ramón y Cajal, who subscribed to the outgrowth theory, and Hans Held, who held onto the protoplasmic bridge theory. Susan Billings explains, "Cajal claimed that *Plasmodesmen* were artifacts of Held's preparative method, while Held, in turn, thought perhaps pale staining or alcohol fixation prevented Cajal from seeing them. The impression one might get from reading their papers is that Cajal and Held were discussing vastly different material."[135] Yet when Cajal visited Held's laboratory in Germany, he said that he "had the pleasure of examining Held's excellent preparations. Just as we expected, they are very successful, and to our great surprise, they show very much the same picture as ours."[136] As with Brownian motion, researchers were unable to agree on what they saw in what were nearly identical preparations. (Indeed, this is another example of the influence of theory on observation.)

The histological technique of slicing, solidifying, staining according to type, and then mounting the preparation on a microscope slide had been the foundation of cell research up through the nineteenth century. It served biology well; as Hannah Landecker notes, "the technique of staining, by suspending the cell in time, freed the experimenter from the temporal exigencies of a living subject."[137] Precisely because the subject did not move, the researcher could examine the preparation at his or her leisure. "Cellular movement, or lack thereof," Landecker continues, "was a matter of inference when using static representations, the hardened moments preserved in histological specimens. It was about inferring what was happening in the spaces between the sequential slices of preserved moments."[138] Indeed, perhaps because this method was so dominant, movement itself was not important to biology's research questions—one could even argue that biologists' questions were often tailored to their technique. However, once questions of movement and growth could not be answered, new techniques were required; alternatively, we could argue that once the new techniques were available, new questions came to the foreground.[139] Physicists had solved this problem of observation and interpretation in the case of Brownian motion by emphasizing the gap or discontinuity, thereby excising from the theory that which was not observable. Biologists could not do this, however, because their ostensible object of study was continuity itself. So they had to develop techniques that would allow them to study cell movement directly.

Harrison believed that "the only hope of settling these problems definitely lies, therefore, in experimentation."[140] In 1905, Braus approached these questions experimentally by transplanting limbs of very young tadpoles to various areas of other tadpoles' bodies. He then examined the nerves in the transplanted limbs, finding evidence in favor of the protoplasmic bridge theory.[141] Harrison repeated these experiments and came to completely different conclusions. In fact, Harrison argued, "The same facts [in Braus] may be interpreted quite readily, if not more so, in accordance with outgrowth theory. The experiments do not approach the problem directly enough to determine questions of histogenesis [the origin of the nerve fiber], and there are too many loopholes left to permit a rigid proof."[142] What was needed, apparently, was a means of isolating the nerve cell itself so that the growth of the nerve fiber could be observed directly. In fact, this is exactly what Harrison did in a now-famous experiment that he first described in a June 1907 announcement:

> The immediate object of the following experiments was to obtain a method by which the end of a growing nerve could be brought under direct observation while alive.... The method employed was to isolate pieces of embryonic tissue, known to give rise to nerve fibers.... The pieces were taken from frog embryos about 3 mm. long.... After carefully dissecting it out, the piece of tissue is removed by a fine pipette to a cover slip upon which is a drop of lymph freshly drawn from one of the lymph sacs of an adult frog. The lymph clots very quickly, holding the tissue in a fixed position. The cover slip is then inverted over a hollow slide and the rim sealed with paraffin. When reasonable aseptic precautions are taken, tissues will live under these conditions for a week and in some cases specimens have been kept alive for nearly four weeks. Such specimens may be readily observed from day to day under highly magnifying powers.[143]

By successfully maintaining a tissue specimen outside of the body (known as "in vitro," whereas the state of tissue inside the body is known as "in vivo"),[144] Harrison initiated the technique of "tissue culture," a biological method whereby fragments of tissue from an organism are transferred to an artificial environment in which they can continue to survive and function in some form. With the cell population thereby isolated, the researcher could better examine and manipulate cell behavior. The histologist could observe the growth of the cell directly under the microscope and even slow down or accelerate that growth by manipulating the environment. Investigators from all quarters immediately took notice. Within the next few years, the literature on tissue culture exploded; its most famous experiments were those conducted by Alexis Carrel and Montrose Burrows at the Rockefeller Institute, where they managed to keep the culture from a chicken embryo alive for decades.[145]

This method also presented a major break with traditional methods of visualization in biology.[146] Not only did the in vitro experimental technique offer a different way of isolating the cells, but this form of representation also replaced the discontinuous form of representation then common to biology (for example, sequential slides, illustrations of stages of development, still photographs) with temporal continuity (movement, growth).[147] Tissue culture allowed the researcher to observe movement in time; there still remained the challenge of representing this movement, of course, given the traditional means (slides at conferences or illustrations in texts and journals). Harrison's experiment also showed fairly conclusively that the nerve fiber grew from a single cell. It demonstrated that the fiber could and did grow without the help of other cells and that nerve growth did not require the presence of preformed protoplasmic bridges. Even so, biologists such as Hans Held challenged Harrison's results for years afterward, maintaining that the nerve fiber grew on its own in vitro but grew with the aid of bridges in vivo.[148]

So in September 1911, Braus's demonstration before the Society of German Natural Scientists and Physicians spoke to both the interest in the new technique and the enduring controversy about nerve fiber development.[149] His demonstration also addressed new challenges in representation in biology, in that he used motion pictures to depict these two issues. At the start of the talk, Braus discussed the still-new technique of tissue culture as Harrison and Carrel had developed it to that point. He then noted, "I thought that many of you would be interested in seeing the phenomena of growth and movement in these cultures, if not in a live culture, at least via cinematographic images. I have films that were made in order to study the more detailed movements in the cultures of my many preparations, in which the transformations were too fast or too slow to be observed in the objects themselves."[150] This offering to his colleagues at the conference deserves further examination, because it indicates two ways in which film had a crucial role in the experiment.

First, Braus used film to *communicate* the results of his work. In this respect, the medium of motion pictures functioned as one of any number of forms of communication that describe or depict experimental procedures so that the (expert) reader can judge the validity and value of the project. Some have called this expert communication and evaluation "virtual witnessing"; in the modern era, when there is no way to inspect or replicate every experiment personally, the significance of such reports for the dissemination and acceptance of new knowledge cannot be underestimated.[151] In this early example, film promoted "virtual witnessing" in an important new way, not only because film made the cultures present to the expert witnesses in a more than virtual manner but because it was portable and repeatable in a way that live cultures were not.

Second, Braus used film as a *substitute* for his object of study (the tissue from a frog's heart). He made the films to study movements that were "too fast or too slow to be observed in the objects themselves"; his films were therefore the object of his analysis. In fact, the capabilities of motion pictures so closely matched the capabilities of tissue culture that they became an experiential substitute for the technique. That is, the projected image moves and can thereby present movement or growth, but film was also, like tissue culture, fragmentary and easily malleable, both temporally and spatially. If the histologist could isolate, dissect, transplant, graft, and retard or accelerate the growth of cells, then the cinematographer could match these actions perfectly by framing, cutting, splicing, editing together, and slowing down or speeding up the film itself. (Of course, motion pictures were quantitatively different than tissue cultures because of the ease with which images could be multiplied or accelerated.) In this respect, Braus's use of film became what we might call a "virtual experiment"—manipulation of the film could isolate and analyze aspects of the tissue that could also be manipulated, isolated, and analyzed in the technique of tissue culture itself.

Braus then described the star attraction of his films: a beating heart extracted from a frog embryo and maintained in culture for days. That achievement was in itself noteworthy, but Braus was interested in proving something more: "As far as I can tell, observations about the *growth* of organs in vitro are not addressed in the literature." That is, he wanted to demonstrate that specimens actually grew in culture, they did not merely survive. So he used staining techniques and time-lapse cinematography: "To prove that the culture, whose pulsation I've shown, is actually growing, I have filmed isolated, stained cells from the culture at greater magnifications. The exposures occurred, on average, every 10 minutes over the course of

10 hours. Here in the projection they have been compressed into just a few minutes, so that we can see the movements of the stained cells unfold rapidly, although in reality they are very slow."[152] The stains allowed Braus to distinguish the movement of the cells and their changing form as they grew and multiplied. His films thereby functioned not just as a substitute for the culture, but as a new form of evidence.

This is a particularly potent example of the power of cinema to present evidence forcefully, because the cinematic record of his experiments with tissue culture caused Braus to reverse his position on the origin of nerve fibers.

> To supplement this, I am showing images of growing nerves, in which—just as with the mesoderm cells—many of these growth phenomena can be observed under the microscope *directly*. The nerves are of particular interest because of the *in vitro* heart movements. At the time of extraction and the end of the experiment, the heart structure had no ganglia cells. . . . But from our experiences with nerves grown *in vitro*, we can rule out the claim that actual nerves originate in any other way than from the central neuroblasts. I want to discuss here objections that could be raised against this conclusion, reached by Harrison based on his preparations, which prompted me to verify his findings. This investigation has persuaded me that Harrison is basically correct.[153]

Having witnessed these processes for himself through his filmic record, he was able to declare that "only the isolation of neuroblasts [the nerve cells] gives total certainty; the nerve fiber grows out of the neuroblast like a mold grows out of an isolated spore."[154] Braus went on to defend Harrison against the objections of other colleagues in the field, with whom he previously agreed.

So Braus also used film as a *confirmation* of Harrison's theory. Braus's motion pictures were more than a demonstration; they *enacted* the technique and Harrison's theory of growth. Harrison posited the nerve cell fiber as a plant-like "outgrowth" rather than a differentiated element of a multiplying mass. He argued that nerve fiber growth moved forward incrementally, outwardly, and linearly. Likewise, Braus's time-lapse technique in this instance—with its statistical sampling of time through regular exposures and its amplification of the forward push of incremental growth—enacted and insisted on an incremental, regular, mathematical, and resolutely *linear* theory of growth. In other words, this set of techniques (the frame's isolation of the tissue, the temporal manipulation of growth, and the magnification and projection of the image) enacted, amplified, and confirmed Harrison's theory of growth and technique of tissue culture with incomparable rhetorical power. While cell growth, when illustrated by sequential images (such as slides), could have been seen as reversible and thus disputable when direction was at issue, the temporal continuity of the cinematic record *pushed* the observer (including Braus) to a particular conclusion—that the growth of the nerve fiber happens in *this* way. If other scientific uses of cinema try to stop time, this application stressed duration and the teleology of growth. As a means of visualization and as an experimental tool, this application of motion pictures precisely matched broader changes in experimental technique and representation in biology at the turn of the century.

As we can see, motion picture technology presented itself as a potential research instrument to biologists involved in tissue culture, but film became a powerful tool for researchers because the research program in biology had changed to accommodate it. That is, Braus came to film not only because it solved a particular problem but because the problem itself was shaped by changes in conceptions of growth and time that motion picture technology could articulate (indeed, that the moving image might have prompted). Of course, the practical benefits outlined at the beginning of this chapter provided compelling reasons to use cinematography. In Braus's case, it offered the opportunity to record his experiments and then to present some of his findings to a wide audience. The photographic or filmic image also projected the impression of objectivity and thereby provided a potent rhetorical weapon in his arguments with the conclusions of other biologists. And film's ability to manipulate time allowed Braus to present quickly evidence that would have taken hours to observe directly. But beyond these obvious advantages, we could argue that the growth Braus documented was researchable only *because of* cinema. The motion picture camera—because it could alter the scale of cellular time through slow-motion or time-lapse cinematography—could capture events that were not otherwise visible. These events were not merely difficult to observe without the camera, in a real sense they did not exist without it. They became, as Walter Benjamin noted, our "optical unconscious," revealed to us by cinematography in the same way that our psychic unconscious is revealed by—*and comes into being by virtue of*—psychoanalysis.[155]

But there is more to the success of cinema in science and its relatively rapid appropriation than even these good (practical) reasons. Not only were filmic techniques increasingly available, they were also adaptable to various scientific enterprises. If Seddig focused on the space between the frames, on cinema's *analytic* ability, Braus similarly used the time between exposures to his advantage. But for Braus the image of the object was paramount. The images made little sense analytically; instead, Braus concentrated on their *synthesis*. For Braus the gap was crucial for the construction of the film, but the ellipsis was elided in favor of a temporal continuity or biological teleology. Growth is not reversible. This is not to say that film's analytic abilities were confined to (or more suitable for) the physical sciences, while its synthetic nature was best suited to the life sciences. The variety of applications in a range of disciplines makes any such generalization rash. In fact, as Latour might argue, cinema's adaptability was no greater or less than that of any technology—from the microscope to the computer. Nor was Braus's "synthetic" use of cinema any less scientific than Seddig's "analytic" emphasis. Braus was doing more than "simply" recording a phenomenon and sharing it with others, even though he did not explicitly extract quantitative data from the images. Braus's use of time-lapse cinematography recorded only certain, regularly paced moments over the continuity of an event, thereby creating, I would argue, *a statistical sampling* of the temporal dimension of the phenomenon. We can therefore see Braus's particular appropriation of motion picture technology as a statistical application of cinematography very much in keeping with the contemporary adoption of statistical thinking for the study of systems in both physics and biology. Indeed, part of the persuasiveness of these records could be found in their "objective," mathematical construction of a representative sample.[156]

Bergson, however, articulated the limits of analysis with respect to dynamic processes, and an extended comparison of the use of motion pictures in physics and biology would expose the paradox of analysis and synthesis as a problem of comprehending continuity through discontinuity. Thinking about Bergson and motion pictures together, in other words, highlights fundamental assumptions in certain scientific approaches about the nature of time and movement. Seddig's use of motion picture technology to study Brownian motion certainly reveals a commitment to interruption, reversibility, and immobility in the study of dynamic systems—a commitment that was obviously present in Einstein's equations as well. The ambidexterity of motion picture technology—its dual nature as an instrument of both still and moving images—not only explains its success in science, but its varied and subtle permutations in scientific application can themselves be understood as a form of commentary on scientific theory and method.

How researchers used motion picture technology in their experiments tells us much about their disciplinary and philosophical investments.

Above all, I want to stress the mutually reinforcing relationship between science and cinema. The problems and solutions that the technology and the research program presented to each other were entwined and dependent. For Braus and others, motion picture technology fit quite well with other elements of this experimental ensemble, in part because certain formal features—its temporal malleability; its ability to frame and isolate; the forward, teleological motion of its projected image; and so on—seemed to match similar features of the technique of tissue culture. There are other, broader homologies as well. The technique of tissue culture extracted life, reproduced it technologically, and prolonged it, allowing researchers to artificially accelerate or slow life down. And what is cinema except a technology that extracts bodies from their natural time and space, reproducing them mechanically, reanimating them repeatedly (faster or slower as needed) long after the host has expired? The idea that tissues, organs, and life itself are separable from the body—and, by implication, that death is foreign to the organism—is, like cinema, an especially modern notion. So we could say that cinema's singular facility with time, space, and life made it an ideal instrument to capture and confront these concepts in modern science. But we need not go so far. In fact, such expansive analogies tempt us to think of cinema and science as single entities, when the point of these case studies has been to demonstrate the opposite: that film is not monolithic and that science, like other broad umbrellas, contains many different disciplinary agendas, each of which has adopted some aspects of film technology while ignoring others. That so many different researchers found something to use in motion pictures indeed indicates a special relationship between scientists and film. Braune and Fischer's deconstruction and reconstruction of the human body, adapting it to modern labor systems; Einstein's and Seddig's flexible notions of time expressed in the invisible, random, and reversible world of molecular physics; Harrison's and Braus's conceptions of life as separable, temporally malleable, and technologically reproducible—in each case, these scientists turned to motion pictures not merely to observe and record phenomena, but because in cinema they found a kindred spirit, an amiable partner that shared their vision, that could look at the world as they did. The cinematograph, then, was not just a handy tool. In more ways than Bergson imagined, the form and application of motion pictures articulated science's modern agendas.

◊ ◊ ◊

In this chapter, we have examined the merger or accommodation of film form and objects, theories, and disciplinary agendas. In each case, film proved its worth because of this correspondence—not exclusively, of course, because the amount of work necessary to adapt film to these aspects of scientific endeavor is prodigious and varied. But we can credit the researchers for recognizing the homology between film and experiment in the first place. The next chapter will examine medical research films and the correspondence between film form and *observational* practices or ideals.

2

BETWEEN OBSERVATION AND SPECTATORSHIP

MEDICINE, MOVIES, AND MASS CULTURE

Stillness is the unattainable value.
— PAUL VALÉRY (1932)[1]

From around 1904 to 1914, motion pictures endured a difficult and very public transition between their good standing as a scientific tool and their growing notoriety as an instrument of mass culture. During the early period in Germany especially, there was a strong contrast between the enthusiasm for motion pictures as a scientific or pedagogical tool and the simultaneous condemnation of its public, commercial incarnation. Physicians in particular wrote many editorials and studies depicting the threat that motion pictures posed to the psychic, physical, and moral health of the nation. To a certain extent, this is unsurprising: the German complaints about cinema follow more or less the same pattern that we can find in the United Kingdom, France, Russia, and the United States at this time.[2] There seems to be no reason to think that the discussions in the scientific community and those in the public sphere were related or exceptional. But if we look closely, we find that the scientific and public debates about cinema were in fact closely related: the reasons scientists and physicians accepted film as a scientific instrument were rooted in the same logic that prompted them and others to reject cinema's public manifestation.

This logic concerned, to put it too simply, the perceived difference between *observation* and *spectatorship*. That is, participants in both discussions seemed to agree on the advantages and dangers of the moving image, but many of these advantages and dangers appear to stem from *different ways of viewing the image*.

If the previous chapter focused primarily on the negotiation between object and apparatus—such as Braune and Fischer's adjustments to their recruit, to their working space, and to their camera to create a scientifically productive image—this chapter emphasizes the relationship between researcher and image. How the researcher *studies* the image may be just as significant as how he or she produces it. Indeed, "observation" is a woefully underused key to understanding medical practice and identity. Many physicians felt that their professional reputation and livelihood depended on carefully cultivated observational skills. No less a personage than the eminent British surgeon Sir James Paget emphasized in 1887 that "becoming scientific in our profession [requires] the training of the mind in the power and habit of accurately observing facts. . . . The main thing for progress and for self-improvement is accurate observation."[3] As Paget indicates, part of this investment in observation stemmed from the contemporary adoption of scientific principles in medical practice, which some contested, but which ultimately bolstered the field's legitimacy.[4] As physicians identified with scientists, "observation" became an important common link. But it is not simply a borrowed scientific practice, it is also key to understanding modern medical logic. As Michel Foucault and others have shown, modern medicine's understanding and approach to disease is inextricably tethered to specific habits of perception.[5] The process of observation was not simply preliminary to diagnosis but actually rehearsed the logic by which medicine grasped its object.

We can see this process especially clearly when physicians encountered the cinematic image. How they used motion pictures and what they said about them can reveal the subtleties and complexities of their observational practice. As I have argued elsewhere, modern medicine's reliance on the correlation between life and death, or between movement and stillness, if you will, makes motion pictures a privileged representational technology for exploring medical hermeneutics.[6] In their attempt to understand and locate pathology, physicians find in the corpse lesions that might be blamed for death and then compare these to signs in the living body that might be harbingers of disease. In the stilled corpse, researchers are able to grasp some elements of disease that are elusive in life. Similarly, physicians take advantage of the duality of motion pictures by going back and forth between the stilled and the moving image to grasp elusive properties of the recorded object or event. Exploring this process in detail reveals not only the doctor's cultural investment in observational practice but also the foundations of modern medical logic. Examining the medical community's reaction to movies, on the other hand, also reveals this investment in expert observation and their characterization of its opposite—spectatorship—as a pathological condition.

Nearly from the beginning, motion pictures were enlisted as an aid to medical research and education.[7] German doctors, in particular, were captivated by the potential of the cinematic image. Certainly, the application of motion picture technology to medicine was not qualitatively different in Germany than in any of the other Western, industrialized countries. Germany, however, did enjoy a quantitative distinction: between 1900 and 1920, Germans wrote far more and far more frequently about medical cinematography than their counterparts in the United Kingdom, France, the United States, and the rest of the world.[8] One could say that Germans wrote more about everything, but Germany's leadership in certain specialties that made extensive use of cinematography, notably radiology and neurology, helps validate this claim.[9] It could also be argued that in the late nineteenth and early twentieth centuries, Germany's institutional research infrastructure was among the best in the world, which allowed its scientists to incorporate new instruments and agendas more quickly than, for example, researchers in the United States at this time.[10] Of course, the number of journal articles praising cinema's potential as a new medical technology should not imply that the general practitioner put it to use on a daily basis. On the contrary, given the expense, skill, and patience required to operate a motion picture camera—much less one adapted for medical use—the average doctor never came in contact with one or even considered the possibility. During this period, the medical application of cinema remained confined primarily to university research laboratories and some medical school lecture halls, with a few privileged physicians leading the way. Still, German researchers enthusiastically applied motion pictures to a range of specialties from ophthalmology to gynecology.[11]

What did cinematography promise the intrepid researcher? This physician waxes poetic about microcinematography, but his claims are representative of all medical and scientific uses of motion pictures:

> Microcinematography captures [*fixiert*] what a scholar has witnessed and researched in his quiet laboratory, so that he can then demonstrate his collected research to a larger group, be it at a conference, before a medical

society, or in an auditorium of students. But microcinematography not only documents movement processes, it is also a helpful tool for research itself. It captures every phase of movement and thereby allows the researcher to examine each image at length and to study the temporal and spatial relationships of the movements carefully and calmly. Microcinematography does not require hasty progress, as living movement often does; it allows the scholar the time he needs to grasp all worthwhile detail and lets him determine when to move to the next image.[12]

Compressed in this paragraph are most of the advantages typically cited in the early literature on the scientific application of motion picture technology. To recap briefly: first, the cinematic image, like the magic lantern slide, was projectable, and therefore could be shared among colleagues and students. This made it ideal for educational purposes, which required an efficient dissemination of information, but also for rhetorical purposes; the scientific film, like all scientific illustration, existed to present evidence meant to persuade others. The size and spectacle of the cinematic image had a powerful persuasive effect in itself, but part of that power came from the photographic character of the image. This speaks to the second advantage, its ability to "capture" or to document moving phenomena. Because its photographic base allowed it to record the variety and randomness of the natural world, the motion picture had an evidentiary status that was very useful for presentations and publications, because it could therefore act as a substitute for the living patient. This also meant that it could function as an object of study itself; it "provides the researcher the opportunity to examine each image at length." This is the third advantage: the spatial and temporal malleability of the motion picture extended the senses and allowed the researcher to explore new domains previously unavailable. In addition, the individual frames of the film effectively managed the data, especially the informational complexity of any kind of movement. Indeed, the most noteworthy aspect of this excerpt is its enthusiasm for the researcher's ability to control the rate of movement of the motion picture. The author lauds the temporal control that came with the moving image and emphasizes the slow pace of scientific study, the almost contemplative stance before the image: "in his quiet laboratory . . . to examine each image at length . . . carefully and calmly . . . the time he needs." The rest of this chapter will explore these advantages in detail.

But for now, this testimonial will stand in contrast to other descriptions of film common in the medical community—those that decried cinema as a threat to public health. As medicine's prestige grew, physicians expanded their expertise to broader social questions and offered their prescriptions for a healthy nation.[13] So alongside the enthusiasm for the scientific potential of motion pictures there was a separate yet related discourse that was not so enthusiastic. Between 1904 and 1914, reformers and physicians in Germany consistently viewed the new medium of motion pictures with suspicion. They not only protested the dank, unventilated storefronts in which many motion pictures were then being screened, but also felt that the motion pictures themselves presented a serious danger to public safety. The sensational subject matter of many of the movies threatened to corrupt the taste of the nation. But a particularly insidious form delivered this content, as this doctor attests:

> The effect of this sensational subject matter is heightened by the *temporal concentration of events*. The cinema concentrates the sensations of a detective story or thick trashy novel into 10 or 15 minutes. The resulting psychological effect is thus completely different. When reading, we can pause at will, critically reflect on the text, relieve ourselves of the internal pressure through contemplation, digest the scary parts. In the cinema, such excitation of the passions is intensified and multiplied by the rapid succession of vivid images that passes before our eyes. There is no time for reflection or release, no time to find an inner balance. These spine-chilling and grotesque "dramas" distress the nervous systems of young and sensitive persons to the point of suffering, but without providing the spectator any of the means with which he usually defends himself against attacks on his nervous system: quiet *contemplation, intellectual assimilation*, and sober criticism are not possible.[14]

The contrast with the previous quotation is instructive. If cinema provided the *researcher* the opportunity "to examine each image at length and to study the temporal and spatial relationships of the movements carefully and leisurely," it did not afford *the average moviegoer* that same luxury—"quiet *contemplation, intellectual assimilation*, and sober criticism are not possible." The difference between the (boring) research film and the (exciting) film drama perhaps played a role in this characterization, but even more important for the purposes of this chapter was cinema's temporal rush, its insistent and impatient movement forward, which this doctor felt could have a lulling, even hypnotic effect. Those without sufficient training or education were susceptible to cinema's lurid charms. But the educated, professional class apparently had the tools to manage this onslaught. What were these tools? Most obviously, researchers, unlike the average moviegoing audience, were in control of the image and its projection—they could literally manipulate their films however they pleased. But this ability to control the rate of projection went hand in hand with a mode of viewing that was very much part of a nineteenth-century researcher's training and identity, which distinguished itself from another mode of viewing attributed to the lay public, especially the cinema audience.

Expert training depends on developing a set of skills, but it also distinguishes itself from those who lack the skills. In the medical profession, this dynamic determined the nature of the doctor–patient relationship, or the nature of internal debates about midwives and quacks, for example; the medical profession relied on the constant maintenance of that hazy line between "normal" and "abnormal," or between "healthy" and "diseased" in the broadest sense. Medical observation similarly depended on its opposite—in this case, spectatorship. The debates about motion pictures are an excellent example of this dynamic. Comparing these two discourses allows us to see quite clearly the criteria for cinema's legitimacy. By understanding what German doctors considered to be the *proper* mode of observation, we come to understand more fully what they thought was an *improper* way of viewing images, and vice versa. We also see that this proper mode was not simply the result of disciplinary training. The "objectivity" of the scientific eye arose not merely out of professionalization but also in contrast to the "subjectivity" of the untrained other. That is, disciplinary modes of viewing relied on *class* distinctions, as well as professional training.

So to appreciate the medical community's diagnosis of film spectatorship as passive, weak-willed self-abandonment to the flow of images—a characterization that was not uniquely made by doctors, but which is best understood with reference to them—we must take into account its investment in observational practice, especially as it relates to experiment and temporality. Physicians and their educated brethren distinguished themselves from the layperson via their viewing practices: by incorporating film into experimental method, for example, they observed and contemplated the image at a leisurely, disinterested pace, while ordinary spectators were swept up and seduced by it. These observational practices were therefore more than cultural capital, more than a sign of expert learning and status—they were a badge of honor that also served as a shield against the encroachment of mass culture and the rush of modern life. How one approached cinema's temporality, it seems, determined whether one was an observer or a spectator.

The rest of this chapter consists of three sections. Because the early history of medical filmmaking is not yet common knowledge in film studies, the first section provides an introductory survey of the early use of motion picture technology in medical research and training, focusing on such work in Germany. This section will examine the multiple functions of film in the medical context by focusing on the needs that motion pictures served in the international medical community. (This discussion will focus on medical research and training films—films by physicians for physicians—rather than medical films intended to educate the general public.) The second section explores how the use of motion pictures corresponded to established principles of observation in medicine. It will also demonstrate the professional investment in this particular mode of viewing. While the first section emphasizes film and the history of medicine, this section focuses on film and the *philosophy* of medicine. It therefore investigates the use of film in relation to medical logic. Finally, the last section takes a closer look at the medical editorials about the dangers of cinema, especially its hypnotic power, in order to illuminate the moral, ethical, and class dimensions of medical observation and cinematic spectatorship.

THE MULTIPLE FUNCTIONS OF THE MEDICAL FILM

Most histories of research films recount individual cases within various disciplines or fields (just as I did in chapter 1), but it would be worthwhile at this point to step back and sketch broader patterns of use across specialties. Describing these patterns in terms of "function" has the advantage of accounting for the full life of a film beyond the immediate task. As we have seen, motion pictures functioned in various ways for researchers; they extended the senses, of course, but they also documented and displayed data. Even though a researcher made a film for a specific purpose—to examine the mechanics of an organ, for example—that same film also functioned as a record of the case and as an educational aid; it was a rare medical film indeed that had only one use. I would argue that there are three main functions of the research film: *exploratory*, *documentary*, and *educational*. These broad functions could be applied to much nonfiction filmmaking as well, but as a triad they are especially applicable to research films of all sorts, medical or scientific. Generally speaking, research films are made to explore unknown domains or relationships; to document and display results; and to train students, teachers, or, sometimes, the general public. These functions are by no means mutually exclusive; any film can have multiple, even simultaneous, functions. In fact, understanding how a film functions necessarily entails pinpointing who is using the film, when, and for what purpose.

We can group various cases in the history of the medical film around function, a retrospective critical category, but also around specific, historical needs that motion pictures satisfied for medical research. For example, as we have seen in physiology and biology, researchers at the turn of the century were increasingly interested in process and function over structure and morphology; the discovery of X-rays only heightened the prevailing interest in the mechanics of organ function, for instance. Motion pictures also proved to be a valuable technology for visualizing, documenting, and measuring the functions of the human body. Likewise, just as photography was an important method of documenting cases, motion pictures also proved indispensable for recording, for example, pathological movement. Such documents served as entries in an archive of images that medical professionals recognized to be vital to the health of their discipline. Physicians established image archives in hospitals and other research centers in Europe and the United States, and calls for similar motion picture archives of medical films were common at this time. Physicians recognized the importance of these films for medical education as well. While the profession was very self-conscious of the modernization of medical education that had taken place since the Enlightenment—replacing the medieval, scholastic model of book learning with an emphasis on hands-on training and observation—the practice of live demonstration of patient symptoms in medical lectures was logistically and ethically troublesome. Motion pictures offered a way out of this dilemma and presented a novel solution to issues in medical training. The rhetoric surrounding the use of film as an educational tool mirrored the broader discourse on pedagogical applications of cinema, but the medical training film illustrates especially clearly the contemporary emphasis on efficiency and modernization. So this section presents a brief survey of early medical filmmaking using the three functions as a preliminary heuristic, but it will also chart film's meanings in the context of specific issues (organ function, image archives, live demonstration, and efficiency) important to the medical profession at this historical moment.

THE EXPLORATORY FUNCTION

Of course, the motion picture camera can be used to explore facets of the natural world below or beyond the threshold of human perception, such as the extremely small (via microcinematography) or the extremely slow (via time-lapse cinematography). The various means by which motion picture cameras can manipulate time and space have been very important to research films. For many medical professionals exploring the rhythms and movements of the human body, this ability to manipulate time also contributed to its use as an *experimental* instrument, more than simply as an observational aid or recording device. Since the mid-nineteenth century, branches of the medical profession in the Western world had incorporated a scientific ethos into their research and practice. Even though this trend was often contested, by the beginning of the twentieth century, the role of experiment in medical research was firm, if not universal.[15] The experimental use of motion pictures certainly had precedent in the chronophotographic research in physiology of Marey or of Braune and Fischer, but also in the increasingly frequent use of photography in medicine.[16] How can the motion picture camera function as an experimental tool? For that matter, what does an experiment do? To answer these questions, let us consider a fairly well-known example in the history of medical filmmaking: Austrian cardiologist Ludwig Braun's use of a motion picture camera in 1897 to record the beating heart of a dog.[17] A specialist in cardiac dynamics and mechanics, Braun employed frame-by-frame analysis to gauge changes in the size and position of the heart as it beat by measuring the shape and displacement of shadows and other markings on his filmed images. From these short strips he was able to extract conclusive information about, for example, the nature of the apex beat (the lowermost and outermost prominent cardiac pulsation) (fig. 2.1).

Braun used the camera in a way that instruments are often used in scientific experiments, that is, to test a theory, to measure constants, or to explore new domains.[18] A scientific experiment is usually designed first to isolate and stabilize variables (such as temperature, air pressure, etc.) surrounding an event or phenomenon. An experiment is also designed to be consistently repeatable, not only so that other researchers can check the results for themselves, but, more importantly, so that the event itself can be scrutinized closely. And an experiment rarely stands alone—usually it is one of a series of experiments, so repeatability is crucial for the success of the

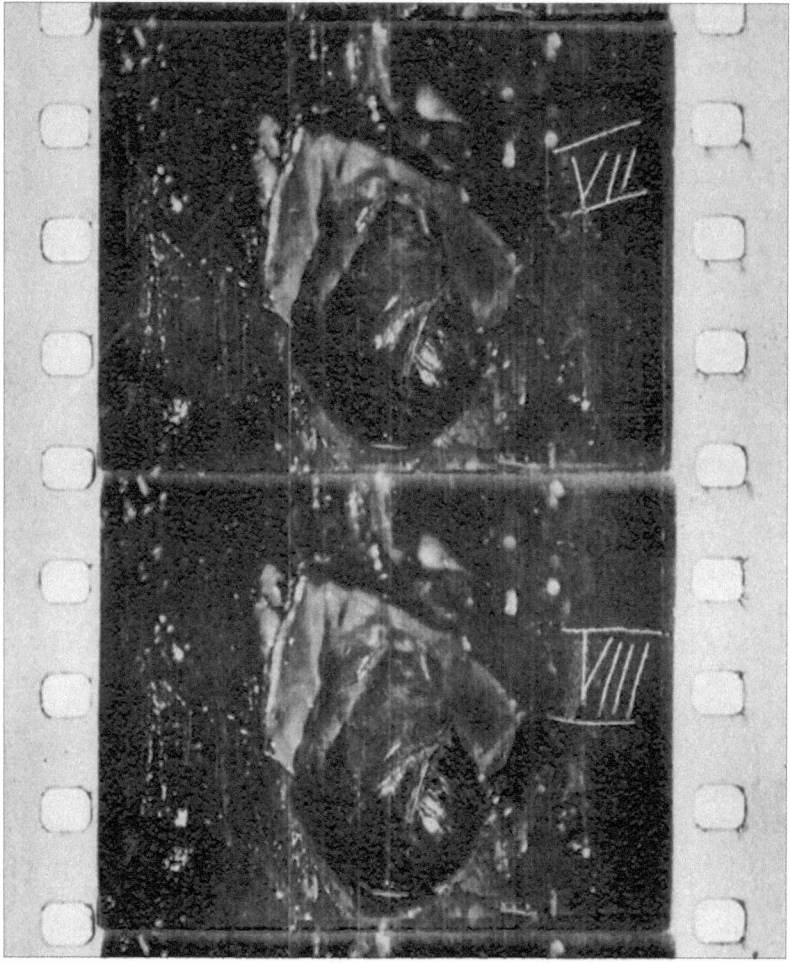

FIGURE 2.1. Frames from Ludwig Braun's film of a dog's beating heart

series as well. Time itself cannot be manipulated, of course, but an experiment allows the duration of an event to be significantly extended through repeated viewings. These three features of an experiment—isolation, stabilization, and repeatability—also allow the researcher to register quantifiable changes in the environment under study. A thermometer is an example of an experimental instrument that isolates and quantifies one particular variable, temperature. Other instruments are used to control or to create an (often artificial) environment for the experiment; the air pump, which creates a useful vacuum, is a good example. Film does not control or alter an environment to a great degree, but it can create something of a *new* environment, which acts as a record of or substitute for the event. Through its framing and focus, for example, film isolated and stabilized the phenomenon for closer study; a beating heart is difficult to examine, so the filmed record of the heart became a "working object" that could be examined repeatedly.[19] The motion picture was the best kind of repeatable experiment: if the record could function as a substitute, it could be endlessly repeated without the work involved in setting up the actual experiment again and again. In this way, as noted in the previous chapter, the motion picture functioned as a "virtual experiment." Researchers also employed motion pictures to register and measure changes by using the constant frame size and frame rate of the film to calculate changes in time and space. Braun used the displacement between frames to measure the size and position of the heart. Finally, the success or legitimacy of an experiment is enhanced if its results can be somehow inscribed (via a published report or visual demonstration) and easily presented or exchanged. In the absence of the actual experiment, the researcher's report of the experiment allows readers a kind of "virtual witnessing," as Steven Shapin calls it.[20] Film, of course, facilitates this "witnessing" in an extremely persuasive way. All these features contributed to the motion picture camera's success as an experimental instrument.

Braun's work also exemplified long-standing medical attention to the mechanics of organ function, which was only intensified with the discovery of X-rays and the development of motion pictures. As we know, the discovery of X-rays generated incredible enthusiasm—in 1896 alone there were more than 1,000 articles published on the phenomenon.[21] Much of this excitement came from the medical community, which immediately recognized benefits; engineers and experimenters sought to combine X-rays and cinematography.[22] In 1897, Scottish physician John Macintyre used a cinematograph to take several X-ray exposures of a frog's leg on motion picture film to create the first moving X-ray image.[23] Others continued to work on the combination to visualize organ function. For example, as early as 1903, Dutch physician P. H. Eykman used X-ray cinematography to study the swallowing mechanism in humans.[24] Similarly, between 1903 and 1906, Joachim-Léon Carvallo at the Institut Marey made a series of X-ray films of swallowing and digestion in small animals.[25] Starting around 1909, the German radiological team of Kästle, Rieder, and Rosenthal made X-ray films of joint, lung, and peristaltic movements.[26] And American Lewis Gregory Cole likewise made films of gastric phenomena.[27] But the

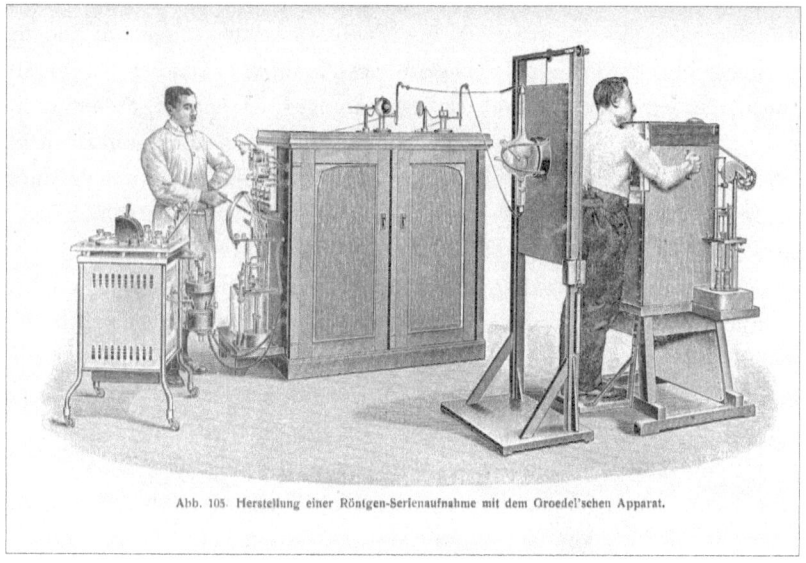

Abb. 105. Herstellung einer Röntgen-Serienaufnahme mit dem Groedel'schen Apparat.

FIGURE 2.2. Groedel's serial cassette X-ray apparatus

combination was always plagued by difficulties: the low X-ray-tube power and slow emulsions of the early days resulted in weak exposures or overlong exposure times, on one hand, or the mechanical problems involved in capturing the image, either directly with large sheets of X-ray film (known as the "direct method") or indirectly through fluoroscopic screens (the "indirect method"), led to unstable images, on the other. Franz Groedel was the acknowledged leader in X-ray cinematography; between 1909 and 1915 he worked tirelessly to develop a viable direct method, including a mechanism consisting of a series of falling cassettes (fig. 2.2). His commitment to the idea of moving X-ray images never wavered: "In spite of [the] present inadequate apparatus," he said in 1915, "important questions have already been solved, or the solution thereof brought nearer."[28]

Many physicians also hoped that cinematography, and especially X-ray cinematography, could be used diagnostically, that is, as a way of searching for clues that aid the identification of a disease or injury, usually via the correlation of visible symptoms with known disease entities or classifications.[29] But even as Groedel and others lauded its potential, the actual diagnostic value of X-rays and X-ray cinematography remained in doubt for many years.[30] In 1912, one prominent leader in radiology said of X-ray cinematography, "Of that, I don't expect too much. It will be didactically valuable, explain some processes, but it will not be able to serve diagnostic purposes in every instance."[31] Nevertheless, researchers and engineers put considerable energy into crafting X-ray images that moved. French radiologist André Lomon collaborated in 1911 with scientific filmmaker Jean Comandon, known best for his masterful microcinematography, to produce moderately successful results with indirect X-ray cinematography.[32] They worked on this approach up to the outbreak of World War I, which "relegated all roentgen cinematographs to storerooms and museums."[33] Groedel, Lomon, and others continued to work on the technical problems after the war, but it was not until the 1930s that the direct method proved regularly feasible, and it took until the 1950s for the indirect method to have daily clinical potential. Despite the inherent difficulties, X-ray cinematography promised at the very least the possibility of viewing hidden processes and movements that would help in understanding organ function and perhaps in diagnosing illness and injury. Indeed, what distinguished the exploratory function, whether manifested in experiment or diagnosis, was the possibility of *discovery*—the image presented a view that revealed something hidden from human perception. Yet to see something hidden, researchers often limited some variables and expanded others, a process made more convenient through motion picture technology, which could isolate or enlarge spatially (through lens or framing choices) or temporally (through recording or projection speed choices), or, as in the case of X-ray cinematography, film could be combined with other technologies that offered new ways of seeing. So film's exploratory function emphasized discovery, but it also underlined and depended upon film technology's formal analog to the functions of experimental restraint and modularity.

THE DOCUMENTARY FUNCTION

If the exploratory function charts new domains and reveals hidden spaces in order to help to understand processes or to identify disease, at some point the domain has been identified and the diagnosis agreed upon. At that point, film serves as an illustration of that which is already known; it becomes equivalent to the statement "Here is an example of X." This is the documentary function of medical cinema. For anything to function as a document, it must have some evidentiary value; the evidential authority of the motion picture derives primarily from its photographic base, which has a well-known rhetorical power. Such is the nature of this image that it served meaningfully as a substitute for the object filmed, especially if the phenomenon was rare or difficult to reproduce in demonstration. This would explain the popularity of motion pictures among neurologists at the turn of the century, who used film to document examples of pathological movement associated with neurological disorders.[34] For example, in 1897, neurologist Paul Schuster utilized the resources of Berlin clinics to create a series of short films of patients with a variety of neurological diseases, including Parkinson's, multiple sclerosis, myoclonus, hemichorea, and ataxia. These single-shot films, between three and ten seconds each, emphasized pathological movement and were designed primarily to illustrate Schuster's lectures without having to rely on live demonstrations of the patients. Schuster also hoped that these films and others like it would someday form an archive of material for medical educators.[35] There were many other physicians and neurologists in Europe and the United States who saw the same potential: Gheorghe Marinescu in Bucharest,[36] Paul Richer at La Salpetrière in Paris,[37] Walter Chase in Boston,[38] Arthur Van Gehuchten in Louvain,[39] Emil Kraepelin in Munich,[40] Hans Hennes in Bonn,[41] Osvaldo Polimanti in Naples and Perugia,[42] and Theodore Weisenburg in Philadelphia[43]—all used motion pictures to document nervous disorders.

These films were especially helpful in medical demonstrations, for which the use of live patients was always troublesome.[44] While the live demonstration was a major advance in medical education—a huge step from the centuries-long, scholastic tradition of learning medicine only from ancient texts—it had its own set of challenges. According to an American student taking classes at the Allgemeines Krankenhaus in Vienna in 1865, each professor was provided with a lecture room near his ward: "At the time of lecture this room is filled in with 'specimens' in the shape of men and women who are transferred from the other wards for the occasion. These patients are looked upon and spoken of as 'material' for the medical instruction and as such are submitted to examination by the students without much reference to any feelings which they as men and women may have on the subject."[45] Patients did not submit gladly, apparently. In another letter, the student draws a sketch of his routine at the hospital, which includes "scolding and pitching into the patients for coming late (wh. they always do in Vienna)."[46] Motion pictures promised an alternative, as an obviously frustrated Hans Hennes noted in 1910:

> How often does it happen to the professor that a patient fails during a lecture, that a manic suddenly changes his mood, a catatonic suddenly fails to

perform his stereotyped movements. Although he executed his pathological movements without disturbance on the ward, the changed environment of the lecture hall has the effect of not letting him produce his peculiarities—so that he does not display precisely what the professor wanted him to demonstrate. Other patients show their interesting oddities "maliciously," only when there are no lectures, continuing education courses, and so on. Such occurrences, which are frequently disturbing to the clinical lecturer, are almost completely corrected by the cinematograph. The person doing the filming is in a position to wait calmly for the *best possible moment* to make the recording. Once the filming is done, the pictures are available for reproduction at any moment. Film is always "in the mood." There are no failures.[47]

Hennes seemed not to be bothered by any ethical dilemmas; his concerns were purely practical. Motion pictures might have accidentally solved some ethical predicaments, but as Lisa Cartwright has demonstrated, filming patients often led to others.[48] Film technology, in this respect, followed a pattern in medical history in which new instruments create a psychic and physical distance between doctor and patient and thereby further tip the power dynamic in the physician's favor.[49] Physicians were already pleased to substitute photographs and slides for patients in lecture, if not to alleviate the obvious ethical concerns, then at least to present all the students with a larger, projected view.[50] Schuster intended his films to substitute for live demonstrations as well: "We wanted to use recordings of typical movement complexes to allow students to see the theoretically oriented lecture illustrated with motion pictures, regardless of the clinic's available material [i.e., patients]."[51]

Neurologists such as Schuster also found motion pictures especially exciting, because the records could be used for further analysis of pathological movement. (This dual use is an example of the function of the film changing according to who is looking at it, students or researchers.) It was also not uncommon to use photography and motion pictures to document intervention outcomes. Orthopedic surgeons, for example, used motion pictures to film "before and after" records of corrective surgeries.[52] These films were presented at congresses to persuade other physicians of a given diagnosis and therapy, or they were presented to students as examples of a particular surgical strategy. Even Comandon's motion pictures of microorganisms could be classified under this general rubric of "demonstration."[53]

Medical films, like medical photographs, also appealed to researchers and educators because they could be part of a disciplinary archive of images. Hospitals such as the Saint-Louis in Paris, Bellevue in New York, and the Charité in Berlin established photographic departments for just this purpose. A report from Bellevue in 1869 indicates that a photographic archive and department could be a magnet for the discipline: "Members of the medical profession begin to visit the Department periodically, for the purpose of obtaining such photographs as pertain to each one's more especial class of investigation. Many interesting cases of skin disease, fractures, and results of important surgical operations have been fully illustrated by series of photographs, which give opportunity for comparison and study not offered by any other means."[54] Motion pictures promised the same opportunity for comparison "fully illustrated by a series of photographs," so to speak. Calls for an archive of film records were common. In his important essay on the use of motion pictures in neurology, Hennes saw the immediate value of a permanent storehouse of medical films: "If one were to collect the most interesting cases in this way with universal participation and support, a cinematic archive, analogous to the phonographic one, could be created, the lasting value of which could certainly not be denied."[55] Franz Goerke went beyond medical cinema in his call for an archive:

> Is it not therefore obvious that we should take up the idea of creating a central collection place, a state archive for the deposit of motion pictures and so save them from destruction? Just as museums are collection places for works in the fine arts, just as the developments in the sciences have their place in city and state collections, just as libraries are protecting the intellectual achievements of mankind, and just as documents and government papers are kept in state archives, so equally important is a state archive for motion pictures, which will allow researchers in future with the help of other sources and sharply drawn conclusions to reconstruct a visual image of former times.[56]

For researchers of all sorts, not just physicians, the idea of motion pictures as documents led inevitably to the impulse to collect and store them. As we shall see, this impulse was not exclusive but was especially pertinent to medical thinking. Indeed, of all the possible hopes the medical community had for photography and film, the dream of a universal and portable archive of cases was the most persistent.

THE EDUCATIONAL FUNCTION

Yet showing films was not as easy as it might seem. Despite the obvious suitability of motion pictures for demonstration or reference, film was simply not used as often as magic lantern slides in the lecture hall during the early period. Most universities lacked projection equipment, finding it expensive, cumbersome, and difficult to master. The films themselves, despite dreams of a ready archive, lacked reliable distribution outlets; other than the lists of films available from the major producers (Pathé, Gaumont, Eclipse, Urban), there was no systematic information about available films on medical topics. The large firms generally distributed only the films they produced, so researchers such as Hennes had few options for distributing their films; procuring copies of someone else's usually involved a personal connection of some sort. Or, often, the films just did not exist anymore. Goerke complains, "For a learned institute it is nearly impossible to obtain a film for an important learned lecture: we know of the film's existence, but in the meantime it has become a victim of destruction or oblivion."[57] This is why the calls for an archive were so common: educators just wanted a reliable source of scientific and medical cinema. So if the use of film as an experimental tool or as a document attracted the interest of just a few dedicated (and historiographically visible) pioneers—usually attached to well-funded research institutions in centers such as Berlin and Paris—we cannot say the same for the clinical or educational applications of medical film, which were scarce, geographically scattered, and sparsely documented.

Still, most medical filmmakers at this time cite as their motivation a desire to improve teaching. In fin de siècle Paris, Eugène Louis Doyen, a maverick surgeon known for his innovative techniques and disdain for the academy, wrote, "It has been with the object of completing our means of teaching the art of surgery that I have been led to study and employ the cinematograph."[58] Doyen hired two cameramen to film his surgeries. Three of these films—depicting a hysterectomy and a craniectomy—were first presented at the July 1898 meeting of the British Medical Association in Edinburgh, where they were enthusiastically received.[59] Despite this successful screening in the United Kingdom, the French Academy of Medicine refused to show Doyen's films to the membership;[60] he had already run afoul of traditional medicine with his unusual (some would say risky) techniques. His colleagues were not about to let him have a forum. Doyen eventually made dozens of films, including one depicting his surgical separation of conjoined twins Radica and Doodica in 1902.[61] During this time, one of his cameramen, Ambroise-François Parnaland, tried to sell his prints of Doyen's films to Pathé and other companies, but it is not clear whether they were actually widely screened to the public at this early date. In any case, Doyen's controversial techniques and the graphic quality of the films further deteriorated his relationship with academic medicine.⁶² Doyen insisted that his films were didactically valuable; they were meant primarily to illustrate and publicize Doyen's tools and techniques, but they were also intended to serve as training films for surgeons.⁶³

In Germany, the watershed moment came at a February 1910 demonstration of "Film in the Service of Medicine," in which the "service" was primarily educational. Representatives from the Berlin medical establishment were so tightly packed into the Kaiserin-Friedrich-Haus that the organizers had to turn people away.⁶⁴ Among the films shown were two surgical films made by Doyen, the first featuring famed German surgeon Ernst von Bergmann and the second focused on Doyen himself. The principle organizer, Robert Kutner from Berlin, also showed his film demonstrating CPR techniques, which drew praise for their potentially wide appeal.⁶⁵ Other films included James Fränkel's results of corrective orthopedic surgery,⁶⁶ Braun's heart film, X-ray films by Carvallo and others, and Karl Reicher's microcinematographic records of single-celled organisms.⁶⁷ While it was not the first time that films such as these had been shown in Germany, the event received much attention from the national and international medical community and helped to focus awareness on the educational potential of medical film.⁶⁸ In a later speech, Kutner sang cinema's praises:

> And how convenient, how effortless! . . . The evidentiary power of [cinema] is more persuasive than that of any other document, exceeding even the most vivid description. . . . The motion picture projector demonstrates its most spectacular educational applications in auditorium demonstrations of microscopic or macroscopic images of movement. In a normal lecture-room demonstration of movement, especially that of small objects (think, for example, of a frog's beating heart), only a small part of the audience really sees anything, while everyone present can observe a film presentation equally well. Without the assistance of the motion picture projector, almost all X-ray motion pictures and certainly all motion pictures taken from a microscope could only be shown to a small group or even just to one person at a time.⁶⁹

Kutner described the pedagogical advantages of motion pictures in a language common to advocates of the educational film at the time, a rhetoric found in Germany but also elsewhere in Europe and in the United States. First, the rhetoric emphasized the "vividness" of the motion picture, usually in opposition to "description" or, more pointedly, book learning. What exactly is this "vividness"? The clarity, texture, and abundant detail of the photographic image combine with projected movement to give the image a *presence* unlike any previous representational form. Its level of detail allows the photographic image to reproduce patterns of texture and variation, hence to represent the structure and randomness of the natural world, while the movement of the image presents this world in real time in a particularly striking way. The object "lives" onscreen. This is perhaps obvious, but it is all to say that "vividness" referred to the sense of presence that the moving image evokes. For early advocates of educational film, it was as if the thing itself were there in the room, available to direct perception.[70] Film thereby functioned as an object lesson, an acceptable substitute for the thing itself (in chapter 3 I will discuss the educational implications of motion pictures as an "object lesson" more thoroughly).

This permitted Doyen to extol the virtues of the motion picture over books and corpses. Complaining about the inadequacy of "surgery of the dead"—the practice of rehearsing operative techniques on cadavers—to confer a sense of the surgical experience, Doyen asked, "Do our books fill the gap thus left? Certainly not. The most detailed descriptions, the best diagrams or photographs of the various steps of an operation are inadequate. . . . It is not sufficient to follow the operation, as it were, secondhand; rather, the author of the technique, the master himself, must be seen at work. The surgeon is judged by his work, and no text-books, however well-illustrated, can sufficiently express his personality."[71] In motion pictures, on the other hand, Doyen found a perfect medium to express vividly the personality of "the master himself." Movies were not "secondhand"; they allowed Doyen to be "present" to the students. Especially noteworthy is Doyen's concept of "personality." Doyen was not publicity shy, by any means, but he was not concerned here with conveying via a medical film his charisma and good looks, or not only those things. Primarily, his films were meant to present his technique and his tools. More specifically, they demonstrated how Doyen held himself and how he moved to accomplish his task. Film provided, better than any other previous medium, a demonstration of the actual movements required in surgery. Doyen's "personality" is his "posture" or "attitude"—his *embodied* technique. To convey that "personality" was to presume that the student would copy it, that while the student watched the film, a kind of kinesthetic empathy took place whereby the movements seen were somehow felt or incorporated into the student's own body. This is the mimetic presumption of most training films, it seems; many training films expect that we will copy their movements and that the student will take on the "personality" or "attitude" of the master.

Second, Kutner emphasized the *efficiency* of the moving image for the pedagogical task. "Efficiency" was a mantra repeated by all sorts of social engineers in Europe and the United States around 1900; the medical community also chanted its name, especially but not exclusively in the United States.[72] Efficiency was a key concept in transforming the turn-of-the-century hospital from "a well of sorrow and charity" into a "workplace for the production of health."[73] In the United States, for example, from around 1900 to 1920, health officials were increasingly dissatisfied with the duplication of services, the lack of coordination of units, and the general low level of effectiveness in patient care among clinics, dispensaries, and hospitals nationwide. "Efficiency" became an institutional logic to promote standardization of facilities, services, and administration. In fact, in the United States at least, efficiency was the rubric through which the modern hospital adopted business practices to establish itself as a desirable place for treatment and to attract paying patients.[74] *Modern Hospital*, the organ of the American Hospital Association, devoted itself to promoting economy and efficiency in hospital management, while the American College of Surgeons was established initially to focus on the standardization of tools and techniques within surgical practice.[75]

Kutner's praise of film also relied on this concept of efficiency. Here he refers, of course, to economies of scale: the simple claim that more people could see a large projected image than could see a small image. As medical school enrollments grew steadily toward the turn of the century, this claim gained traction—lecturers used projected images more and more from the 1870s onward. But Kutner had in mind another form of efficiency; when he wrote "how convenient, how effortless!," he was probably not referring to the motion picture apparatus, which was definitely not convenient and effortless. Instead, he was referring to the efficiency of the image itself. It had a "persuasive evidentiary power beyond that of any other document, beyond even the most vivid description." For Kutner and others, that power came naturally to the image, especially to the photographic image. It worked quickly and effortlessly on the spectator. When Doyen insisted that "with the cinematograph we can make hundreds of people follow in one minute what a whole lecture could not make clear to a limited number of students," he made a similar claim about cinema's efficiency: not simply about numbers of students but about the immediacy of the image versus the indirectness of the spoken word. If the image was direct, instantaneous, vivid, and penetrating, then the written or spoken description was aloof, dull, and circuitous. In a way, Kutner's and Doyen's preferences echo a bias common in modern medical education. The nineteenth century continued a long transformation in medical education (and education in general) that emphasized direct perception of the objects of study over their mediated presentation in books. The discussion of film as an educational tool made this bias even more explicit. The direct perception of objects was seen as a much more effective and efficient mode of learning. The image was perceived as efficient because it could affect the viewer immediately, like a drug, or a blow to the head, whereas the word must be read or spoken and then processed cognitively, all of which takes time. The image was understood as physical and immediate, while the word was intellectual.

Precisely these perceived qualities of the filmic image—its presence, its directness, its immediacy, and its ability to prompt mimesis—made it an especially efficient training tool in the eyes of physicians and others. Protected by the twin firewalls of expertise and closed community, medical professionals could extol the virtues of film for educational goals. Once films such as these escaped these protections, however, they had the potential to become scandalous, as Doyen's example demonstrates. Furthermore, these cinematic properties *themselves* were condemned as part of a system of commercial amusements. As we shall see, the objection to cinematic spectatorship revolved around precisely the same constellation of properties; the difference was not only the type of films (fiction versus nonfiction) but also the protections in place for regulating cinematic projection. Under the watchful eye of the educator, film's immediacy could be mediated, paradoxically, by the written or spoken word. The educator provided an expert framework, such as a lecture and projection control, that corralled and attenuated cinema's more vivid effects. In the movie theater, however, the content and its immediacy were left unchecked, which often led, these experts complained, to mimesis of a criminal sort. The paradox of the rhetoric of the educational film was that even though the moving image exemplified the very image of efficiency, that efficiency cut both ways, for good and for ill. The difference between these two visions of film was also the difference between observation and spectatorship, which is the subject of the following sections.

MOTION PICTURES AND MEDICAL OBSERVATION

The exploratory, documentary, and educational functions of the medical film give some indication of how motion pictures were and are used by the medical community. But they do not, by themselves, present a clear picture of how motion pictures fit established, disciplinary observational models, nor do they explain the shrill and regular objections to movies in the theaters. The next section will examine those protests in detail, but the argument of this section is that to understand those complaints, we must first understand the researcher's investment in a particular mode of viewing related to medical and scientific observational practices. The condemnation of "movies" in the public sphere and the praise of "motion pictures" in the scientific context have in common a stance about the proper way to process visual information. This section outlines that position by exploring how researchers processed data in the motion pictures they used for medical purposes. On one hand, as we have already seen, medical researchers found the ability to analyze movement frame by frame to be an enormous advantage of motion pictures. On the other hand, the uses for motion pictures were not exclusively analytic—the moving image itself was also very important for medical purposes; the efforts to perfect cineradiography are a good example of this application. In either case, however, the patterns of use reveal a need to regulate the insistent temporal push of the moving image. The patterns also correspond to certain common features of medical observation, especially the importance of the series to medical logic and the claim that observational accuracy depends on the observer's willingness to be methodical and leisurely.

In other words, I argue for a homology between observation and film; more precisely, there are certain features of film—namely, its repeatability, its photographic detail, and its ambidexterity between movement and stillness—that correspond to features of observational practice emphasized in the medical literature. Moreover, certain aspects of observation, such as correlation and description, could be easily adapted to film technology. Selected observational practices appear to have found purchase in the cinematic image. This section will chart these correspondences.[76]

My approach to the argument will be unorthodox, however. Rather than survey the history of medical observation and compare it with early uses of motion pictures in medicine, I will take one statement as representative of all. In an 1897 presentation to the Society of German Natural Scientists and Physicians (Gesellschaft deutscher Naturforscher und Ärzte), Ludwig Braun listed six characteristics of cinematography most important for his work.[77] Braun provided, at the very beginning of the medical community's interest in the technology, an articulate and representative example of the discourse on medical film. Taken collectively, all six theses recognized and applauded the exploratory, documentary, and educational functions of motion pictures in the medical arena. The film provided a "good, detailed" document that could be used as a working object or substitute for the heart itself. The nature of the medium allowed frame-by-frame exploration of the movement of the heart. And the film could be presented in an educational setting, where everyone could see the movement clearly. Throughout the history of medical film, these have been the primary advantages that advocates have listed in favor of using this technology. Braun's list can therefore function synecdochially for the larger discourse on scientific and medical film. But it also demonstrates quite clearly the elements of medical observation most germane to his use of film. So in what follows, I examine each of his theses as a pretext to an explanation of medical logic, the relationship between analysis and synthesis in medical observation, and the temporality of the medical gaze.

1. *The cinematograph delivers a long series of chronophotographic images. Each individual image has the virtues of a good, detailed photograph. The film can record the movements of a normal-colored, physiologically moving heart just as well as those of a heart that has been purposefully manipulated in various ways that serve the specific experiment.*

At first, the last sentence appears out of place; its discussion of film's ability to record a manipulated organ equally as well as one that has not been altered seems both obvious and odd, given the emphasis on the photographic character of the medium in the previous sentences. But it is precisely the history of the use of photography in medicine to this point that prompts the statement. Despite the popular conception of photography as a reliable document, in 1897 there was a wide gap between what one saw in an organ and what one got in a photograph. The extreme variability in film emulsions, for instance, along with equal variability in photographic expertise meant that "standardization was not one of photography's strong points" in the nineteenth and early twentieth centuries.[78] For example, until the introduction of orthochromatic and panchromatic photographic emulsions in 1884 and 1905, respectively, the reddish-yellow color of anatomical specimens came out too dark in photographs unless special precautions were taken.[79] So if a researcher were interested in the color or texture of an organ, some sort of manipulation was required to create a good photograph. Braun emphasized here, however, that when using motion pictures, movement is the object of study, so one does not *need* to make these adjustments, although that option is still available. For the purposes of recording a "physiologically *moving* heart," film's problems with color and texture were less relevant, thereby giving motion pictures a hidden advantage over the "good, detailed photograph."

Nevertheless, we cannot ignore the importance of the "good, detailed photograph" in Braun's scheme. The photographic character of the image was the foundation upon which Braun built his argument for motion pictures. But the most important aspect of the image was its *repeatability*: that "each individual image" could be examined as a photograph and that there were a series of them available for that purpose. This statement is one of the clearest indications that, as Lisa Cartwright has noted, medical researchers at this time used film primarily as series photography.[80] But why was series photography interesting for physicians? How did it fit their disciplinary needs? Series photography presented, in a representational form, what we might call the "statistical" aspect of medical logic. Ludwik Fleck, in a 1927 article on "Some Specific Features of the Medical Way of Thinking," argued that the fundamental cognitive task of medicine is to find a law for irregular phenomena. Medicine attends to the human body, which is a wildly variable and stubbornly individual object. Finding some sort of consistency in this variability, especially when focusing on atypical, morbid phenomena associated with disease, requires the examination of many, many cases. The clinic was the prototypical site where this observation occurred; physicians used "statistical juxtaposition and comparison of many such phenomena" to arrive at an "ideal" picture of the disease. Hence, "the role of statistics in medicine is immense. It is only numerous, very numerous observations that eliminate the individual character of the morbid element" and allow a regular "law" to emerge, to the extent that this is possible. Medical logic relies on series of cases found in clinics to come to an understanding of the "facts" or symptoms of a particular disease. Through the repetition of these facts and their variation, the researcher begins to see the pattern that becomes the ideal picture of the disease.[81]

Certainly, observing a patient in a clinic is much different than looking at a photograph or a film. The physical presence of the patient calls for not just a visual examination but palpation, percussion, auscultation, and other physical diagnostic techniques, not to mention the active questioning of the patient, none of which can be done to a photograph. But as Bernike Pasveer explained, part of the process of making images diagnostically viable involved encoding the images with information gathered by physical examination. That is, a successful X-ray image of a patient with pulmonary tuberculosis was able to show information (such as position of the lesions) that one could render or confirm from auscultation, for example. Learning to read the image required correlating sight and sound. The image thereby became a visual analog of the sounds and sensations obtained by physical examination. At that point, the photograph became a legitimate working object and could be compared with *other* photographs. Photographs could be compared like cases, but the photo could even become the basis of comparison of these cases, in that the image eventually standardized a vocabulary and a way of looking at the disease (and the patient).[82]

So series photography acted as a formal analog to the series of cases in a clinic, without having either the clinic or the actual cases present. Photography functioned "statistically" by presenting a series of images of different cases to be compared, or different views of the same case, either spatially (for example, different angles of view) or temporally (for example, tracking the stages of a disease as it develops).[83] However, it is very important to emphasize that at this point in the development of medical cinema, the diagnostic value of photography, series photography, or cinematography was very limited.[84] (Even the diagnostic value of X-ray images at this time was restricted.)[85] Usually the image functioned as a document of an already established diagnosis or as the presentation of a question for others to answer.[86] But the image itself did very little work in the detection required to establish an etiology of disease. Yes, the archive of images did help in the recognition of *patterns* of disease—a series of photographs might declare, "Here is the course smallpox takes over nine days," for example. But the cause of the disease remained hidden to the camera. In this respect, the value of series photography and motion pictures for Braun and others was more experimental than diagnostic; the images functioned to control variables, to manipulate time, and to measure phenomena more than they detected causes or invisible relationships. It was only with the development of radiography and cineradiography that this possibility was broached, but the true diagnostic potential of the moving medical image was still decades away.

Series photography was much different than still photography in that the series was capable of depicting the duration and transformation of disease.[87] But series photography and Braun's use of cinematography were similar to still photography in one clinic-like respect: they all functioned as an aid to *correlation*, by which I mean establishing a mutual or reciprocal relationship between objects or events, such as correlating lesions in a cadaver to disease elements in a living patient. The clinic, as a collection of similar cases in a single location, functioned as an archive of the common symptoms and development of a disease. Series photography worked in that way as well. Photography's ability to isolate, frame, and repeat similar cases was a powerful aid in the standardization and multiplication of observational views. In addition, the arrangement of photographs in a series allowed not only their sequential organization but also their *simultaneous* presentation. Georges Didi-Huberman has argued that Charcot's arrangement of his patients into living tableaux functioned like tables of data by organizing their signs into simultaneous events.[88] Photography allowed this same organization, and much more easily. In series photography, the sequence was important, because it suggested a causal order or chronology, but the simultaneous display of images was arguably equally essential to the process of comparison and correlation. Series photography, as a research tool that allowed both sequential analysis and simultaneous display, thereby succinctly articulated the ideals of medical observation and logic.[89] Braun's use of motion pictures as series photography also matched this logic, especially in the way he compared the images, as the next theses demonstrate.

2. *The study of the resulting—especially the enlarged—images allows the analysis of movement, the recognition of every intermediate state and every phase of the process, and with that, a more exact assessment of the resulting transitional steps than was possible before.*

3. *The individual photograms are very uniform. If one lays two successive images on top of each other, those parts that remain motionless match perfectly, while the moving parts show positions that correspond to the differences in their movement. From this, one can perceive the spatial displacements, judge them better than before, and to a certain extent measure them and calculate the speed of the displacement in space from the time of exposure and the number of exposed frames.*

Clearly, medical observation is not simply passive looking. Observation calls on more than vision alone—it requires all of the senses combined with an extensive linguistic and logical apparatus. Foucault described "the clinical gaze" as a perceptual act sustained by logical operations,[90] and correlation is one of its primary logical maneuvers. Medical photography and film functioned in many ways, but they operated primarily, I argue, as an aid to correlation. In fact, with his detailed description in thesis three of how he compares individual images, Braun offered an example of how correlation works. He laid two images on top of each other, aligning their similarities; from this alignment, the salient differences emerged. Likewise in the clinic, the repetition of cases over time created a cumulative, collective "image" of the disease, in which the more or less constant aspects of the entity were stabilized and the variations were set aside for further study, then reintegrated into the "picture" of the malady.[91] Medical photography abstracted and accelerated this process; the number of photos of a particular affliction accumulated and created something of a "virtual" clinic, an archive of cases to be aligned and compared. Series photography worked the same way on a smaller scale by providing images to align and compare so that variations in the single case could be correlated. Cinematography worked on an even smaller scale: it created images not just of a single case but of a single moment. But the point here is that one of the key features of motion pictures in this mode was not necessarily the depiction of movement but *the generation of images for comparison*. That is, for Braun's minute research on the mechanics of the heart, ever more images were needed to make meaningful correlations. Cinema was useful in this case not because it depicted movement but because it functioned as an "instant archive" of images for comparison.

Braun was mostly interested in measurement; he was working toward a precise tabulation of the dynamics of the heart. So his second thesis emphasized cinema's ability to aid the "analysis of movement." The comparison of images that he outlined was designed, like Braune and Fischer's apparatus discussed in chapter 1, to extract data for the measurement of movements. The paradox, of course, is that his analysis of movement entailed, as we learned from Bergson in chapter 1, the comparison and measurement of different states of rest. When Braun emphasized "the analysis of movement," he focused on various stages of motion or "transitional steps." Those intermediate stages, however, were actually states of rest, or at least they functioned that way as still images. The analysis of movement was therefore a carefully designed comparison of various states of stillness. Just as paradoxical is the equation of "intermediate states" of movement with the film frame. That is, if analysis implied the "recognition of every intermediate state," then it is not going too far to suggest that those states, those "resulting transitional steps," to some extent resulted from the process of analysis itself. As the heart beat, the film frames divided and presented its natural articulations. The smallest possible "step" was between two frames; another step might have been between several. But in every case, the frame was the measure of the transition; the "event" was equated with the temporal boundaries implied by the spatial boundaries of the frame. Chronophotography, for example, was essentially a record of successive phases of motion: in deciding the time between exposures, the researcher already conceded that X amount of time would equal a single "phase." This is as clear an indication as any that the technology the researcher used helped to shape his or her conception of the object. Braun's understanding of movement depended at least partly on his understanding of motion pictures. Indeed, it also appears at first glance that

Braun's understanding of movement and continuity depended entirely on a method based on stillness and discontinuity, à la Bergson. This, however, was not always the case, as the next three theses indicate.

4. *The cinematographic shot can be used to present to the observer any movement of the heart synthetically, at will, and even decelerated to a rather great extent without impairing the clarity of the images.*

When Braun wrote that cinematography can present "any movement of the heart synthetically," he meant it can show actual movement not the decomposed movement of analysis. If analysis breaks down movement, synthesis is the reconstitution of movement, the projection of a moving image. We could argue that the relationship between film's still images and their movement in projection corresponds to the relationship between analysis and synthesis in scientific method. In that tradition, synthesis functioned as an important check on analysis; analysis decomposed a phenomenon into its constituent parts, and synthesis reconstituted those parts to verify that the analysis was correct. Synthesis appears therefore to be only a supplement to the more valuable labor of analysis. This was how Marey used cinematography: as a way of training expert vision by decomposing movement and building it slowly back up through synthesis, which taught the viewer what to look for when the movement was encountered again. If he was dismissive of the moving image, having no interest in it other than as a verification of his analytic results, historians of the scientific film have often taken that dismissal to be definitive. But that ignores the real scientific value—not to mention the pedagogical value—of the moving image. Many other researchers, including Braun, have noted the importance of the moving image for understanding their objects of study.[92] Thesis four gives an indication of that value, which in this case was primarily correlative. As we saw in the second and third theses, motion picture technology generated an "instant archive" of images for comparison; in frame-by-frame analysis, this comparison was used most often for measurement: the images were used to track spatial displacement. Braun used moving images not for measurement but for correlation of a different sort. The temporal malleability of film—it could be "decelerated to a rather great extent"—created a new set of views for comparison. That is, just as the proliferation of images generated many *spatial* displacements to be compared, so the variability of projection speed generated many *temporal* "views" to be evaluated. Different projection speeds revealed different aspects of the movement, which were collected and compared for a better understanding of the whole. This is a rather common procedure in the scientific appraisal of research films, which demonstrates that while frame-by-frame analysis may be the best way to derive quantitative data from film, it is clearly not the only kind of information that researchers get from motion pictures. Synthesis was not merely a supplement to analysis, but an equal partner.

5. *The reproduction of movement in slow motion grants the observer more time to recognize significant features, to evaluate precisely the elements of movement, to determine whether or not all the parts of the heart start moving at the same time, and to compare the contraction process of the various parts of the heart, especially of both ventricles.*

In thesis 5, Braun goes on to describe exactly how slow-motion projection of the images helped his project. It allowed him to recognize, compare, evaluate, and decide on various features of the heart's actions. How did it do this? The ability to manipulate the speed of the presentation gave him "more time" (*längere Zeiträume*). This is arguably the key feature of scientific and medical filmmaking, the reason any researcher chooses motion pictures over other representational technologies. To comprehend—from the Latin "to grasp"—any complex action requires that we slow it down and bring it within reach of our senses and limited processing abilities. Motion picture technology allowed that to happen. It gave the researcher "more time" to contemplate the details of the actions under study, especially when he or she exercised his or her exclusive privilege to control the pace of projection. In scientific filmmaking, as in experiment, the inexorable forward push of time and of the motion picture can be manipulated, slowed, stopped, accelerated. This was also true for narrative or any other kind of filmmaking, but the difference rested not in the technique but in the ability of the scientist to control the image, an ability not available to the spectator in the audience. I will have more to say later in the chapter about how this scientific approach to time becomes normative when applied to movies in the theater, but for now we can simply note the importance Braun and others give to cinema's ability to make "more time."[93]

This ability to control or to extend the time for examination of the event was a crucial advantage for researchers. Before exploring its manifestations in the medical research film, I want to point out how this ability related to established practices of medical observation. I am especially interested in the temporality of observation, which is addressed in historical discussions about proper method in medical observation, in Foucault's discussion of "the clinical gaze," and in Michael Hau's concept of "the holistic gaze." First, method: throughout the nineteenth century, physicians insisted on the difference between careful and careless observation, which depended on the ability to diligently apply a prescribed method. Textbooks at the time sought to outline this method for students and junior practitioners. British physiologist Thomas Laycock wrote one such text, in which he made it clear: "The foundation of medical experience is observation of disease, and the requisites to successful observation are minuteness and accuracy. The clinical student must therefore make up his mind to be sedulously minute and carefully accurate in investigating the cases under his notice."[94] German pathologist Rudolph Virchow described in detail his method of performing autopsies. Autopsies conducted in a haphazard way promoted interpretive error, he argued, so he "drew particular attention to the necessity of insisting—in autopsies for medico-legal purposes, as in everything else now—upon completeness of examination and exactness of method, both in the investigation and in notetaking, so that it might be decided subsequently, though not in anticipation, whether there was any significance or importance in what was observed, or whether it was accidental and unessential."[95] Virchow was particularly careful to describe exactly what should be observed in an autopsy and the order in which observations should be made. Method therefore precluded haste. An observation could be made quickly, to be sure, but over and over physicians warned their students against the dangers of hasty observation. Instead, observation was to be practiced carefully, leisurely, with an eye to detail (at least until one was trained).

Indeed, detail itself prohibited hasty observation; if one properly attended to and assimilated the details, it took time to do so. Foucault argued that "the clinical gaze" emerged from precisely this intersection of vision, detail, and language. With the nineteenth century, according to Foucault, "Rational discourse is based less on the geometry of light than on the insistent, impenetrable density of the object."[96] That is, if Western thought since Descartes equated truth with light and light with an abstract ideal, then during the nineteenth century "the solidity, obscurity, the density of things closed in upon themselves, have powers of truth that they owe not to light, but to *the slowness of the gaze* that passes over them, around them, and gradually into them, bringing them nothing more than its own light."[97] The clinical gaze cast its own light, searching across the dense landscape of information in the clinic and separating the essential from the inessential. So for Foucault, this careful gaze was inseparable from the process of analysis. Foucault characterized the analytic aspect of medical perception as a process of simultaneously recognizing, separating, naming, and acting upon some disease element.[98] The clinical gaze was a process of seeing, thinking, and naming, which therefore became a deeply contemplative maneuver that took its time. German surgeon and professor Theodor Billroth emphasized the imaginative element in this process:

"Solitary, meditative observation is the first step in the poetry of research, in the formation of scientific phantasies, the reality of which we then test with the tools of logic, mathematics, physics and chemistry. Our tests will be the more successful the better we have learned to handle these tools. The diseased organism, the patient, must be observed in just this way, thoughtfully, and in a state of mental solitude and meditation."[99] The abundance of detail, the richness of the information—what Foucault calls "the plentitude of concrete things"[100]—was at least partly the reason for this emphasis on a contemplative, solitary, slow, and steady gaze.

But physicians characterized medical observation not only in these terms. Michael Hau has described medical observation as an active, gestalt-like process of holistic apprehension; he argues that many German physicians at the turn of the century objected to the analytic, overly scientific approach to observation that had become fashionable in the medical community. Threatened by the new technologies and methods common to laboratory science, many physicians took refuge in an explicitly artistic approach to observation, emphasizing their "ability to 'see' and to synthesize seemingly fragmented characteristics of a human body into an aesthetic, coherent whole."[101] By making proclamations about the relationship between health and beauty, these physicians stressed an aesthetic approach to the whole body to differentiate themselves from the more specialized and decidedly more fragmented approach of, say, bacteriologists. If the "holistic" gaze was more immediate and intuitive, that might distinguish it from the contemplative and correlative gaze described earlier, but it could also describe the brand of medical perception Foucault calls "the glance." Foucault explains the difference between the clinical, analytic gaze and the more synthetic glance: "The gaze implies an open field, and its essential activity is of the successive order of reading.... The glance, on the other hand, does not scan a field: it strikes at one point, which is central or decisive; the gaze is endlessly modulated, the glance goes straight to its object. The glance chooses a line that instantly distinguishes the essential."[102] This form of the expert eye is similar to what Lorraine Daston describes as "all-at-onceness" in scientific observation: an intuitive, immediate understanding of the relationships presented by an image or by phenomenon.[103] This distinction between types of medical perception is useful for our purposes, because it highlights the duration or temporality of observation: the difference between the searching, contemplative gaze of the explorer and the immediate, intuitive glance of the connoisseur.[104] With regard to motion pictures, Braun found the first mode of observation especially valuable, because it corresponded with film's ability to make "more time" to recognize, compare, evaluate, and decide. But neither Marey nor Braun dismissed the glance or the synthesis of images entirely.

6. *Finally, the cinematograph can project single images to a large auditorium, and can also present animated images as "living photography" by showing the entire chain of images synthetically.*

While they perhaps do not immediately correspond to the experience of motion pictures, Foucault's "gaze" and "glance" remind us that medical (and scientific) observation was not only analytic but also encompassed a more synthetic, holistic, or intuitive element. Likewise, Braun's use of motion pictures both as still photographs and as moving images should stress the equal value of both for medical cinema. The individual frame was very helpful for measurement or correlation, while the projected image had important rhetorical, educational, and even documentary uses. It makes little sense to separate them; we should, in fact, consider the relationship between the still image and the moving image as a vital dialectic for medical understanding.[105] Physicians went back and forth between the two forms of the medical film in a productive, yet prescribed way that preserved and proved their expertise. Indeed, Braun's sixth thesis is all about expertise, however obliquely. Motion pictures could be used in the auditorium, where they could instruct and dazzle the audience with the latest in medical knowledge and technology. Imagine Braun (or others) giving a medical lecture on heart function, in which he has control of the film image. Braun could present his material one frame at a time, as a slide show, carefully delineating for his audience the precise difference between one phase and another. He could take as much time as he needed, with the image still projected, to describe the salient features of the organ's mechanism. But Braun could also capitalize on the rhetorical power of motion pictures by turning, with a flourish, the single image into "living photography."[106] This provided the medical students with another kind of information, while at the same time powerfully persuading them of Braun's argument. But even the spectacular aspect of the moving image was attenuated and contained by two barriers mentioned earlier in this chapter: expertise ("we know what we are looking for in the image") and closed community ("this knowledge is exclusive to us").[107] That is, the seriousness of purpose implied by the specialized educational setting contained the full power of the moving image. Even so, presenting "the animated image as 'living photography'" implied an almost Doyen-like commitment to showmanship, mastery, and self-promotion. Through the alternation of still and moving images, of analytic and synthetic approaches, surgeon became magician and back again, experts both.

It is also worthwhile to note the cultural capital invested in the gaze and the glance. In textbooks and treatises on medical observation, authors stressed that the expert eye did not come easily: "Aspiring physicians will never be able to develop an eye for diagnostics from books. That can only be learned through careful observation and constant contact with his patients."[108] Here the author emphasized a truism of medical education in the nineteenth century: direct observation was preferable to, even precluded, the scholastic, bookish approach that ruled medicine for so long. But he also confirmed several aspects of the clinical gaze: it was contemplative, it relied on the accumulation of cases, it was learned only through vigorous training. On the other hand, another author described the glance: "It is that piercing gaze, which unconsciously separates the essential from the unessential and penetrates to the bottom of things with ease and certainty, without requiring difficult research, which is easily misled by irrelevancies that are easily apparent, while important things that elude perception are overlooked."[109] The glance was decisive, confident, second nature. Like the gaze, the glance was acquired only after long training and experience. While Foucault characterized the distinction between the gaze and the glance as a historical progression from one mode of observation to another, and Hau described it as a cultural or disciplinary differentiation, I would suggest that the duo, like stillness and movement, constituted a set of alternatives or a dialectic, each always part of the larger category called observation. But the character of the two does not ultimately concern us here, nor even their relationship to each other, as much as their function as badges of professional expertise. The cultural investment in these modes of observation is palpable, as indicated by the numerous texts on observational method or by the number of debates about the specificity of clinical observation, especially when threatened by new technology (the sphygmomanometer, for example).[110] Observation, in whatever form, was central to the German physician's professional and class identity. The explorer and the connoisseur may have had different skills, but they invariably came from the same place and were quite comfortable in each other's company (or even in the same skin).

This exercise in textual analysis has been designed to highlight the three-way correspondence between certain formal features of motion pictures, particular techniques used to examine them, and specific principles of observation. The way Braun used individual film frames, for example, illustrates and corresponds to the principle of correlation in observation. These correspondences demonstrate that the acceptance and legitimacy of motion pictures in science and medicine depended not merely on the availability of the technology but also on its ability to conform to established disciplinary imperatives. At this historical moment in the development of Western medicine, motion picture technology partly addressed a number of representational and observational issues, including the increased demand for rigorous observation in medicine and the rising interest in the documentation and analysis of complex movements. (So it is perhaps not surprising that the number of research projects on pathological movement, for example, increased as film was accepted as a tool for studying it.) Beyond the question of legitimacy, these correspondences also demonstrate, I hope, that observation is a complex operation and that we can learn much about it by studying how film is used. Specifically, it is clear that for nineteenth- and twentieth-century scientists, medical researchers, and scholars, "observation" was considered to be primarily *a self-disciplinary method of ordering thought*. As a means of protecting the community against the chaos of irrational and biased subjectivity, observation was closely aligned with experiment as a universally adopted method of enforcing professional standards. Observation was a prescribed method of organizing information through such means as stabilizing and isolating variables; measuring, describing, and classifying objects and events; and arranging and correlating these items or events to arrive at verifiable conclusions. This process is what it meant for physicians to be "scientific" as Paget so forcefully insisted: "In our calling careful practice and scientific study should be inseparable"[111] (which also emphasizes the futility of separating experiment and observation).

Film worked as a research instrument to the extent that it could conform to or inform this method. In particular, film was an aid to correlation. The motion picture camera was a tool for generating images or views for comparison, either comparison of the individual frames or comparison of different temporal "views" via slow-motion or time-lapse cinematography. Alongside this feature, film was praised as something that gives the user "more time," either by storing material for convenient retrieval, by reproducing and repeating events, or by offering variable filming and projection speeds. One expert applauded film's utility in the medical classroom:

> Cinematography could help with that. The images can be shot with all the time in the world, using specimens that have been prepared with the utmost care, and are then permanently available. Instead of the uncertain outcomes of fleeting experiments, the serial images can be shown repeatedly, which not

only makes a greater impression on students, but by presenting the frames individually, can also provide an analysis of the phenomenon.[112]

Motion pictures gave more time to the researcher and the student alike—more time to prepare, more time to analyze and comprehend. Especially important was the flexibility of film with regard to the dialectic between analysis and synthesis; if researchers have always had analysis and synthesis, only with motion pictures did they have that choice in the same representational system.[113] Alternating back and forth between the still image and the moving image, researchers took advantage of cinema's ambidexterity to present an "impressive" yet scientific picture of the phenomenon. Yet this ambidexterity was not merely an advantage of film—it signaled a hermeneutic strategy that in part became standard procedure because of it. The cognitive gains that resulted from going back and forth between the still and moving image—which have been known ever since a mental image could be sketched—were codified into experimental and observational method with the advent of motion pictures. But more to the point, the scientific legitimacy of this alternation—that stillness and movement could be *equated with* analysis and synthesis—was to some extent due to the availability of motion pictures as a research tool that could offer both options. So film was not simply a useful tool for understanding complex events: the use of film actually made manifest a mode of understanding. The ambidexterity of film operationalized a way of thinking and made it visible to the community in both images and practice. The motion picture camera was not just an instrument, but a theoretical approach.[114]

But I want to return to the researcher's obvious glee, which the quotation above by K. W. Wolf-Czapek vividly demonstrates, in being able to manage the flow of time by controlling motion pictures. Wolf-Czapek's cheerleading betrays a certain emotional investment in this potential, as if it were crucial to the lecturer's pedagogical success or the physician's identity as a researcher. Controlling the passage of time almost became the very definition of "scientific" because of all it implied about careful preparation, leisurely observation, accurate description, and deliberate correlation. But as we saw with Braun earlier, the giddiness that the researcher felt toward the moving image was not only about the control of time, but the submission to it as well.[115] Or more precisely, the moving image was thrilling in itself: its invitation to mimesis, its temporal rush, its presence. Controlling the moving image completely would also make it less spectacular, less effective as a persuasive representation. So the researcher had to negotiate this duality by deciding how and when to use the motion picture in its two forms. Various techniques in the laboratory and the lecture hall—such as frame-by-frame analysis, slow-motion and time-lapse cinematography, varying projection speeds, looping a single shot so that it repeats, or tracing an image to abstract and manage its detail—help to manage the thrill and temporality of the moving image (even if these techniques come with their own sensory pleasure). But even when the image was projected in all its glory as a thrilling example of "living photography," the conventions of the lecture attenuate that potential. And it goes without saying that a film about pathological movement did not have the same immersive potential as a feature narrative. Physicians recognized the full impact of the motion picture, to be sure, but they did their best to manage that impact in the laboratory, the classroom, and the professional meeting. Was it simply a generic demand? That is, was it simply the case that any representational form would have the same restrictions? Other media did have similar boundaries of decorum and professionalization. But cinema was a special case unrelated to its content—it *moved*, and this meant that there was a little bit of carnival even in the most staid presentation or most boring research film. With the possibility of this kind of presentation, what signified "scientific" or "professional" depended especially *on how one managed the passage of time*. If scientific experiment always implied some control of the temporal variable, with motion pictures that control became explicit and essential to scientific identity, especially in the face of its mass culture incarnation. Hence the medical community's swift and vicious reaction to "the movies."

TIME, SPECTATORSHIP, AND THE WILL

Movies were only one of many ailments of modernity, of course, and in the scheme of things, they took up no more editorial space—probably considerably less—than complaints about trashy literature, modern art, lurid advertising, trolleys, trade unions, or other by-products of industrialized life in the metropolis. But with their forceful entry into public consciousness, starting around 1907, motion pictures touched a very particular nerve among the educated elite. The previous section exposed one tendril of that nerve: the professional and emotional investment in methodical, leisurely, observational practice attached to scientific method. The pace of motion pictures threatened that practice; more precisely, for most commentators the (perceived) quickness of movies endangered a cultural tradition of learning, expertise, and status that invested heavily in studious attention, methodical description, and logical reasoning. Modernity jeopardized that investment in a variety of ways, but motion pictures best represented the problem of social acceleration: movies, like change itself, moved too quickly for comfort. They therefore put attention and reason at risk and, in that way, exemplified a problem of national importance. In their editorials about movies, physicians discussed the problem of cinema's pace in the context of this investment in attention and reasoning. They also addressed the objectionable subject matter of many films and the unhygienic conditions of many movie theaters. Physicians joined reformers in their protest against the number of "trashy films" (*Schundfilme*) that exposed themselves to a dazed public in seedy storefront cinemas (*Ladenkinos*). This final section will examine some of these complaints about the physical and psychic or moral threat of motion pictures. I want to provide a survey of the variety of complaints, but I also want to emphasize *pace* as the primary source of nagging discomfort with movies felt by doctors and other members of the educated middle-class (*Bildungsbürgertum*). This section, then, serves as a segue between the focus on film as a research instrument in this and the previous chapter and the exploration of film as a social tool and phenomenon in the following chapters. Let us start this examination, however, with a brief check of credentials: How did German physicians position themselves as experts on movies, of all things?

As late as the 1850s, academically trained physicians in Germany were not considered experts of much. Their therapies were, by and large, no more effective than those available to the lay public; barber-surgeons, midwives, and folk medicine presented plenty of competition; and a physician's social standing was usually lower than the upper classes he served, which meant that medical advice was routinely questioned and rejected. Claudia Huerkamp has demonstrated that the turnaround was a result of the Prussian state bureaucracy's policy to take control of all guild organizations, to standardize medical education, to license medical practice, to thereby suppress folk alternatives, and to prioritize hygienic practices among the population.[116] These policy decisions effectively eliminated the competition and, especially through the establishment of health insurance in the 1880s, expanded the clientele for academic physicians, particularly into the working classes. As doctors took on more patients from the working classes, their social standing was above the majority of their patients, thereby reversing the previous patient–doctor power dynamic. In addition, the state directives promoted a general "medicalization" of society, in which academic concepts of health and disease were adopted by society at large; social problems that were not previously seen as medical problems, such as alcoholism, were gradually treated as diseases.[117] Furthermore, the academic medical community stabilized and enforced professional standards, especially by adopting scientific methods and techniques, which led to new approaches to diagnostics, anesthetics, and surgery, for example. As their expertise and their association with the craft of science grew, so did their authority.[118] So state policies, the expansion of clientele, the growing professionalization of medical ranks, and the association with scientific principles, all combined not only to create a medicalization of society but to place academically trained physicians in position as its chief diagnosticians.

By the 1890s this authority began to consolidate, and by 1914 it seemed perfectly natural to view many social problems within a medical framework and equally natural that physicians were qualified to comment on them. Doctors examined and prescribed for the body politic as well as for the individual. They saw themselves not merely as professionals or experts, but as *Kulturträger* (bearers of culture) who had a responsibility to support, carry on, and protect the nation's heritage.[119] This intervention into the social sphere manifested itself in two ways: more doctors became involved in social medicine or public health, and more doctors wrote cultural critiques from a medical standpoint. For Munich psychiatrist Emil Kraepelin, it was simply a matter of applying one's skills in diagnosis, prophylactics, and therapy to the growing need for public medicine. Such public health issues as alcoholism, tuberculosis, and sexually transmitted diseases increasingly occupied the expertise of academically trained physicians in Germany at the turn of the century. While Germany lacked a national ministry of health until after World War I, individual doctors were very much involved in national campaigns.[120] Kraepelin, for example, was committed to the temperance movement of the day, a typical example of the growing medicalization of social issues in industrialized nations.[121] Likewise, motion pictures were swept up in this trend as certain aspects of the cinematic experience came to be regarded as a public health threat.[122]

Kraepelin's protégé, Robert Gaupp, also directed his skills toward public health issues, such as child psychology and welfare, but he combined his specialized training with his cultural capital and status as a University of Tübingen professor to bring less obviously medical issues, such as motion pictures and pulp fiction (*Schundliteratur*), under the rubric of medicine.[123] Perhaps the most famous example of the medical diagnosis of culture, however, is Max Nordau's *Degeneration* (1892). With chapters titled "The Symptoms," "Diagnosis," "Etiology," and "Prognosis," and various case studies of modernist endeavor, from the symbolists to Leo Tolstoy, Henrik Ibsen, Richard Wagner, and Friedrich Nietzsche, *Degeneration* argued that cultural products of the day displayed classic symptoms of degeneration, decadence, and hysteria, which were "the consequences of the excessive organic wear and tear suffered by the nations through the immense demands on their activity, and through the rank growth of large towns."[124] Drawing in part on the contemporary discourse of "nervousness," Nordau (fig. 2.3) found that the excesses of modernity had physical consequences, the symptoms of which could be read in cultural products.[125] Just as Italian psychiatrist Cesare Lombroso theorized that social deviance was based on biological retrogression—criminality, he surmised, must be due to some primitive remainder or reversal in the normal progression of the species—so Nordau argued that the cultural avant-garde was not progressive and forward-thinking but instead atavistic and regressive, displaying traits common to deviants and hysterics.[126] Using medical norms to evaluate literature and the arts, Nordau combined physiological theories of social deviance, Darwinian evolutionary models, and liberal ideals of ordered progress and rationality into an old-fashioned, curmudgeonly disapproval of modernism.[127] Nordau's work was extremely visible—it was one of Europe's ten best-selling books of the 1890s[128]—but certainly not unique. Darwinian theories of heredity and degeneration; medical or psychological investigations into hysteria, nervousness, and hypnotism; social ferment due to the rise of social democracy, industrialization, and modernization; all combined to create an uneasy mood of cultural pessimism among the educated middle classes of Wilhelmine Germany, which the commentaries of Gaupp and Nordau exemplify.[129]

So the metaphorical extension of "health" and "disease" to all aspects of social life became more than a metaphor for doctors and their readers in imperial Germany. Likewise, the moral and physical threats that motion pictures and other cultural forms presented were also very real for these commentators. One physician's complaint about the dangers of representing the erotic and sexual in film presents a common stance:

> The arrival of the cinematograph brought much damage in its wake. . . . While the danger, which undermines serious moral principles and hence the moral foundations of a man, must be fought, it is, of course, only a threat to the masses who don't think for themselves, for the minors, the young and

FIGURE 2.3. Max Nordau

easily impressionable, who are not used to having their own opinion about things or probing things with critical method. And because these masses make up the majority of our nation, this public, easily accessible sexual fare, which is apt to damage the sexual apparatus [*sexuellen Tractus*], represents a serious *sexual danger*.[130]

Of course, the idea that sexually arousing images or narratives might damage one's moral fiber was not new. But this complaint is especially representative and remarkable for its insistence that "critical method" (*kritischer Methode*) was the difference between the protected and the unprotected. Without it, apparently the masses were at modernity's mercy. But it is not simply that the masses lacked the tools to defend themselves; again and again, medical and lay pundits claimed that movies corroded "critical method" itself.

We find this idea even in editorials that outlined the mere physical threat that motion pictures presented to the average viewer. A Dr. Paul Schenk, for instance, discussed the flicker effect of early cinema, warning teachers who hoped to use motion pictures for educational purposes:

> From the standpoint of visual hygiene, the usefulness of cinema as an educational tool remains dubious. Modern man systematically ruins his eyes. We suffer from an excess of visual stimuli. . . . The much-maligned "flicker" of the cinematic image is a malaise that presently, and probably forever, deprives the cinema of the claim to be a "hygienic" means of instruction. . . . This impression is further strengthened by the unnaturally fast changes of scenery. Our eyes cling intently to the screen, where, in addition to the flicker of the images, a change of scenery takes place almost every minute within a time period that lasts 8 to 12 or up to 15 minutes.[131]

Cinema's educational utility, according to Schenk, was hampered by the threat its "flicker" presented to student health. On one hand, this flicker supposedly damaged the eyes; there was apparently a physical connection between the flicker and nerve damage of some sort.[132] On the other hand, the *pace* of the film intensified this damage. In fact, this physical damage acted as a metaphor or outward symptom of a deeper, psychic damage caused by the temporal push of cinema. Reformers or *Kinogegner* (enemies of the cinema) also took up this diagnosis in a general medicalization of cinema in the discourse of the time.[133] Albert Hellwig, not a doctor but one of Germany's most prolific *Kinogegner*, relied on medical terminology and cited an Italian doctor's discussion of cinema's assault on the sensitive mind: "Of a group of neurasthenics, d'Abundo observed that frequenting cinematographic presentations brought about all sorts of ailments, especially insomnia. It was not so much the influence of the contents of the presentation, but rather an effect of the rapid, constantly moving action and attendant flickering."[134] Another nonphysician, O. Götze, put a finer point on it: "The hasty tempo of the images, and especially the Leipziger-medley-like program [i.e., one resembling a mixed vegetable dish], seduce the child into superficial *viewing and observation* habits. Even the images that could offer intellectual enrichment go by so quickly that it is impossible to form clear impressions."[135] With this observation, Götze articulated the position of the medical community and reformers with regard to the movies. Medically speaking, the cramped, oppressive atmosphere of the theaters, together with the flicker effect, presented a threat to physical health, while the immoral content of the films and their delivery by means of a relentless temporality presented psychological dangers. Cinema's quick pace threatened reason itself.[136]

This complaint about tempo corresponded to a broader contemporary discussion about modernity and "social acceleration."[137] We could trace this feeling that "the world is moving faster" back to the eighteenth century, of course, but objections piled up toward the mid-nineteenth, especially at the fin de siècle. Nordau complained about a lot, but close examination of *Degeneration* reveals that the root cause of the state of things was, for him, the heightened pace of modern life:

> Its own new discoveries and progress have taken civilized humanity by surprise. It has had no time to adapt itself to its changed conditions of life. We know that our organs acquire by exercise an ever greater functional capacity, that they develop by their own activity, and can respond to nearly every demand made upon them; but only under one condition—that this occurs gradually, that time be allowed them. If they are obliged to fulfill, without transition, a multiple of their usual task, they soon give out entirely. . . . To speak without metaphor, statistics indicate in what measure the sum of work of civilized humanity has increased during the half-century. It had not quite grown to this increased effort. It grew fatigued and exhausted, and this fatigue and exhaustion showed themselves in the first generation, under the form of acquired hysteria; in the second, as hereditary hysteria.[138]

So, according to Nordau, the human body was not able to keep up with the increased demands and pace of modern life and grew fatigued, which led to a corrosion of the nerves and, subsequently, hysteria and degeneration, which showed itself in daily life and in modern art. "All the symptoms enumerated are the consequences of states of fatigue and exhaustion, and these, again, are the effect of contemporary civilization, of the vertigo and whirl of our frenzied life, the vastly increased number of sense impressions and organic reactions, and therefore of perceptions, judgments, and motor impulses, which at present are forced into a given unity of time," writes Nordau.[139] Too much information compressed into too little time was Nordau's recipe for decline.

Nordau's antidote for sensory overload was not withdrawal but "attention." Attention was the savior, the critical faculty that brought order to chaos:

> Thus we see it is only through attention that the faculty of association becomes a property advantageous to the organism, and attention is nothing but the faculty of the will to determine the emergence, degree of clearness, duration and extinction of presentations in consciousness. . . . Culture and command over the powers of nature are solely the result of attention; all errors, all superstition, the consequence of defective attention. False ideas of the connection between phenomena arise through defective observation of them, and will be rectified by a more exact observation. Now, to observe means nothing else than to convey deliberately determined sense-impressions to the brain, and thereby raise a group of presentations to such clearness and intensity that it can acquire preponderance in consciousness, arouse through association its allied memory-images, and suppress such as are incompatible with itself. Observation, which lies at the root of all progress, is thus the adaptation through attention of the sense-organs and their centers of perception to a presentation or group of presentations predominating in consciousness.[140]

As Jonathan Crary has demonstrated, Nordau's views here represented a vast literature in experimental psychology and social commentary focused on attention.[141] Crary has covered this ground well, but Nordau's remarks nevertheless bring up two issues especially salient to this chapter. First, we should note the absolute centrality of attention for Nordau's Enlightenment project: "Culture and command over the powers of nature are solely the result of attention; all errors, all superstition, the consequence of defective attention." Attention was not simply the ability to focus, it was the motor of human progress. The stakes were high, precisely because Nordau, as others before and after, tied attention to human volition.[142] "Attention is nothing but the faculty of the will," which meant that human action relied almost entirely on attention in Nordau's scheme. Attention, as choice, expressed free will; to focus is to choose. But more than that, attention, as Crary points out, is also inhibition; it is the suppression of stimuli and the ordering of information. It is not simply choice that makes us human, according to Nordau, but the kind of choices we make for the good of progress. Attention, then, had a powerful ethical charge. Without the proper exercise of attention, according to Nordau and this tradition, we would be automatons and decadents, abandoning our future to the caprice of chance, fate, or worse, the flow of modernity.

What did the proper exercise of attention look like? For Nordau, it looked exactly like "observation," which brings us to the second issue: observation as the practical manifestation of attention and volition. According to Nordau, attention determined "the emergence, degree of clearness, duration and extinction" of stimuli, thus controlling the flow of information to consciousness. Attention was an important gatekeeper against the rush of modern life; significantly, it could control the *duration* of the presentation. Attention, or its equivalent in observational practice, controlled the rate at which consciousness is exposed to stimuli, just as a scientific experiment isolates and controls variables and the rate at which the researcher receives information. The difference between overstimulation and an unacceptably quick pace is sometimes difficult to distinguish, but if we think of observation and experiment as a set of practices that manage the flow of data, then it really does not matter. In fact, attention itself, as Crary emphasizes, was a notoriously slippery concept; not only did it bleed easily into distraction, but psychologists more or less failed to locate or define it, except as an "imprecise way of designating the relative capacity of a subject to selectively isolate certain contents of a sensory field at the expense of others in the interests of maintaining an orderly and productive world."[143] Attention still received significant play, but precisely because of its ambiguity it makes good sense now to focus instead on observation. In the context of the hailstorm of modern life, attention functioned for Nordau and others as a "stimulus shield," as well as a metaphor for bourgeois values of order, rationality, and discipline. But most of all, it functioned as an abbreviation for a set of observational practices derived, at least in part, from scientific method. These practices, which I outlined in the previous section, could be learned; the medical student or the young scientist was constantly required to adopt and perfect this mode of viewing. Once incorporated, this training became second nature, of course, but more importantly, as we see in Nordau and others, it carried an intense moral responsibility: "Observation . . . lies at the root of all progress." For Nordau and the *Bildungsbürgertum*, that which threatened these practices, or risked the possibility of passing these practices to another generation, attacked the very foundation of human endeavor.

So it is easy to see why physicians and other members of the educated elite greeted cinema's temporal rush, emblematic of the incessant pace of modern life, with something less than enthusiasm. Stuck in a theater seat, with no way to slow or stop its continuously unfurling movement, they could only helplessly watch film corrode their (and their children's) "critical faculties." We have already seen how psychiatrist Robert Gaupp, along with his colleague at Tübingen, art historian Konrad Lange, condemned "the temporal concentration" of cinema: "In the cinema . . . excitation of the passions is intensified and multiplied by the rapid succession of vivid images that passes before our eyes. There is no time for reflection or release, no time to find an inner balance . . . quiet *contemplation, intellectual assimilation*, and sober criticism are not possible."[144] Others echoed his medical opinion that the rush of images was too much for the average viewer: "All eyes greedily fix on the living picture as it flits by. And the mind must hurriedly link together the connections between images that were only briefly glimpsed. . . . In the theater I must *mentally exert myself* more throughout [than in the cinema]. I've got to pay attention to the words and thoughts, which often appear in a very terse, sharply focused form. Such a tight line of thought is not necessary in the movies."[145] For some commentators, cinema's rush of images threatened reason itself:

> The mere habituation to the darting, convulsive, twitching images of the flickering screen slowly and surely corrodes man's mental and, ultimately, moral strength. First, one gets used to switching quickly and abruptly from one impression to the next; one loses the slow continuity of the succession of ideas, the ability to grasp, which is a precondition for all sound judgment. Second, one becomes accustomed to yielding to and unthinkingly following its random string of images and ideas; one no longer misses the logical structure of an overarching thought, which is indeed what binds individual impressions together into what is generally referred to as "thought." . . . Third, as a result of the rapid passing of images, one gets used to absorbing only approximations of an impression; one does not grasp the image clearly and consciously, in all its details.[146]

For these writers, cinema's threat was ultimately a question of mental hygiene: methodical observation as taught by the scientific method was one good practice, akin to flossing, that would help insure the health of the psychic mechanism. It was only one of many good observational practices—aesthetic contemplation was another—that reflected a well-ordered mind. Motion pictures, on the other hand, encouraged sloppy habits of thought. Instead of the active self-control that these hygienic practices implied, motion pictures suggested disrepair and self-abandonment.

We can see these implications especially clearly in the characterization of cinema as a hypnotic agent. The ground between cinema and hypnotism has been well tilled,[147] but I return to it because hypnotism provides not only a succinct and well-known model of spectatorship, but because, surprisingly enough, it also offers a model of observation. That is, if we consider the diagnostic value of hypnotism, or its function as an experimental tool, the double-edged value of this trope will become especially clear. So let us first briefly explore the discussion of hypnosis as a metaphor for film spectatorship. As Stefan Andriopolous has shown, the comparison of cinema with hypnosis tapped into contemporary debates about hypnotic crime that flourished during the initial heyday of hypnosis, between 1885 and 1900. In the clinical literature, researchers such as Hippolyte Bernheim in France and Auguste Forel in Switzerland wondered whether the power the hypnotist had over the subject could extend to forcing the hypnotized person to commit crimes against his or her will. To investigate this phenomenon, they even staged fake crimes, in which the hypnotized subject was asked to shoot or stab another person with blank bullets or wooden daggers. If the results were inconclusive, the "belief in the possibility of perfectly camouflaged suggestions produced the paranoia that there might be an unlimited number of unknown hypnotic crimes."[148] The idea that an agent could plant a suggestion in a hypnotized subject to be carried out later excited the already overwrought imaginations of many.

For many physicians and reformers, cinema acted as a hypnotist, sending impressionable subjects to the streets with powerful, posthypnotic suggestions to commit crimes of all varieties. Or it functioned as a "trigger" for latent psychopathies. One physician cited a 1911 case in Frankfurt, in which a twelve-year-old boy stole a purse. When interrogated, the boy confessed, "I admit to having carried out the purse-snatching: I by chance ended up in a crowd where Frau K. stood with her purse. I had once seen the presentation of a purse-snatching in a movie theater. I was thereby compelled to try something like that, too."[149] This criminal sounds rather literate for a twelve-year-old, but nevertheless the cinematic depiction of a crime apparently presented a suggestion too powerful to resist. The doctor commented, "The effect of the dramatic, moving images on the child's psyche must have been especially strong because the purse-snatching took place on a crowded street in the middle of Frankfurt."[150] The medical literature on cinema during this time was littered with such cases in which a youngster who has committed a crime admits to visiting the movie theater at least once a week or even to learning the technique at the movies.[151] Apparently, like Drs. Caligari and Mabuse, cinema invisibly enlisted an army of impressionable youth to do its bidding.

How was this accomplished? Gaupp explained the connection between cinema and hypnotism:

> Add to this the well-known psychological fact that, when hearing or reading about exciting events, few people have sufficient imagination to visualize the events graphically before the mind's eye. Cinema, however, brings everything right before our eyes in embodied form, and does so under psychological conditions that are conducive to deep and often lasting suggestive effect: the darkened room, the monotonous sound, the forcefulness of exciting scenes following each other beat by beat lull every critical faculty to sleep in impressionable souls, and thus, not infrequently the content of the drama becomes a fateful suggestion for the complaisant youthful mind. We know that *all suggestions adhere more strongly when the critical faculties sleep*.[152]

Gaupp was not the first or the last to note the structural similarity between cinema and hypnosis; as Rae Beth Gordon notes, critics from Ycham to Raymond Bellour have been fascinated by the parallels: the darkened room and the luminous screen of the theater mimic or encourage the intense focus and narrowed consciousness of the hypnotized subject; the monotonous noise of projection and the "scenes following each other beat by beat" are like the lulling rhythms of the hypnotist. Everyone from Hugo Münsterberg to Jean Epstein has remarked on film's suggestive power.[153] Noteworthy, however, is the image that emerges from this varied commentary of the spectator as suggestible, childlike, and feminized—a characterization that was fairly commonplace, especially in social psychology. The idea of "suggestibility," especially, was a lynchpin that connected the trope of "the child" to other groups, namely women and crowds. That is, reformers and physicians made a conceptual leap from children to crowds (and to the cinema audience) via the idea of suggestibility, which had become an important concept in crowd psychology. For example, Gustave Le Bon, the most well-known popularizer of nineteenth-century crowd psychology, characterized the masses as "an enraged child." Furthermore, according to Le Bon,

> It will be remarked that among the special characteristics of crowds there are several—such as impulsiveness, irritability, incapacity to reason, the absence of judgment and of the critical spirit, the exaggeration of the sentiments, and others besides—which are almost always observed in beings belonging to inferior forms of evolution—in women, savages, and children, for instance.[154]

Darwin's evolutionary scheme provided a quasi-scientific basis for comparing crowds to children, but "suggestibility" was even more significant for this comparison. Le Bon devoted a chapter to "the suggestibility and credulity of crowds," arguing that the crowd is "perpetually hovering on the borderland of unconsciousness, readily yielding to all suggestions,"[155] a mental state most commonly found in women, children, and the hypnotized subject.[156]

Via Charcot, hypnotism had already come to be associated with female hysteria. Under his supervision, female patients (and even some male patients) fell into a deep trance and, through the power of suggestion, disassociated themselves from their bodies, allowing Charcot to practically sculpt them into various positions.[157] Likewise, in Le Bon's crowd psychology, this pliability and impressionability was a distinctly feminine trait. Those susceptible to suggestion—or more precisely, those unable to withstand suggestion—were feminized, namely women, children, savages, crowds, and weak-willed men (fig. 2.4). This is also why hypnosis was such a powerful trope for characterizing the cinema audience: for Gaupp and others, the viewer was in a state of self-abandonment and pliability, swept away by the moment like a hysteric or hypnotized subject. Film viewers—whether actual children or merely uneducated and therefore childlike—lacked the will to distance themselves from the psychological and physiological effects of suggestion. Precisely this lack of will was at issue; *willenlos* is the word Gaupp uses to describe impressionable youth. His diagnosis described both pathology and morality; the lack of will was a moral weakness that manifested itself in mental impressionability and physical mimesis. Those who *could* resist were, of course, male experts. Against the feminized subject, physicians such as Charcot, Le Bon, and Gaupp pitted the moral strength of the masculine professional.

FIGURE 2.4. A hypnotist practicing his craft, France, circa 1900

This image of the film spectator is a familiar picture. But what of the hypnotist himself? That is, if hypnotism were a valuable trope for describing the spectator, could it be equally valuable to describe the observer? What properties of hypnotism expressed the hypnotist's stake in the process, especially with regard to his mode of viewing? What was the medical purpose of hypnosis? Psychoanalysis and the medical community abandoned serious work in hypnosis early in the twentieth century for a variety of reasons; not only was it becoming too much of a sensational sideshow, but the power dynamic it implied was indefensible within humanist inquiry.[158] Nevertheless, before this crisis and in its modern form, hypnosis held therapeutic potential, because it allowed the investigator to probe deeply into the patient's psyche. It had the effect of anesthesia—it allowed the physician to explore the psyche without the interference of a "live" consciousness. It was considered a form of "psychic dissection." In this respect, it *suspended time*. More precisely, it functioned *experimentally*: like an experiment, it controlled variables, stabilized the environment, and provided enough time to examine. Berlin psychiatrist Albert Moll declared, "Hypnotism is a mine for the psychological investigator, for hypnosis is nothing but a mental state. When we think that psychologists have always used dreams so much in their investigations of mental life, and that experiments can be better made in hypnosis than in ordinary sleep, *because it can be regulated at pleasure*, we cannot deny the value of hypnosis to psychology."[159] One prominent French psychiatrist put it this way:

> Through numerous examples, I will try to show that with hypnotic processes we may practice, if I may express it in this way, an actual *moral vivisection* (if the reader is not too frightened by the word) and witness with our own eyes and *make* function the intellectual mechanism just as the physiologist sees and *makes* function the organic machine.[160]

When did a physiologist "make" a body function like a machine? In the course of an experiment, of course. Similarly, hypnotism was a process that stabilized the subject; isolated important variables; and allowed the time to inventory, describe, and classify those variables in a scientific way. Like a filmstrip, the subject's consciousness under hypnotism could be paused, rewound, slowed down, or inspected bit by bit. It gave the researcher time to explore, to accumulate data, to contemplate the details, and to correlate their patterns. In this way, the comparison between cinema and hypnosis goes further than what we find in the discourse itself: it is not just that cinema was a hypnotist or even that hypnotism was a constant theme in the films themselves, but that hypnotism and cinema were functionally equivalent in the eyes of the researcher. Both were experimental apparatuses that had a flexible, controllable regulatory mechanism in the hands of the scientist, and both had, in relation to the layperson, a potentially dangerous pace precisely because it could not be controlled.

Hence, proper observational method and experimental control functioned as mental prophylactics against the sweep of time. "Will" was the name they often gave to the power to resist, but if "will" ever existed beyond the realm of bourgeois fantasy, it existed as ingrained, scientific, observational method. For example, hypnosis was often characterized as an extreme "narrowing" of consciousness with an attenuation of peripheral awareness; the paradox, of course, being that those most capable of intense focus and attention were most easily hypnotized. Scientific observation, on the other hand, was emphatically *not* considered a "narrowing" of consciousness; this was not what Nordau or others meant when they lauded "attention." Indeed, given the emphasis on logical operations such as correlation, observation was nothing if not an "expansion" of awareness; the researcher constantly mentally compared the object under study to other objects in past and present experience. So the prevailing view of observation held that logical operations attached to a viewing protocol—the finest aspect of "attention" as the educated elite understood it—were the best defense against slipping into a dangerous immersion.

The masculine, medical professional was distinguished not only by his will, but by his training in scientific observation. Indeed, this training was his primary defense against the onslaught and seductive power of (moving) images. The scientific appropriation of motion pictures depended on the ability of the researcher to halt this onslaught, to forestall its assault and to read *slowly*. It must be so: if the researcher were to correlate the data, he had to be able to control its flow. Ultimately, then, the negative comparison between cinema and hypnosis was about mastery: mastery not only of the hypnotized subject but of the ability to hypnotize. Hypnotism and motion pictures were perfectly acceptable scientific tools, as long as they were in the qualified hands of (male) professionals. Hypnotism, for example, was condoned (and still is) as a legitimate procedure for psychic exploration. Likewise, physicians in Germany used motion pictures to help diagnose or visualize disease. Like motion pictures, hypnosis was used to isolate, stabilize, and present—in this case, psychic or somatic trauma. Hypnosis was a tool for acquiring the distance and time necessary to observe and to diagnose. Physicians used motion pictures to record the human body, but then they slowed the images down or stopped them to examine the phenomenon frame by frame, also giving them the time and distance they needed to master the event. In the same way, hypnosis allows access to different temporal registers. In both cases, the scientific use of hypnosis or motion pictures implied a mastery of time and of the human body. Film's popular incarnation, however, implied a lack of mastery of both.

In his essay "What Is Enlightenment?" (1784), Immanuel Kant made a compelling argument for self-mastery: "Enlightenment is man's release from his self-incurred tutelage [*selbstverschuldete Unmündigkeit*]. Tutelage is man's inability to make use of his understanding without direction from another. It is self-incurred when its cause lies not in lack of understanding but in lack of resolution and courage to use it without direction from another."[161] For Kant, true freedom meant being able to think (and speak) for oneself. Those who did not or could not were condemned to political and social infancy. "Tutelage" is an important term, in this respect; literally translated,

Unmündigkeit, means "minority" in the sense of not yet being "of age." (*Mündigkeit* means "majority," as in "the age at which full civil rights are accorded.") Kant basically argued that those who refused to assume this responsibility to think for themselves were politically no more than children. So in this essay he gave license to society's "enlightened" to lead the others. Hence the importance of education, especially *Bildung* (self-cultivation), for attaining this state of political and social responsibility. The emphasis we have seen so far on observation and attention reveals that *educating the eye* was absolutely central to discharging this ethical duty. Self-cultivation was not limited to training in literature, music, or history; training *vision* was also very much a part of this process. Scientific or medical observation was only one element of a range of viewing practices that could be considered *Bildung*-worthy. Pedagogical practices of visual education and practices of aesthetic contemplation, which the next two chapters respectively explore, were two others. Their opposite number—what we call "spectatorship"—can be fully appreciated only by examining the infantilization of the "unenlightened," which is the task of chapter 3.

3

THE TASTE OF A NATION

EDUCATING THE SENSES AND SENSIBILITIES OF FILM SPECTATORS

I step in; intermission has just begun. An oppressive, damp draft hits me, even though the doors are open. The spacious room (500 capacity) is filled to the last seat with children. There is an incredible ruckus: running, yelling, shrieking, laughing, chatting. Boys scuffle. Orange peels and empty bon-bon boxes fly through the air. The floor is studded with candy wrappers. Along the windowsills and radiators young toughs roughhouse. Girls and boys sit together, densely packed. Fourteen-year-old boys and girls with hot, excited faces tease each other in unchildlike ways. . . . Children of all ages, even two- and three-year-olds, sit there with glistening cheeks. Women walk among them selling sweets. Many children sneak candy and drink soda [Brause]; young boys smoke furtively.

Then the movie begins.

—A SCHOOLTEACHER FROM BREMEN (1913)[1]

For the cinema reformers of imperial Germany, this was a scene from hell. This is what German *Kultur* had come to, what modernity had wrought: children melting and spoiled like day-old candy on the floor of a movie theater. Like Professor Rath of *Der Blaue Engel* (The Blue Angel, 1930), who follows his students into a seedy nightclub and is initially shocked by the sexuality and degeneracy within, this schoolteacher from Bremen walked into a matinee and was horrified by what she saw. Her emphasis on the corporeality of the scene, on the *body* of the audience, so to speak—fighting, eating, sweating, and awakening sexuality—attests to the perception that cinema presented a grave danger to the children's emotional and moral fitness and especially their physical health. Given the traditional bourgeois association of sensuality with the lowly masses, this scene also represented a threat to the well-being of the nation, of the body politic.[2] As this chapter will demonstrate, "children" and "the masses" were often interchangeable concepts. Indeed, while cinema was often portrayed as a gaping Moloch devouring innocent children in some pagan ritual, the children depicted in this passage are far from sacrificial lambs. They present something of a veiled threat to the narrator, as if she had entered a strange, chaotic culture. In the contemporary literature on children and cinema, scenes like this one provoked twin paternal urges: to protect and to control.

"Women walk among them selling sweets." Along with the concern for sensuality (and its tacit partner, capitalism), the numerous references to sweets stand out in this description. Implicit was the assumption that cinema was spoiling the "taste" of the children for financial gain. Reformers complained constantly about the "tastelessness" of both the theater atmosphere and the films themselves. Figuratively speaking, the concession candy that ate at the children's teeth was also rotting their aesthetic sensibilities. Konrad Lange, a noted art historian at the University of Tübingen and a ferocious enemy of film, put it more bluntly: "If one were to judge the artistic understanding of our good, middle-class citizens, one would have to say that their taste is rotten to the core. They have a morbid taste for the slick, the effeminate, the sentimental, and the sugary. They display a demoralizing aversion to the healthy dark bread of true art."[3] Lange made the relationship between taste, class, and the body as explicit as possible. Taste, as the simultaneous expression of individual judgment and social distinction, serves as a connection between the private and the public spheres. It is, as Pierre Bourdieu notes, "a class culture turned into nature, that is, *embodied*."[4] It therefore provides a link between individual consumption and a national agenda. Protective of the masculine ideals at the heart of that agenda, Lange condemned the feminization of culture accompanying the onslaught of modernity; he was not alone: most middle-class males of the Western world seemed to share his concerns.[5]

Lange's solution, like that of many teachers and educators involved in cinema reform, stressed the education of children and adults. The abiding faith in the ability of education to overcome social ills and promote social progress was a fundamental plank in the platform of many fin de siècle movements, from the socialists to the progressives. But the tradition of *aesthetic* education, with its promise to harmonize the senses and sharpen judgment, offered a quintessentially German solution.[6] By pointing the way to a "true" and "pure" aesthetic experience, aesthetic education also pledged to counteract the corrupting influence of the cinema. While many reformers, such as Lange, steadfastly refused to be seduced by cinema's charms, some flirted with this particular Lola, courting her in hopes of making her an honest woman by giving her an aesthetic education (or an education in aesthetics). Even while the film industry used the reformers for its own ends, it revealed the contradictions of their ideology. Just as Professor Rath's affair with Lola reveals the indefensibility of his position—a teacher who, ultimately, does not have the best interests of his students at heart—so, too, the reformers' involvement with film shows that there were larger issues at stake than the health of the children.

Even if Professor Rath portrays the hypocrisy of his class, I see him as a distinctly modern character caught between—and profoundly ambivalent about—the solemn textbooks of his classroom and the cunning spectacle of the nightclub, or more broadly, between *Kultur* and *Zivilisation*. As Norbert Elias and others have noted, these terms played an important function in German self-identity; by the beginning of the twentieth century, they formed a contrasting pair. If *Kultur* implied "inwardness, depth of feeling, immersion in books, development of the individual personality," or the general cultivation of one's inner life, then *Zivilisation* implied "superficiality, ceremony, formal conversation."[7] Or, as Raymond Geuss puts it, *Zivilisation* "has a mildly pejorative connotation and was used to refer to the external trappings, artifacts, and amenities of an industrially highly advanced society and also to the overly formalistic and calculating habits and attitudes that were thought to be characteristic of such societies."[8] By the 1910s and 1920s, *Zivilisation* came to be associated with the primary burdens of industrialized society: capitalism and technology, or the utilitarian, material, even decadent world of commercial interests. By contrast, *Kultur* was associated with not only inner, spiritual values but a specifically German configuration of ethical and moral ideals.[9] Professor Rath is caught between his allegiance to learning and teaching the subtleties of the German cultural heritage, to cultivating his soul and those of his students to become good Germans, on one hand, and his attraction to the superficial, voyeuristic pleasures of a crass milieu that combines the worst of sensuality and capitalism, on the other. His downward spiral, of course, begins with his choice between them.

German educators interested in reforming film and its theaters felt they faced the same dilemma. But unlike Rath, they did not approach it as an either/or choice. Instead, the appropriation of motion pictures for educational purposes was a negotiation between aesthetic and moral values, on one hand, and modern technology or commercial interests, on the other. Such negotiations necessarily implied that commercial cinema was not viewed negatively across the board. Just like the reform movements in general, film reform was nothing if not multiple, consisting of many different voices expressing the full range between progressive optimism and cultural pessimism.[10] Yet a picture emerges of a well-meaning but deeply ambivalent group committed to preserving traditional ideals while modernizing the social curriculum. Motion pictures, in this case, were for many not merely a problem to be solved but perhaps also a solution in themselves; if they could be reformed, or "ennobled" as reformers often put it, then they might serve as a means to reconcile *Zivilisation* and *Kultur*, a goal shared by most of the reformers of imperial Germany, no matter what their chosen object of improvement.[11] The educational use of motion picture technology reveals this attempt at reconciliation, especially in the way that films were made to fit established pedagogical practices. As reformers pressed film into an educational mold, their work also highlighted the ideological and practical contours of that framework.

Specifically, the use of motion pictures in education followed the contours of a pedagogical trend or approach known as *Anschauungsunterricht*, or "visually based means of instruction." Popularized in the nineteenth century, this method exemplified the reformist impulse in education, in that it attempted to counter the perceived overrationalization and rote memorization of the traditional pedagogical approach with a self-consciously modern strategy that emphasized the immediate visual perception of things themselves, as opposed to their description in books. Known as "the object lesson" in English-speaking countries, the approach asked teachers to present an object, such as a seashell, and ask the students a series of simple questions about it. The questions were designed to lead the students from concrete description to a higher, abstract understanding of the object in relation to its environment and to other things. Many claimed that motion pictures could function as outstanding representations of objects that could not be brought into the classroom, if and only if film could be made to conform to this and other methods of visual training. This chapter, then, continues some of the concerns traced in the previous chapter, especially the investment of the expert class in specific methods of processing visual information. At stake here is not merely the type of viewing, such as medical or scientific observation, but the culture's interest in *what counts as learning*. *Anschauungsunterricht*, in this respect, was not just one more pedagogical method but a way of acquiring *Kultur*; it was a strong component of *Bildung*, or the process of self-cultivation. What counted as learning, in this case, was a way of forging *vision* and *thought* into *taste*; if "taste" were the sign of cultivation, then "vision" was the means by which taste was trained.

This chapter argues that the goal of reform was not merely to align film and its theaters with standards of taste and morality but to conform motion pictures to specific modes of viewing; *Anschauungsunterricht* serves as an obvious and convenient method of training vision to which reformers and educators adapted motion pictures and their projection. While *Anschauungsunterricht* was explicitly a method of training observation in school-age children, the ideal pertained to adults as well, as we see in all sorts of adult education programs at the time, especially those concerned with aesthetic education. Hence, the image of spectatorship or of ordinary movie audiences at this time must be understood in contrast to this attempt to assimilate motion pictures into a larger ideological arena, signaled most strongly by the desire to train audiences to a particular mode of viewing associated with *Anschauungsunterricht* and aesthetic education.

So this chapter charts the work involved in fitting motion pictures to educational and reformist agendas; it also surveys the cultural and ideological landscape on which these efforts took place.[12] Three sections alternate between practical efforts and historical and theoretical context. The first section discusses the general contours of reform in Wilhelmine Germany before outlining some of the concrete steps reformers took to protect child audiences from the hazards of cinema. Censorship, taxes, and child protection laws were accompanied by attempts to create an alternative film system by controlling means of production, distribution, and exhibition. I will also describe reformers' efforts to persuade production companies and theater owners to support reform films and exhibition values, which led to the creation of reform theaters and community cinemas. The second section examines some of the guiding principles or assumptions behind these efforts, including *Anschauungsunterricht*, aesthetic education, and the discourse on "the child" and "the masses." The urge to protect children from the "degeneracy" of mass entertainments went hand in hand with the desire to educate the general public. Both concerns drew life from child and crowd psychology popularized at the turn of the century. Reformers saw aesthetic education and *Anschauungsunterricht*, entwined in theory and practice, as potential solutions to the problematic picture of spectators the experts painted. The third section looks closely at the work of two reformers who set out to create alternative, educational, and edifying exhibition spaces in response to the perceptions and problems outlined in the previous section. Educator and reformer Hermann Lemke articulated principles for the educational use of film for school-age children, especially at commercial theaters that arranged special screenings for elementary schools. Another reformer, Hermann Häfker, hoped to offer educational or edifying film presentations for adults. Häfker's attempts to create "model presentations" (*Mustervorstellungen*) exemplify the use of film as an instrument of ideological solidarity. Increasingly worried about the "bad taste" of mass entertainments, Häfker enlisted film in a program of aesthetic education designed to raise the sensibility of the people to a unified, national level. Taking his cue from a long tradition of aesthetic education dating back to Schiller, as well as the art education movement then taking place, Häfker wanted to use motion pictures to train the tastes of the nation. Finally, this section takes a longer look at Schiller's ideas about aesthetic education to lay bare Germany's long-standing ideological investment in the relationships between vision, taste, pedagogy, and nation.

CINEMA AND THE SPIRIT OF REFORM

In their desire to make a change for the better, the men and women involved in film reform were part of a much larger set of movements sweeping the industrialized world around the turn of the century. As increased industrialization and urbanization brought one social upheaval after another, "reform" expressed the mood of the times in a variety of ways throughout Europe and the United States. In the United Kingdom, constitutional reforms swept through Parliament as groups demanded suffrage throughout the last third of the century. In the United States, the agrarian Populist movement of the 1890s and the Progressive movement of the 1900s reflected a broad impulse toward criticism and change. Progressivism, in particular, captured the spirit of reform through its outrage over the excesses of capitalism, its faith in progress, and its interventionist policies. During the 1880s, the pressures of industrialization and democracy prompted the French parliament to create the only state-run, compulsory, secular primary school system in the world. The growing confrontations between the forces of labor and capital also prodded republican politicians to campaign for social legislation, such as regulation of working conditions, to ensure social peace.[13]

Germany, in particular, was deluged by swelling transformations in the public sphere provoked by rapidly changing demographics. The Industrial

Revolution and national unification came relatively late to Germany and accelerated very quickly. The resulting discord between the classes and between rural and urban lifestyles seemed especially acute.[14] During the high tide of these changes, which occurred from around 1890 to 1920, the concept of "reform" as an expression of the sense of transition and as a plan for managing it took on special significance for self-understanding. Germany's groundbreaking legislation providing for compulsory insurance for workers' sickness, workplace accidents, and retirement pensions became an influential model for the United Kingdom, the United States, and France. These measures, dealing in some form with physical conditions and consequences of the workplace, illustrate the strong connection between class and somatic issues in reform movements during the late nineteenth century. Reform manifested itself in everything from *Jugendstil* decor, to a new, more "natural" style in women's clothing (*Reformkleidung*), to nutrition reform (*Ernährungsreform*).[15] "Reform" implied a battle against tradition, against perceived cultural and social stagnation; as such, it provided a plan for the formation of new, more "authentic" concepts for living. In fact, the connections between the reform movements and the more general tradition of *Kulturkritik* are very strong; from Jean-Jacques Rousseau and Johann Heinrich Pestalozzi in the eighteenth century to Friedrich Nietzsche and Paul de Lagarde in the nineteenth, the critique of society paved the way for a general re-evaluation of values in the twentieth.[16]

Very often, the critique of society focused on the educational system. Education reform was among the first movements to sweep across Germany. The kaiser himself had set the agenda on a December morning in 1890 while addressing a congress of educators in Berlin; Wilhelm II claimed he grasped "the spirit of an expiring century" with his calls for school reform. In answering the question of how the German schools of the nineteenth century could be reshaped to meet the needs of the twentieth, the kaiser echoed sentiments that had been expressed throughout Europe during the often rocky transition from the Victorian age to the modern. Specifically, he voiced his fears that the *Gymnasium* (high school) failed to train its students adequately for the requirements of Germany's rapid industrialization. Second, he complained about the "excess of mental work" in the schools, arguing that such "overburdening" was threatening the physical health of Germany's youth. Finally, he insisted that German schools devote more time and energy to fostering specifically national values, thus turning away from the traditional emphasis on the classics: "We must make German the basis of the *Gymnasium*; we should raise young Germans, not young Greeks and Romans."[17]

The kaiser's concerns about "the modern," "the healthy," and "the national" reflected and reinforced similar fixations of the European elite. Like them, he found the educational system to be both the problem and the solution to perceived crises. By inadequately preparing the nation's children for the demands of the future, the system risked irrelevance, according to reformers of the time. Swift reform promised both a brighter future and a greater measure of control over the rapid changes taking place. Among the different examples of education reform were the country boarding schools (*Landerziehungsheime*), which were experimental schools located in the countryside as an explicit rejection of urban culture.[18] Their emphasis on physical education echoed the hopes of the youth movement for a spiritually renewing combination of countryside, fresh air, and *Volk*. Likewise, work schools (*Arbeitsschule*) hoped to renew the creative (and ethical, hence political) spirit through manual labor, such as gardening, and handicrafts, such as wood sculpting or leatherwork.[19] The art education movement (*Kunsterziehungsbewegung*) similarly stressed the creative capacities of children, advocating renewal through art and education of the aesthetic sensibility.

Film reformers shared the kaiser's interest in "the modern," "the healthy," and "the national."[20] At the center of their concerns lay motion pictures, which they also found to be both scourge and cure. An emblem of modernity, cinema represented a plague, especially toxic to children, and proper education of the public was the only hope to halt the epidemic. At the same time, cinema was emerging as the most powerful instrument of mass education and therefore potentially provided the surest treatment for whatever ills modernity had spread. Before treatment could begin, however, commercial interests had to be persuaded to participate in this remedy. Moreover, film needed a stamp of legitimacy to have any authority in this rescue mission. Film reform was the process by which these goals were attempted, if not completely achieved. It shared roots, objectives, and ideology with other reform movements of the day, especially, and not surprisingly, educational reform.

It does seem rather unusual that the head of the German empire, almost by definition the representative of a conservative status quo, would come out so strongly in favor of reform. A mixture of progressive reforms and reactionary politics indicated the ambivalent attitude of the bourgeoisie toward the troubling issues of the day.[21] All reform movements revealed, in one way or another, the fundamental irony of the kaiser's position. The calls for clothing reform in Germany exemplified this contradiction. In his 1901 book, *The Culture of the Female Body as a Foundation for Women's*

Clothing, Paul Schultze-Naumberg made an extensive study of the debilitating physical effects resulting from methods of forcing the female form into an aesthetic ideal. In a graphic and impassioned plea to eliminate the corset, in particular, Schultze-Naumberg demonstrated how its use eventually caused deformation of the muscles, bones, and internal organs. He called for a more functional clothing in keeping with "a new concept of corporeality."[22] While consistent with similar efforts by women's movements to liberate themselves from the pressures of social constraints, Schultze-Naumberg's "new concept" of more "natural" bodies included only those from healthy German stock. Carl Heinrich Stratz, another strong advocate of clothing reform, took a similar approach in his book *Women's Clothing and Its Natural Development*, grounding his arguments for the elimination of the corset on the conclusion that it threatened the racial superiority of European women.[23] Schultze-Naumberg and Stratz, whose worries about the integrity of the Fatherland were cloaked in concerns for the health of women, are excellent examples of the fusion of progressive goals of more liberal movements with reactionary, nationalist politics.

Clearly, Schultze-Naumberg and Stratz shared with their fellow guardians of culture a rather paternalistic attitude toward the role of women in the changing public sphere of modernizing Germany. Women in general, and female sexuality in particular, were often targets of public disapproval about modern life, as in Stratz's work above, or in the complaints about Asta Nielsen's "unladylike" behavior in her films, or in the concerns about the number of women in cinema audiences. For some, these complaints are signs of a deeper anxiety about the increasing role of women in public life in Germany; the women's movement or the youth movement in their various manifestations are often cited as prime motivators in the perceived decline of male cultural authority.[24] According to this version, the women's movement exemplified a menace that appeared to surround German intellectuals; actions such as the youth movements seemed to threaten paternal credibility even in the home.[25] Faced with such massive structural changes, the cultural elite embraced reform as a way of coping with, and controlling, modernization. While it is true that most voices that survive in print are male, we must not forget that women were not simply the objects but the agents of reform. In fact, late nineteenth-century imperial Germany saw a proliferation of volunteer organizations—instigating reform, helping the poor, and exchanging information—that provided structure to a society in transition. Women entered the public sphere not only as consumers but also through volunteer organizations that functioned in tandem with, even as a substitute for, the state and provided steady and urgent pressure for changes in social policy. So social policy in imperial Germany was not simply a top-down affair instigated by male experts but grew out of women's culture.[26] Likewise, while Germany's "crisis in culture" might have been acutely felt by male intellectuals as a decline of male authority,[27] this perception must be tempered by the knowledge that the social mobility that women and workers gained from modernization also benefited males and the middle class not only through the greater economic power that came with a consumer society but the political power that came with the alignment of state policy with middle-class goals—an alignment that occurred partly as a result of the intervention of women's and volunteer organizations.[28] So while the hyperbole of public debate sometimes provides an easy example of male, middle-class "anxiety," we should not take it entirely at face value.

The *Schund* campaigns of late nineteenth-century Germany are a great example of the battle between modern consumer culture and the goals of middle-class volunteer organizations. *Schund* is usually translated as "trash" or "rubbish," but reformers of the late nineteenth century applied it to serialized novels and pamphlets purveyed by colporteurs or kiosks, especially literature perceived to lack any redeeming value yet not obscene. While the serialized novel emerged in the mid-nineteenth century, the market grew exponentially thereafter; by 1890 it was so popular that *Schundliteratur* accounted for fully two-thirds of all German literary sales.[29] By the time of the Weimar Republic, *Schund* referred in general to any thin, mass-market paperback novel sold by the millions at kiosks and stationery shops.[30]

Educators and school administrators were the first to launch campaigns against this form of entertainment; according to Corey Ross, "the roots of the campaign against *Schund* can indeed be traced back to the efforts of elementary school teachers in the 1870s and 1880s to influence the reading habits of their pupils by drawing up lists of recommended titles."[31] Teachers continued to be the primary force behind these campaigns, even as they comprised a diverse group of interested parties, including temperance groups, religious associations, women's commissions, leagues against public vice, and local police forces.[32] Yet, as Kara Ritzheimer has argued, the rhetoric of reform unified these groups, despite their diverse goals and methods. Ritzheimer paraphrases, for example, a professor who warned that

> children who read excessively were likely to develop a lust for reading (*Lesewut*) that might "effeminize the body," "cause the senses to lose their acuteness,"

"weaken the memory," "over-excite the imagination," and "destroy a will to pay attention to serious matters." Furthermore, this "reading lust" was capable of breeding indolence (*Schlaffheit*), indifference (*Gleichgültigkeit*), absentmindedness (*Zerfahrenheit*), mental laziness (*Denkfaulheit*), extravagance (*Überspanntheit*), and slack behavior when it came to work and play (*Unlust zu Arbeit und Spiel sich einstellen*).[33]

This rhetoric of sensuality, addiction, feminization, and weakness was a portrait of media effects that transformed apparently normal children into the worst caricature of the lower class. Beyond the vocabulary of media effects, these groups also promoted *children* as the primary focus of their efforts, thereby unifying their protectionist rhetoric around a class-neutral issue, rather than waging a campaign in the name of *adults*, which could barely evade paternalistic connotations, if not all-out class warfare.[34] In any case, the rhetoric was incredibly pliable, in that it was applied equally vociferously to books and motion pictures.

Cinema reform not only patterned itself after the educational reform movements, the instigators were often hardened veterans of the anti-*Schund* campaigns. Karl Brunner, for example, was a *Gymnasium* professor and a leading anti-*Schund* campaigner who eventually became the head film censor in Berlin.[35] The Hamburg commission discussed later started as a response to *Schund*. Most of the reformers were teachers and educators, so their close ties to the education reform groups of the period were also formative. Hermann Lemke, a *Gymnasium* professor from Storkow and one of the founders of *Kinoreform*, was well connected to the Society for the Dissemination of Popular Education (Gesellschaft zur Verbreitung von Volksbildung), the leading educational organization in Germany. Hermann Häfker, the most articulate representative of film reform and arguably Germany's first film theorist, was a journalist and writer who was also close to the leaders of the art education movement. Konrad Lange, one of the leading voices of the art education movement, taught art history at the University of Tübingen.

Despite their similar backgrounds, these reformers were not all of one mind. The disparate views and priorities of all involved, as well as the absence of a central organization or platform, make it difficult even to characterize *Kinoreform* as a movement. Scattered around mostly northern and small-town Germany, the representatives worked primarily at the local level, trying to coordinate national efforts through friendly trade journals, such as *Der Kinematograph* (1907–1935) out of Düsseldorf and especially *Bild und Film* (1912–1914) out of Mönchen-Gladbach. The birth of trade magazines devoted exclusively to film coincides with the birth of the reform movement in 1907; during its earliest years, *Der Kinematograph* acted as a willing partner in *Kinoreform*.[36] The range of viewpoints in its pages, and in the other magazines that followed shortly thereafter, testifies to the difficulty the reformers had in choosing the most effective course of action.

If they did not agree on methods, their approaches at least reflected the experience and infrastructure already gained in the work against *Schund*. Basically, the efforts of the film reform movement overall can be divided into "negative" and "positive" solutions, in the parlance of the day: those that emphasized regulation, taxes, police enforcement, and censorship, and those that offered alternatives to the objectionable fare; this chapter will focus on the latter, "positive" approach. We can also divide these approaches further based on the object of reform. *Filmreform*, for example, expressed an emphasis on the *content* of films, with an accompanying strategy that focused on production. *Kinoreform*, on the other hand, expressed an emphasis on the *space* of film exhibition, and with it a strategy to clean up and "uplift" film theaters (fig. 3.1). *Filmreformers* such as Brunner and

FIGURE 3.1. A typical storefront movie theater (*Ladenkino*) from the pre–World War I era
Courtesy Deutsches Institut für Filmkunde, Wiesbaden

Lange spent their energy devising negative methods to control film production and reception, while Lemke and Häfker developed positive strategies for both *Filmreform* and *Kinoreform*.

Yet these reformers had a set of common objectives framed in ways similar to those in the *Schund* campaigns. First and foremost, they felt compelled to protect children from what they perceived to be the dangerous effects of cinema. This was first explicitly stated in 1907, when a teacher's group in Hamburg, the Society of Friends of the Schools and Instruction for the Fatherland (Gesellschaft der Freunde des vaterländischen Schul- und Erziehungswesens), formed a commission to study the effects of cinema on schoolchildren. Its conclusions were predictable and familiar: both the films themselves and the theaters produced physical and moral side effects in school-age children. The combination of the "flicker effect" and the lack of adequate ventilation in theaters caused "eyestrain, nausea, and vomiting," according to the commission. Emphasizing the connection between the body and ethical judgment, the symptoms were a sign of a deeper moral sickness, manifested in school by "apathy for learning, carelessness, and a tendency to daydream."[37] Jurist Albert Hellwig, certainly the most prolific reformer who advocated the negative approach, echoed these concerns in 1914, when he argued that "a promotion of a certain superficiality and inattentiveness, as well as a retardation of concentration and aesthetic cultivation" could be counted among the psychological dangers to young moviegoers—a diagnosis familiar from the discourse examined in the previous chapter.[38]

Second, the reformers made it clear from the very beginning that they hoped to use cinema for educational purposes. In this and many other ways, the German reformers were very similar to their American counterparts, who also took it upon themselves to "uplift" both the theaters and the films for the good of the masses.[39] The Hamburg commission concluded its study with the recommendation that

> Technically and thematically impeccable cinematic presentations can be an outstanding instrument for education and entertainment. Pedagogically and artistically minded groups must advocate for better, nobler uses of the cinematograph by encouraging the big corporations in this industry to present good, child-oriented productions in special screenings for children.[40]

Hermann Lemke answered this call to arms independently in the summer of 1907, when he persuaded a Friedenau cinema theater owner to host Germany's first "reform theater," which was likely merely a "film reform" night or series at a commercial theater, given that it did not last very long in this form. Lemke gave the opening address, making the goals of cinema reform clear to the mostly middle-class audience of teachers, press, and community leaders. He charged that the current state of cinema had caused the aesthetic sensibilities of the people to regress. Calling on the combined power of educators and the press, he maintained that "when the taste of the people is so backward, it's the duty of the intellectual [*geistigen*] leaders to influence them and put their aesthetic taste back on the right track." Cleaning up the cinema theaters was the first order of business in this project, making it one of the earliest examples of *Kinoreform*. Lemke applauded the improvements the theater owner had already made:

> This reformation is already apparent in the way this auditorium has been given a worthy interior decoration. Gone is the small, narrow room where everyone is crammed and squeezed together; in its place we find a larger, airier hall . . . so that the patron's sense that he is in a second-rate establishment vanishes all on its own. Good ventilation has been provided in order to reduce health risks.[41]

Lemke's concerns demonstrate how closely "taste" and "respectability" were tied to "the body," and especially the body of "the masses." He was preaching to the converted, however. *Der Kinematograph* later reported, "it seems that the middle class is more interested than the working class in the direction the reform theater is taking. While the seats in the third section show hardly any patrons, the first section (50 cents admission) is mostly sold out."[42] Still, Lemke was sufficiently encouraged to organize a Cinema Reform Association the following autumn.[43] Represented by teachers, members of the press, theater owners, and production companies, the association was one of many throughout Germany that hoped to coordinate efforts from these quarters toward their educational goals. Indeed, composed of businessmen, teachers, council members, and theater owners in the community, local *Kino-Kommissions* such as this were the primary means through which reformers organized their efforts, disseminated their results, and created larger networks.[44] Lemke's society received contributions from such firms as the German branches of Eclipse and Gaumont.[45] While cleaning up the *Kinos*, the reformers turned their attention to the films themselves.

Enjoying an easy fraternity with producers during the early years, film reformers hoped to capitalize upon their good relations with the motion picture companies to increase the number and availability of reform-type productions. In a particularly idealistic move indicative of the "positive" approach, Lemke in 1908 suggested that his reform association act as a "Film-Idea-Central," a clearinghouse of sorts for reform-minded scripts. Members of the association could submit ideas for scenarios, and the society would negotiate with the studios on the writers' behalf. Lemke explained, "Because we're in constant contact with the manufacturers, such an exchange will be relatively easy to arrange, particularly because we know what is being demanded. We would provide distribution free of charge and only require that those who use it become members. In this way, we may be able to bring the film companies to the cutting edge of this moment and also have a productive influence internationally."[46] Unfortunately, while the members of the movement might have held some early enthusiasm for this plan, the film companies themselves apparently did not take to it; the idea never went beyond the initial stages, and no further mention is made of the Film-Idea-Central in the trade press or reform publications.

The failure of the Film-Idea-Central and the film reform theater in Friedenau established something of a frustrating pattern for the reformers. Film companies and exhibitors expressed early interest in reform projects, even going so far as to sponsor events, but eventually refused more meaningful and lasting support. The end of 1908 saw the opening of Germany's first film trade show/exhibition at Berlin's Zoological Gardens. Jointly sponsored by Lemke's reform party and the leading film companies at the time, and with the rather obvious motto of Refining the Cinema (*Veredelung des Kinos*), it was nonetheless heavily criticized even by friendly periodicals for its lack of organization.[47] Exhibitors, manufacturers, and production companies declined the reformers' help for the next exhibit in 1912.[48] Likewise, when Lemke and Häfker attempted to muster support for their special exhibitions, the film companies were initially supportive but lost interest fairly quickly. Realizing that domestic companies could not or would not produce sufficient numbers and variety of educational films, Häfker went so far as visiting foreign film companies, such as the Charles Urban Trading Company in London and Eclipse in Paris, to find suitable nature films for his exhibitions.[49] Lemke even went to England and wrote film treatments to jump-start some sort of interest in his program.[50] Very early on, it was quite clear that the production companies were cautious about backing the reformers and their schemes.

This did not mean, however, that the film companies wanted nothing to do with the reform movement. They were certainly willing to use the reform movement for their own ends; despite their difficulties, the reformers were still a legitimating presence—they were, after all, educators, clergy, journalists, and otherwise pillars of their respective communities. Film companies were eager to cash in on this allegiance. Advertising trumpeted this relationship, even if the reformers had nothing to do with the making of the film. A 1912 Italian film distributed by the German company PAGU, *Die Irrfahrten des Odysseus* (The Wanderings of Odysseus), was rather disingenuously labeled a "Reformfilm" and carried this blurb: "From a special press screening, which was attended by the most respected Berlin literary figures and art critics, came the unanimous decision: 'This film signals the long-awaited reform of cinema'" (fig. 3.2).[51] Aware of the potential directions cinema could take, the film companies initially went along with the reformers, especially if they could be used as a selling point. But as soon as it became apparent that the vast majority of the viewing public was more interested in narrative entertainment free from "ennobling" connotations, the companies brushed off the reform societies' efforts to influence the product directly.

The extreme positions of some reformers did little to help the overall cause with the production companies. Lange and Brunner, for example, were steadfast in refusing film any legitimacy whatsoever as a medium of entertainment. Their regular denunciations of "film drama" (*Kinodrama*) merely antagonized an industry leaning heavily toward narrative films. This prejudice against narrative films often disguised stronger rhetoric against international domination of the German film market. "In the international film drama, the wildest passions of all nations come together for a gruesome rendezvous," clergyman Paul Samuleit charged.[52] Likewise, their complaints about capitalist interests tainting cinema's potential were thinly veiled laments about the presence of *foreign* capital. Some reformers, such as Häfker and Lange, dismissed film drama because of aesthetic concerns. It did not offend their sensibilities because of sloppy production qualities, although these did attract attention. Rather, the filmed drama betrayed what they saw to be cinema's primary mission: to record movement and "real life." The argument for filmic realism, of course, coincided with their desire to use cinema for educational purposes, as we will see in the next section. As Sabine Hake notes, they did not dismiss the possibility of story elements in their educational films, but the excesses of the "trashy film" so contradicted their stated ideals that many rallied against film drama altogether, for both political and aesthetic reasons.[53]

FIGURE 3.2. *The Wanderings of Odysseus* (L'Odissea, Italy, Milo, 1911), touted to be a "Reformfilm"

Lemke hoped to reform the cinema through example, stressing cooperation with and from the industry, and to rally schools together to create a distribution system.[54] Others were not so willing to rely on this teamwork. One faction of the reform movement, led by Albert Hellwig and Karl Brunner, saw censorship and regulation (the "negative" approach) as the best way to combat the onslaught of *Schundfilme*. Both Hellwig and Brunner advocated a series of legal restrictions on the cinema, including censorship, entertainment taxes (*Lustbarkeitsteuer*), poster censorship, safety regulations, and child protection laws (*Kinderschutz*).[55] Authorities tried to maintain some control over child audiences (and, consequently, theaters) by restricting their visits to specific hours of the day, regulating the length of the matinees, and requiring that they be accompanied by an adult, that police should have unlimited access to the theater during the matinees, that the day's program must be given prior approval, or that a "suitable pause" separate the films.[56]

Reformers recognized early on the importance of establishing a distribution network for their educational films. For this task, Lemke and his circle enlisted the help of the Society for the Dissemination of Adult Education (Gesellschaft zur Verbreitung von Volksbildung, hereafter referred to as the GVV). An umbrella organization for more than 8,000 local education groups, clubs, associations, and societies, it was a formidable partner in *Kinoreform*. *Bildungs-Verein*, the house publication, had a circulation of 13,000—many times that of any film trade magazine. Yet the GVV leadership was hesitant about cinema's importance as an educational tool. Even though Johannes Tews, the director of the GVV and editor of *Bildungs-Verein*, had attended the opening of Lemke's Friedenau reform theater, he still considered cinema to be of minor significance.[57] The GVV resisted involvement with cinema until 1912, when it established a film distribution center of 180 films in 16 categories, from history of the Fatherland to educational films on biology.[58]

The reformers found a more willing and beneficial partner in the Lichtbilderei, established in 1909 as a foundation of the Association for Catholic Germany (Volksverein für das katholische Deutschland). The Lichtbilderei was Germany's largest educational film institute before World War I, with an extensive catalog of titles. It began as a rental source for magic lantern slides, which could be used for public lectures, but started collecting films as well after 1911. By the end of 1912, it reportedly had around 900 titles and was collecting more at about 30 films per week, and by 1913 offered 400 slide series and 1,400 film titles.[59] The Lichtbilderei was not limited to providing films for schools, churches, and clubs; it also provided programming for many commercial theaters. Approximately 40 weekly theaters and 50 to 60 Sunday *Kinos* showed Lichtbilderei films regularly.[60] The Lichtbilderei was also involved in the distribution of more commercial dramas, actually acquiring "monopoly" rights over such established hits as *Quo Vadis?* (Italy,

1913), *Giovanna d'Arco* (The Maid from Orleans, 1913), and *Tirol in Waffen* (Tirol in Arms, 1914).[61] From 1912 to 1915, the Lichtbilderei was something of an organizational center for the cinema reform movement.[62] Its stock of films gave life to the community cinemas and private *Reformkinos*, and its publications—the periodical *Bild und Film* (Image and Film) and the series of books from the association's Volksvereins publishing company—were the principal forum for the discussion of *Kinoreform* issues after 1912.

In 1912, the GVV, in association with the Lichtbilderei, established the funds for two educationally oriented *Wanderkino*. These traveling cinemas toured from town to town, playing for four to six weeks in each place, in an effort to offset the influence of commercial cinemas and unify aesthetic and educational standards across the nation. Showing between nine and eleven films an evening, accompanied by lectures concerning such topics as "A Modern Factory," the enterprise was basically modeled after the GVV's successful *Wandertheater* and public lecture series. Between the fall of 1912 and the outbreak of war, the *Wanderkinos* offered a total of 1,279 such evenings.[63]

Reformers had most success with their exhibition experiments. In addition to the reform theaters and *Wanderkinos* already mentioned, a number of communities established their own public cinemas. The first was founded in the town of Eickel at a cost to the community of 14,000 German marks. Others opened soon afterward, in such towns as Altona, Wiesbaden, Osterfeld, Frankfurt (Oder), Gleiwitz, and Stettin.[64] These cinemas became the center of local reform activity and provided the precedent for the state-run cinemas of modern Germany, which continue to illustrate the relation between taste and nation. The proclamations of the early *kommunale Kinos* articulated this relationship and the goals of the reform movement in general:

> To oppose, for aesthetic, cultural, and patriotic reasons, the trash that is generally offered in the private theaters; to replace it with films of scientific, entertaining, and educational value; to exert, in association with established institutions with similar principles, a gradual influence on the film market, which is currently almost entirely dependent on foreign countries, and thereby keep here the millions that are flowing out of the country. Finally, to enlist cinema in the service of youth organizations and colleges by providing suitable presentations.[65]

To the modern observer, the cinema reformers of imperial Germany might seem a bit quixotic. Tilting their lances to such impassive windmills as capitalism and narrative, they only reluctantly and belatedly conceded that they were charging against the wind of public opinion. As the movies became more popular and an evening's entertainment began to look less and less like a lecture series, instead relying more heavily on *Kinodrama*, the reformers began to look more and more irrelevant. Their own Dulcinea—the children of the nation—seemed oblivious to their activities. Even those sympathetic to their cause, like this reviewer of a book on cinema and theater reform, found their efforts somewhat naive.

> [The author] is certainly entitled to his opinion in this terribly serious matter. However, he will surely also understand the skepticism of those who do not believe at all in the "ennobling" of films towards a literary bent [*literarischen Seite*], because they see completely heterogeneous things being forced into an unhappy marriage. The idea that benevolent corporations will free the theaters of commercial interests altogether is too pretty to be given much credence.[66]

Others were not so kind. Speaking on behalf of the industry in 1911, a trade journal editor pointedly replied to the reformers, "We ourselves know what we need, and we don't need tutelage."[67] One theater owner from the 1920s remembered them as "sanctimonious folks and hypocrites, morality sleuths in male and female guise."[68] Film histories, until recently, have been equally dismissive. Siegfried Kracauer charged simply that, with their zealous efforts to defend the literary canon of the nation, "they yielded to the truly German desire to serve the established powers."[69] Even if a bit condescending, Kracauer was not far off the mark. While the proclamations of the various *Kinoreformers* embraced a wide range of opinion, they never strayed far from the status quo. As Sabine Hake noted, "In sharp contrast to the intellectuals, the reformers aligned themselves openly with the existing power structure."[70] We must not, however, underestimate the reformers' contribution to German culture. In trying to sway what Kracauer called "the salutary indifference of the masses," the reformers succeeded in dominating the discourse on cinema in the years before World War I. In addition to the permanent impression they left on German film culture, German mass communication research owes them an especially heavy debt: their focus on media effects had a lasting influence on the vocabulary and goals of modern mass media studies in Germany.[71]

Ultimately, of course, cinema reform was not completely successful. The reformers failed to meet their stated goals and, considering the extreme position of many reformers, this is perhaps all for the best. World War I abruptly changed the nation's priorities, and even though the calls for reform were heard again through the Weimar years, the urgency of the moment had passed. In 1913, lances heavy with disappointment, the movement clearly appeared to be running out of breath. Sighed Lemke, "I was always hoping that someone would take over the chairmanship from me, assist me, and further expand the [Cinema Reform] Association, but no one was willing to do so and the result was that the Association remained in its infancy [*in seinen Kinderschuhen stecken blieb*, or literally, "stuck in its children's shoes"]."[72]

CHILDREN, CROWDS, AND THE EDUCATION OF VISION AND TASTE

Lemke's metaphor was apt, because it reveals the extent to which the reformers thought about the cinema (and themselves) through the metaphor of "the child." Because they were educators and teachers, this is perhaps to be expected. It is noteworthy, however, that they applied this trope to adult audiences. References to their audiences as "children," especially in connection with mention of "the masses," are scattered throughout the discourse.[73] One reformer, looking for the underlying causes of cinematic drama's continued popularity, maintained that "just as much blame belongs to the audience, the people [*das Volk*], this 'big child,' whose alarmingly spoiled taste craves for cinema's dramatic trash and silly comedies, practically forcing the theater owners to present them with aesthetic and moral duds week after week."[74] Even Georg Lukács thought about cinema spectatorship in terms of children: "In the 'cinema' we should forget these heights [of great drama] and become irresponsible. The *child* in every individual is set free and becomes lord of the psyche of the spectator."[75] Whether Lukács's "inner child" was inherently good or evil depends upon one's viewpoint, of course, and there were many to choose from at this time.

This section will survey some of the prevailing assumptions about child psychology and pedagogy to clarify the underlying ideological presumptions about child and adult moviegoers. "Suggestibility" was the common denominator linking children and crowds; studies of children were even the source of theories about social psychology. So this section will demonstrate how the portrait of (film) spectatorship usually derived from expert analyses of children and crowds. At the same time, I will show that symptoms of this spectatorship were problems to be solved by training in observational methods, specifically aesthetic education exemplified by the kind of programs promoted by Hamburg museum director Alfred Lichtwark, who was the driving force behind the art education movement of the time. His approach was popular and well known to film reformers—especially because its nationalist flavor appealed to the taste of many experts of the day—but it was also familiar because it exemplified the observational approach to general education known as *Anschauungsunterricht*. Looking closely at this approach or trend in pedagogy reveals it was a self-conscious response to the perceived disorder and quickened pace of modern life; observational training was a way of ordering thought that countered pace and disorder by emphasizing "dwelling" and correlation. This section thereby connects psychology, reform movements, and pedagogy to explain the ideological and practical emphasis on expert modes of viewing as a solution to the multiple problems spectatorship posed to film reform.

Child psychology of the period provides a partial explanation for the equation of children and the masses. Swedish author Ellen Key's *Century of the Child*, an enormously popular children's rights manifesto published originally in 1900, advocated a reassessment of the prevailing view that children were inherently evil. Summing up a trend in child psychology that emphasized the creative nature of the child, it called for new teaching methods to correspond to the new century, leaving behind the authoritarian methods of the old school and reassessing pedagogy "from the child outward" (*vom Kinde aus*). If adult society, utilitarianism, and the demands of the "real world" had determined the standards of pedagogy before, now attention turned to the child's needs and inner nature. Whereas the old pedagogy might have emphasized uniformity, now the child could expect to be treated as "the measure of itself."[76] As Key insisted, "instruction should only cultivate the child's own individual nature," which Key and others assumed to be creative, good, and even wise.[77]

Freud, however, was less optimistic about the life of the child. His essay on "Infantile Sexuality," published in his 1905 *Three Essays on Sexuality*, painted a darker picture of childhood as a "hothouse of nascent psychopathology."[78] His explanation of the importance of the child's body—describing the oral, anal, and phallic stages—on mental development was groundbreaking. Its lasting contribution is manifold, but most immediately it underlined the influence of childhood development on adult mental life. While there is little indication that Freud's theories were wholeheartedly accepted by garden-variety reformers, the new child psychology of both Key and Freud provides a clue to the urgency reformers felt when they argued for aesthetic cultivation and against the influence of sexually charged *Kinodramen*.

Despite Sigmund Freud's seminal contributions, Darwin's evolutionary theories of child development still had a strong grip on the public imagination during this period. In particular, Darwin argued that child development recapitulated the mental evolution of the species. Accordingly, the maturing child was expected to exhibit mental characteristics of subhuman species. In *The Descent of Man*, Darwin observed, "We daily see these faculties developing in every infant; and we may trace a perfect gradation from the mind of an utter idiot, lower than that of an animal low in the scale, to the mind of a Newton."[79] Discussions of crowd psychology latched onto this teleological comparison between children and primitive mentalities.[80] Gustave Le Bon, the most well known popularizer of nineteenth-century crowd psychology, characterized the masses as "an enraged child." Furthermore, according to Le Bon,

> It will be remarked that among the special characteristics of crowds there are several—such as impulsiveness, irritability, incapacity to reason, the absence of judgment and of the critical spirit, the exaggeration of the sentiments, and others besides—which are almost always observed in beings belonging to inferior forms of evolution—in women, savages, and children, for instance.[81]

Darwin's evolutionary scheme provided a quasi-scientific basis for comparing crowds with children, but even more significant for this comparison was the concept of "suggestibility." Le Bon devoted a chapter to "the suggestibility and credulity of crowds," arguing that the crowd is "perpetually hovering on the borderland of unconsciousness, readily yielding to all suggestions" (21), a mental state most commonly found in women and children (29). Most serious psychologists of his time dismissed Le Bon's rather crude arguments, but the metaphorical connection between children and the masses was still quite powerful for researchers. In fact, one could argue that social psychology has its roots in child study. Alfred Binet, a disciple of La Salpêtrière's Charcot and one of the founders of experimental social psychology, used the observational opportunities provided by public school classrooms to test his evolving theories of suggestibility. His conclusions about children and suggestibility worked their way into his formative studies of social behavior, which had a profound impact on the direction of modern social psychology.[82]

Reformers borrowed the concept of "suggestibility" as they described the cinema audiences and their scopophilia or *Schaulust*.[83] The Hamburg commission noted this condition in their report, complaining that

> many cinema presentations endanger children morally as well. Let's assume, for example, that a young boy with a tendency towards thievery were to see crimes presented with elegance and brilliant success. Wouldn't that arouse his imitative instinct? A young girl could easily learn how to get easy money and enjoy an apparently carefree and, in her eyes, wonderful life by selling her honor. When she needs to earn a living later in life, she might ask herself: "why work at a sewing machine for 10 pfennigs an hour, why work at a factory for 10 marks a week?"[84]

Why, indeed? These remarks prefigure persistent themes in the discourse on cinema during this period, especially the preoccupations with suggestibility, crime, and female sexuality. Emilie Altenloh, author of the first sociological study of cinema, even found parts of this equation in the nature of female spectatorship:

> The female sex, of which it is generally said that it always purely and emotionally absorbs an impression in its entirety, must be particularly receptive to filmic presentation. By contrast, it seems very difficult for people who are highly developed intellectually to project themselves into the sequences of events, which are often strung together haphazardly. On various occasions people who were used to grasping things on a purely intellectual level said that it was extremely hard for them to comprehend the logic of a movie plot.[85]

Altenloh equated holistic or synthetic approaches to the image with female spectatorship, and analytic approaches with expert or educated observation. She suggested that, on one hand, this open or holistic approach to the image enabled an empathetic projection that is unavailable to those who approach the film analytically. On the other hand, this empathetic mode of viewing left the spectator vulnerable to suggestion, and in this step she equated feminine and childlike modes of spectatorship. She further maintained that, in the absence of a strong family life or education, cinema held a mesmerizing influence on its young patrons, especially young male workers: "It is undeniable that the cinema's lack of all higher interests has a certain influence on the entire way of thinking and living for these unstable people [*ungefestigen Menschen*]," she concluded. "From the lives of outlaws, the morals of Apaches, and the fearlessness of heroes in cowboy films they take a philosophy of life that forces them into trajectories similar to that of their celebrated idols."[86]

Albert Hellwig also wrote often on the suggestive power of cinema and its dangers for the criminally inclined or morally weak. In one article, he described a neurasthenic woman's response to a night at the movies. In the film, a postal clerk dreams that he is attacked by robbers: "there appear a series of threatening faces and ghostlike hands, which reach out to others in their sleep." This made such an impression on the young lady that she began to see hallucinations of these hands day and night. "The rather intelligent lady was initially fully aware that it was merely a hallucination, a product of her imagination. She was nonetheless quite upset because she saw this group of gigantic hands appear suddenly and in a variety of circumstances."[87] Hellwig implied that the cause of the woman's hallucinations was a combination of cinema's suggestive power and the woman's pathological condition, neurasthenia, a vague, yet debilitating nervous condition in vogue during this time. It left its victims incapable of work and inflicted upon them a dazzling array of symptoms, including headaches, the fear of responsibility, graying hair, and insomnia. According to Anson Rabinbach, "neurasthenics were identifiable by their impoverished energy and by the excessive intrusion of modern urban society on their physical and mental organization."[88] It was a form of mental fatigue that left its victims unable to resist the stimuli of the modern world; it was characterized, in short, as a weakness of the will, as moral exhaustion.

The combination of pathology and morality is significant, because the concept of "moral weakness" metaphorically connected judgment and physical strength. The reformers' focus on both the unhealthy atmosphere of the nickelodeons and suggestive power of film reveals an underlying concern for both the bodies of the audiences and their moral judgment. This concern manifested itself as a problem of "taste"—taste lies between the realms of sensuality and reason. As with the question of the nature of the child, reformers were divided over the nature of the masses, especially their judgment. Against those who argued that the masses were not ready for reform, that they were not interested in what interested the educated classes, Hermann Lemke argued, "I for one cannot imagine that the general population has such bad taste; and even if the people were not yet mature enough for it [cinema's reform], they would have to be educated. But one must never indulge their lowest instincts—that is harmful to the community and must be prevented."[89] Hellwig was less willing to entertain the idea that the masses were inherently good: "It is the bad taste of the audience that ultimately makes the trashy film."[90] The solution to this problem of taste and, by extension, the crisis of moral judgment, was aesthetic education.

Since Schiller, aesthetic education has offered a solution to the twin problems of sensuality and suggestibility. That is, Schiller suggested the category of the aesthetic as a medium between alienated Nature and Reason. In an alienated world, the aesthetic provided Schiller with hope for reintegration and, hence, social harmony. The aesthetic category acted as a corridor between raw nature and a higher morality. "In a word," Schiller wrote, "there is no other way of making sensuous man rational except by first making him aesthetic."[91] The reformers were very interested in making "sensuous man rational." Schiller's importance for the reformist agenda is illustrated by an editorial in the trade periodical, *Lichtbild-Bühne* (fig. 3.3). The headline reads, "The Cultural Work of the Cinema Theater: Thoughts from the Year 1784, by Friedrich von Schiller."[92] The essay invoked Schiller's "The Stage Considered as a Moral Institution" to argue that cinema could function in the same manner. The aesthetic, however, was a precondition to the moral, and cinema must first go through that transformation. An illustration from a 1918 reform pamphlet illustrates the axiomatic nature of this relationship between the aesthetic and the moral (fig. 3.4).[93] The upper-left sphere represents "entertainment with immoral effect" and "morally unobjectionable entertainment," while the upper-right sphere signifies "art" and "non-art." A transubstantiation occurs when the rather plain problems of morality (left) and aesthetics (right) are superimposed to reveal the nature

Die Kultur-Arbeit des Kinematographen-Theaters.
Gedanken aus dem Jahre 1784
von
Friedrich von Schiller
(† 9. Mai 1805).

FIGURE 3.3. "The Cultural Work of the Film Theater: Thoughts from the Year 1784 by Friedrich von Schiller"

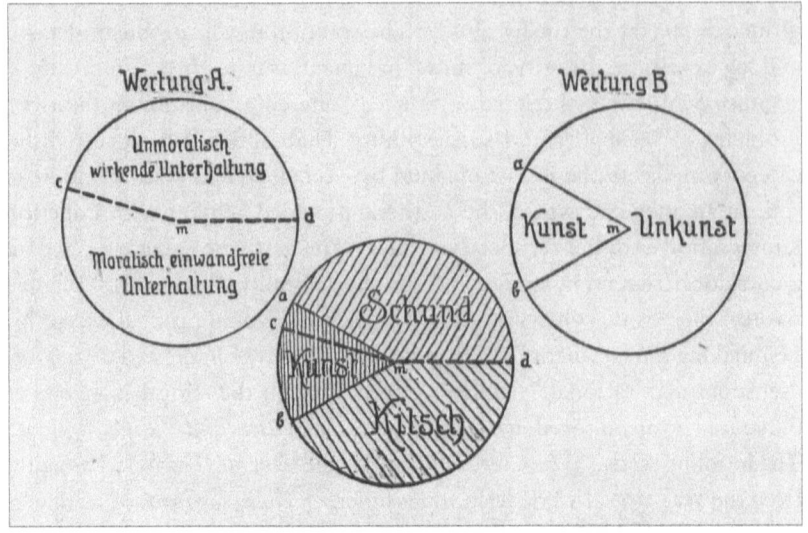

FIGURE 3.4. The geometry of taste: the superimposition of morality ("entertainment with immoral effect" [*Unmoralisch wirkende Unterhaltung*] and "morally unobjectionable entertainment" [*Moralisch einwandfrei Unterhaltung*]) and aesthetics ("art" [*Kunst*] and "non-art" [*Unkunst*]) reveals the nature of taste ("art" [*Kunst*], "trash" [*Schund*], and *Kitsch*.)

and proportion of "art," "trash," and *Kitsch*. We could say that this new sphere represents the issue of "taste."

Schiller represents the beginning of a long tradition of aesthetic education in Germany, one that eventually became grafted onto questions of nationalism. The most famous, or infamous, example of this development was August Julius Langbehn's *Rembrandt as Educator*, first published anonymously and with enormous success in 1890.[94] Like Schiller and de Lagarde before him, Langbehn reacted against the perceived excessive rationalization of the Enlightenment. The preoccupation with systemization, objectivity, and book learning had, in his opinion, brought about "the decline of the spiritual life of the German people" (7). Specialization, he complained, precluded exercise of creative power: "one thirsts for synthesis" in overeducated Germany, he wrote, and so "one turns to art!" (8). The German people could be rescued from this "systematic, scholarly, cultured barbarism" by "going back to their original source of power, their individualism" (9). Furthermore, if individualism was the root of all art, and he claimed it was, and if education should correspond to the nature of its students, then art education would be the most effective and natural form of instruction. Rembrandt, even though he was Dutch, was for Langbehn "the most individual of all German artists" (11). "The scholar is characteristically international, the artist national," he wrote, underlining the difference between science and art, word and image (11). The goal of art education, as Langbehn saw it, was to effect a spiritual regeneration of the German people by reacquainting them with their own inner nature as it was exemplified by the masterpieces of national art. But if this program sounds reasonable, most of Langbehn's essay was noxiously antimodern, antiliberal, and even anti-Semitic, which unfortunately struck a chord among the reading public and resounded throughout German culture of the time.[95] He advocated approaches to aesthetic education rooted in national (as opposed to international/foreign) sources, but he understood the local and national in biological, racial, and ethnic terms. Progressive agendas had no place in his conception of aesthetic education.

Alfred Lichtwark, generally recognized to be the driving force behind the art education movement, also valued local artistic traditions in his address to the 1901 art education conference in Dresden. "Our education still lacks a *firm national foundation*," he declared.[96] The basis for a national culture, as with Langbehn, could be found in German art. "Up to now," Lichtwark said, "the schools have not considered it their task to acquaint youth not only with the names, but the works of the great artists who express the German character" (104). And he blamed this lack of attention to "national art" for the lack of "formative power" in German culture. But Lichtwark's idea of *Heimat* was not rooted in blood and soil, like Langbehn's, but rather in the historical and cultural environment in which talent and cultural forms develop; Lichtwark's more historically contingent conception of *Heimat* therefore led to a more liberal and modernist understanding of the relationship between local and national culture.[97] Yet like Langbehn, Lichtwark held that "the challenge of art education" was inseparable from "a moral renewal of our life" (99). This hope was certainly not limited to Lichtwark; most representatives of the art education movement held it as their ultimate goal.

But Konrad Lange was cautious of such sanguine hopes, asking at that same conference whether "with '*Kunsterziehung*' [art education] we've actually found the magic word to solve all social questions."[98] If he seemed less concerned about the spiritual state of the people, he was, like Lichtwark, still very anxious about the state of German art. He acknowledged that "we actually have masters of the first order in all the areas of the fine arts, men who are living proof that the creative German spirit is not yet dead," but claimed that this was not enough. For this relative good health to survive, it needed good soil in which to grow. "And this soil can only be the people's understanding of art."[99] Worried that the elements of modern urban life could undermine children's sense of culture, Lange advocated leading them to art to maintain a sense of artistic tradition, to bring "the artistic education of our youth . . . in closer connection to the living, creative art of the present."[100]

The education of taste was also very important to Lichtwark. To his contemporaries, he was even better known as the director of the Hamburg *Kunsthalle*, which came into international prominence during his tenure. There he was instrumental in organizing groundbreaking exhibits of amateur and artistic photography, promoting local artists, and rediscovering forgotten local talents, such as the romantic painter Philipp Otto Runge. "We do not want a museum that simply stands and waits," proclaimed Lichtwark upon assuming the directorship of the Hamburg *Kunsthalle* in 1886. "Rather, we want an institution that actually works for the aesthetic education of our population."[101] Aesthetic education, for Lichtwark, was about individual self-cultivation, of course, but it also had value for the community. Summarizing Lichtwark, Eckard Schaar writes, "Artistic cultivation is not an innate ability . . . but rather a participation in a national collective property that, carried by the spirit of the people, influences the soul of the individual."[102] Lichtwark envisioned his museum as an educational center for the artistic life of the region that would focus and direct this process of individual and communal aesthetic development. It would be a clearinghouse of taste, wherein exhibits of art from around the world would help raise the sensibilities of the general public and teach new techniques to local artists. For instance, in his introduction to the first exhibit of amateur photography in 1893, Lichtwark stated that the show's goal was to "raise the artistic taste of the public and stir interest" in the new art.[103] The development of a national art depended upon the aesthetic education of both the public and the artists—his museum would take up that task.[104]

Lichtwark's influence on actual educational practices came through one of his most popular books, *Exercises in the Contemplation of Art Works*. The drills consisted of Aristotelian question-and-answer sessions between teacher and student, demonstrating by example how the child's inherent aesthetic taste could be cultivated and guided to acceptable standards. The student would gaze upon a painting and answer the teacher's questions about its form and content until the work's meaning would reveal itself to the child. For this process to be successful, Lichtwark stressed the importance of extended contemplation of single artworks in a quiet, conducive environment. Consistent with the *vom Kinde aus* philosophy mentioned earlier, this method of aesthetic training soon gained wide favor among German art educators. Lichtwark's system also confronted the important issue of national taste.

> The typical modern German has a weakness in the area of aesthetic education. He lacks an external refinement and solidity of form as well as an inner connection to the visual arts. He has no desire for artistic pleasures that require an education of the eye and heart. His eyes see poorly and his soul not at all. For the preservation of our nation as well as our national economy, these inadequacies must be forcefully redressed.[105]

Lichtwark designed his *Exercises* to provide a training program for children and others who were "aesthetically weak." By teaching youngsters how to look, gaze, and, ultimately, *see*, Lichtwark was following a set of presumptions common to aesthetic education: train the eye and the heart follows. For the art education reformers of imperial Germany, then, educating public taste was a project in nation building. Simply, education *through* art was a way of building a distinctly national art, while education *to* art was designed to build consensus and therefore national unity, as well as maintain traditional standards and methods of evaluation. These two directions were and are common for all art education programs from Friedrich Schiller to John Dewey.[106]

Lichtwark's approach caught on. A number of books in the following years staged an encounter between an imaginary viewer, a painting, and a helpful questioner.[107] *Bildbetrachtung* ("image viewing" or "image contemplation") became a common method of teaching art appreciation, which emphasized the contemplative, especially the spiritual side of aesthetic education and experience. Schoolteacher Heinrich Wolgast echoed the views of many in Hamburg and elsewhere who stressed the connection between vision and spiritual development through disciplined viewing exercises: "the child who conquers the world with these more sensitive organs will reap greater rewards than one with untrained eyes," he wrote, and "intellectual understanding of the world based on a deeper grasp of its appearances will guarantee an improved preparedness for life."[108] Reformers such as Wolgast hoped that training the observational acuity of children and adults would result not just in spiritual but ultimately national renewal; the art education movement is therefore a good example of the explicit, parallel investment in visual education and national goals that we see consistently in discussions of positive film reform.[109] Lichtwark's program enjoyed national visibility and increasing popularity, especially among educators interested in current ideas about visual education. The parallels were not lost on film reformers such as Lemke, Sellmann, and Häfker, who had similar educational aspirations for film.

Yet these pedagogical impulses sprang from a broader trend in visual education that suffused nineteenth-century pedagogy, for which Wolgast's formulation could have easily served as the motto. This trend, known generally as *Anschauungsunterricht* ("observational instruction" or "visual means of instruction" or simply "visual education"), started with the educational principles of Swiss pedagogue Johann Heinrich Pestalozzi (1746–1827), who believed that *Anschauung*, or "sense impression" was *"the foundation of all knowledge."*[110] *Anschauung* is a difficult word even for Germans: in Kant's *Critiques* it is usually translated as "intuition," but it also means "sense perception," which are two very different things. In German dictionaries the two meanings of the word, one cognitive, one perceptual, sit side by side, matching the epistemological and phenomenological (or the knowable and visible) sides of the experiential coin. Yet *Anschauugsunterricht* strove to develop the child's innate sense of form (in the Kantian sense) through observational exercises focused on the visible world, so it thereby embraced *Anschauung*'s seemingly opposed connotations. It is perhaps best understood as a process through which the pupil attains an appreciation of both the detail of the individual object and its place in a larger system, whether philosophical, taxonomic, or social. Pestalozzi's program was notable for its then-radical approach to education. Rather than a top-down, authoritarian, deductive, and speculative method that demanded the student listen to the lecturer and recite back what was said or read in books, Pestalozzi advocated a bottom-up, democratic, inductive, and empirical approach to learning that asked the teacher to interact with the student—similar to the process in Lichtwark's *Exercises*—and trust that the student could come to a higher understanding of the material through careful observation of the object in front of him or her. It countered mere book learning and rote memorization with a program that emphasized the pedagogical power of natural objects themselves.

But it was not merely about bringing the child to an object; the method was primarily about teaching the child how to *observe*. Yet as we saw in the previous chapter, observation is above all a complex logical operation, as Clive Ashwin notes about Pestalozzi's *ABC der Anschauung* (1803): "Its content was designed to activate and exercise the child's faculties, first by distinguishing separate entities and isolating them within the perceptual manifold; secondly by enabling the child to observe and note their peculiarities of form; and thirdly by associating each form with its correct name. This integration of number, form, and word led Pestalozzi to put the *ABC der Anschauung* at the centre of his educational scheme."[111] In practice, this involved a specific, step-by-step process. The teacher and the student would look at an object, such as a seashell, and the child would be asked to describe what he or she saw and be encouraged to point out individual characteristics of the object ("distinguishing separate entities and isolating them"). A series of questions asked by the teacher would bring the student to understand the basic form of the object, sometimes by drawing the object or its geometrical shape ("enabling the child to observe and note their peculiarities of form"). Naming each element and its form would be an important part of the exercise ("associating each form with its correct name"). Equally as important, the student would come to an understanding of the salient elements of the object (such as the seashell's function as both protective covering and housing) and its relationship to other such objects or abstract principles. Through this "object lesson," the child would learn from direct perception but would also learn the method for correct observation "in which essential qualities of the object are distinguished from the accidental ones."[112]

Pestalozzi perceived this process as a solution to the problem of rote memorization and excessive book learning, but also to the problem of society's perceived acceleration. He sought to advance learning through direct, visual encounters as opposed to verbal or written encounters, but he also implicitly designed this approach to *decelerate* the learning experience. The entry on Pestalozzi in the *Encyclopedia of Philosophy* summarizes the goal: "According to Pestalozzi, it is the curse of modern civilization that its hasty and primarily verbal education does not give man enough time for the process of *Anschauung*, a term perhaps best translated as 'internalized apperception,' or as dwelling on the meaning and challenge of an impression."[113] Written and spoken information were too abstract for Pestalozzi, but the spoken lecture also expressed the "hasty" pace of modern life. The object lesson allowed the child to dwell on the salient features and relationships, or on *details* and their *correlation* to principle and form. Indeed, Pestalozzi's method embodied the modern educator's contested relationship to detail, which on one hand exemplified the confusion, ornamentation, and overstimulation that many saw as the problem with modernity's visual field. Melanie Keene, in her dissertation on the object lesson in England, discusses the example of palaeontologist Gideon Algernon Mantell, whose distaste for the crowds and hyperstimuli of the 1851 Great Exhibition in London fortified his belief in the value of attention to single, small objects as doorways to general principles.[114] Mantell's motivation for adopting Pestalozzi's pedagogical principles was common; in the profusion of detail, the object lesson commanded *time to dwell* on the form underlying the confusion.

Yet on the other hand, this very detail functioned as the object's ground of authenticity. Not just any object could serve as the focus of a lesson. Preferably, the object came from "nature," a term which for Pestalozzi was synonymous with all that is genuine, authentic, and free from artificiality.[115] A useful contrast in this scheme is the drawing, which was often an important *part* of the lesson—students learned about form by drawing shapes—but which was explicitly eschewed as the *basis* of a lesson. That is, students could draw an object and thereby learn about its form, but they could not use a drawing as the lesson's object itself. Pestalozzi was unhappy with rote memorization of words, but he was "equally dissatisfied with the use of pictures as a substitute for the direct experience of objects."[116] The difference, of course, is in the details, which in this method signified the randomness, contingency, and authenticity of nature; "authentic" meant "free from human influence," or "that which resists us" (one German word for "object" is *Gegenstand*, or "that which stands against"). Pestalozzi and his disciples complained about the overabundance of stimuli and detail in the modern world—almost invariably associated with the man-made world and its objects—but they clung to the detail of nature as the ground for their method.

The *Encyclopedia of Philosophy*'s entry on Pestalozzi translates *Anschauung* as "internalized apperception," which Christopher Ritter helpfully unpacks as "self-directed understanding through which newly observed qualities of an object are related to past experience."[117] The editor and translator of Johann Herbart's 1804 elaboration of Pestalozzi's method agreed that apperception was key: "He [Herbart] teaches that the chief object of instruction is to secure the reaction of the mind upon what is offered to sense-perception. We must understand what we see. We must explain it by what we know already."[118] Looking at *Anschauungsunterricht* in this way, it is clear that Pestalozzi's method signaled a distinctly modern moment when self-identity was aligned with observation, or more precisely, with the logical operation most closely associated with the practice of observation: *correlation*, which we explored in the previous chapter. To relate newly observed qualities to past experience and then to order those qualities and experience into a hierarchy and network of relationships— this was what modern educators and experts defined as *understanding*. And this is what they demanded from their students as they trained their vision to be expert. Observation and correlation ruled as all fields committed to visualization and new forms of seeing developed across the disciplines. Herbert Spencer, who popularized Pestalozzi's method in England, exclaimed, "Of new practices that have grown up during the decline of these old ones, the most important is the systematic culture of observation."[119] So educators versed in *Anschauungsunterricht* understood observation as a twofold operation: to *dwell* on authentic, natural detail and to *correlate* that newly observed detail to past experience with form and function. The object lesson was therefore a reaction to the perceived *quickness* and *disorder* of modern life.

Pestalozzi and his followers soon and quickly disseminated his method throughout Europe, where it found fertile ground.[120] The commitment to visualization in the sciences encouraged visual instruction in education, which was the logical extension of Pestalozzi's approach. And obviously, given its emphasis on nature, *Anschauungsunterricht* was perceived to be an especially effective method for teaching natural history and science.[121] Science educators, like nearly all those interested in visual education whatever their discipline or national origin, expressed the belief that images worked *quickly* and *efficiently* on the viewer's mind.[122] Chapter 2 explored this assumption about the presence and power of images; briefly, the trope presumed a homology between image, idea, and the structure of the human mind. Because of this presumed structural similarity, images were thought to leave a stronger impression than verbal or written descriptions. German physiologist Carl Jacobj put it best:

> Symbolically descriptive word images [*Wortbilder*] of concepts can be replaced by the simultaneously created visual image [*Anschauungsbild*] that represents the factual object of observation immediately and in all its details, so that it [the visual image] is imprinted in a faster, stronger, and more sustainable way on the conscious mind and, as a consequence, on memory.[123]

Even if the image worked as a fine substitute for a wordy description, Jacobj indicated that no educator ever left images to work on their own.

Indeed, without proper guidance, images were almost always considered anathema to proper understanding. So the second major trope or guideline in visual education, beyond the vividness and efficiency of images, was that the real power remained with the instructor's words. But this relationship to words was fraught, as we have seen: a discomfort with words as the basis of a lesson but a recognition that words needed to frame a lesson. As Ashwin notes, "In its most fundamental sense, then, *Anschauung* meant something like 'direct and correct observation': observation which was closely *associated with* language to the extent that the child was given the correct verbal equivalent at the point of making the observation, but which was not *mediated by* language."[124] These dicta remained unchanged as motion picture technology became an option for visual education, as we will see in the next section.

"CINEMATIC LESSON PLANS" IN ELEMENTARY AND ADULT EDUCATION

Yet this tidy portrait of Pestalozzi and *Anschauungsunterricht* should not deceive us into envisioning nineteenth-century pedagogy as an orderly sequence of inheritances. Over the course of the century, these principles were not only diffused so broadly that some did not even know their origin (if they could be said to have one), but there were so many interpreters, objections, and tweaks to theories of educational praxis that average reformers could be forgiven if they were sometimes confused about which practice belonged to which theory. For example, the art education movement positioned itself against what it perceived to be the excessive emphasis on science and mathematics in the approach of Herbart and his disciples, even as it adopted his method for ordering the lesson plan and even though Herbart himself argued against excessive rationalism in education, too. Nor should we be confused about the degree of impact these theories actually had in the authoritarian, state-run classrooms of imperial Germany—which is to say, hardly at all.[125] We should, however, recognize that all of these reform efforts hoped to provide *balance* and *order* to the educational experience, even if the understanding of proper balance and order was constantly in dispute. So when faced with the educational potential of film, which hardly anyone disputed, and the chaotic application of such potential, which everyone noted, advocates such as Lemke and Häfker drew upon the trends in pedagogical theory at the time in order to place film into an orderly and recognizable system of practice. This section will discuss precisely how this was accomplished with regard to elementary education (Lemke) and adult education (Häfker).

Films in the elementary school classroom were very exceptional, if not unheard of, during the early period in Germany. The apparatus was cumbersome and difficult for the untrained, but the expense was prohibitive as well. Because hardly any elementary schools could afford motion picture technology, the only solution, if film were to be a modern addition to their visual instruction agenda, was to take students to the local theaters. These excursions would generally take place during special screening times set aside during classroom hours, when the children and teachers could attend a matinee showing of a program of films deemed suitable by teachers, reformers, and school administrators. Most teachers and reformers, however, balked at the idea, having already decided that the movie theater was inherently corrupting, no matter what the content. Even so, commercial theaters were undaunted, especially after decrees in 1910 and 1912 in Breslau and Prussia, which restricted entry for children to film theaters except for special children's matinees.[126] Most provinces eventually adopted these laws, leaving theaters without an important part of the market and prompting many exhibitors to cooperate more readily with schools and educators. So a proprietor of a film theater in Berlin, in a pamphlet promoting the educational use of (his) commercial theaters, assessed the situation, stressing the importance of film for visual instruction:

> It is hardly necessary to mention the value of cinematic presentations for visual instruction [*Anschauungsunterricht*]. For some time there have been efforts everywhere to use this important element in teaching. Institutions of higher education, which have the necessary resources, have their own screening rooms equipped with cinematographic equipment. Similar plans have repeatedly been made for elementary schools, but they have not yet been carried out, mainly because of high costs.[127]

The word *Anschauungsunterricht* was commonly used in connection to film's potential place in the curriculum. It was especially popular with regard to teaching natural history, for which motion pictures seemed to enjoy a preternatural inclination.[128] Yet almost as often as educators suggested that motion pictures could function well as an object lesson, they questioned the "motion" part of "motion pictures," as in this declaration by a teacher named Rüswald: "Based on my experience, I would only support the use of film in teaching in the following circumstance: namely, that slides and film are used simultaneously. This demand is based on the facts that the impressions of the cinematographic image are too quick and thus too superficial, and that now more than ever a calm, quiet, measured dwelling on a single subject matter of education is bitterly needed."[129] This equation of movement, haste, and superficiality was common, and it was certainly consistent with the principles outlined in the previous section (we will see more of it in chapter 4). But educator Adolf Sellmann would have none of it:

> This objection is unjustified. Does the observation of movements in nature make one superficial? Not at all. On the contrary, this kind of observation can and must lead in many cases to an especially acute attention and thoroughness. If I want to grasp the action precisely, I must look closely, observe keenly, and turn focused attention to the moving process. Motion processes observed in nature can imprint themselves deeply on the soul, so that the observed motion becomes an inner experience. This can, of course, also be the case with motion that is observed on film. If I have focused on it with all my attention and therefore with all my soul, the impression lasts longer. *The living picture can often have a longer-lasting impact than the still picture.* Why should observed movement only fleetingly be remembered? It surely depends entirely on the mind [*Seele*] that looks at it.[130]

Sellmann and other advocates argued that movement actually *sharpened* the attention and that film's ability to replicate that movement—and, crucially, to reveal hidden aspects of it—gave it a privileged role among the *Medienverbund* (media ensemble) of early visual instruction. Educators argued back and forth about the role of filmed movement in the classroom, but their discussion could be distilled to a question that was rarely articulated: What is the role of observed movement in understanding the natural world? Or better: What does it mean to understand movement? Fundamentally, the split in camps corresponded to a choice between discontinuity and continuity, with teachers such as Rüswald advocating the use of slides (and film, in his case, although some argued for slides only), while Lemke, Sellmann, and others supported the use of *motion* pictures. Lobbyists for film saw themselves on the side of modern pedagogy, while critics of film defended against unnecessary and potentially dangerous technology.

Yet the primary justification for the use of film as an object lesson resided not necessarily in its movement but in the ability of the photographic image to replicate in its detail the randomness, variety, and disorder of the natural world. Educators signaled this by consistently calling film's image "faithful to nature" (*Naturgetreu*), which was a term often used by scientists and researchers when justifying their faith in the photographic image. Of course, not every aspect of the photograph is perfectly faithful to nature, but in these discussions, *Naturgetreu* referred most often to the level of detail that allowed the photographic image to reproduce patterns of texture and variation. Physician Richard Kretz wrote, "Photography is perfectly faithful to nature [*Naturgetreu*], that is, the images reproduce . . . all forms and proportions, the distribution of light and shadow in a completely correct manner."[131] Later, a geography instructor declared,

> The cinematograph offers an excellent substitute for student hikes and field trips. It leads the student not only through the wider area of his home province, but also through the most distant latitudes. He gets to know countries and peoples through his own observation [*aus eigener Anschauung*]. The most accurate description of a landscape, the most in-depth, vivid [*anschaulichste*] description of life and the activity in it, the most detailed painting cannot replace what film, with true fidelity to nature [*Naturgetreu*], parades before the eyes of the students.[132]

This "fidelity" referred not to color or depth or emphasis or any of the aspects of nature and observation that many complained photography could not represent well. Instead, it referred to the same qualities that brought scientific curiosity to bear on nature in the first place: the abundant variations on patterns of similarity and difference found in the forms and random textures of natural phenomena. Because photography could replicate these forms and textures with such detail, it could act as a substitute or a representation of the object of study. Like "vividness" (*Anschaulichkeit*), *Naturgetreu* referred to the advantage of images over words. But *Naturgetreu* was finally a stronger justification for film's role in the object lesson. Even if some disagreed on the role of movement in films seen in the classroom, *Naturgetreu* was a description of film that nearly all could agree upon, especially given the importance of nature's details to the principles of *Anschauungsunterricht*, as we have seen.

Nevertheless, if the image itself was more or less pedagogically controversial, educators objected, often rightly, to the lack of films specifically made for their curricula or the lack of rigor in most special school presentations at commercial theaters. Many complained that the *wissenschaftliche* ("scientific" or "academic") presentations programmed by commercial exhibitors were often nothing more than inoffensive nonfiction titles randomly strung together.[133] *Reformkinos*, or theaters especially procured in order to offer more edifying screenings, were sometimes a solution to this problem, but they required financial and institutional support not often forthcoming. In his 1909 pamphlet, *Kinematographie und Schule*, Georg Victor Mendel argued for creating standing theaters devoted to educational films for schoolchildren, a goal later achieved in the Urania theaters and the *kommunales Kinos* discussed earlier.[134] Ernemann, the Dresden-based equipment manufacturer, dedicated in 1909 its standing exhibition space to educational or scientific screenings for children three afternoons per week.[135] Also in Dresden in 1910, civil engineer August Kade funded twice-yearly educational and scientific screenings—with live musical and song accompaniment—in a city-owned exhibition space, which was dubbed the Kosmographia "scientific theater."[136] Overall, at least eight reform-oriented theaters or screening spaces opened in Germany between 1909 and 1914, all of which were touted as alternatives to educational programs at commercial theaters.[137] Even with these efforts, commercial cinemas remained the most widely available option for elementary school educational film screenings (fig. 3.5).

FIGURE 3.5. Children at Luisen-Kino in Berlin, circa 1910

Yet the problem of curricular integration remained. Educators complained that the film programs at these theaters were well intentioned, but offered far too much far too quickly:

> It is precisely the broad scope of the programs, which are all condensed into one to two hours, that must stir the most concern from a pedagogical point of view, as students do not arrive at calm observation or reflection, and none of the images can leave a lasting impression on them. . . . Not until projectors and film are cheaper and each school has its own cinematograph, or a special device can be attached to the projection apparatus that occasionally allows cinematographic images to be shown, which the teacher can present himself and explain as the lesson requires—only then can film have a profitable application in the school.[138]

From the typical teacher's point of view, school screenings at commercial theaters fundamentally lacked the *control* they required: motion pictures moved at a pace that could not be controlled, but also the program itself was out of their hands. According to this teacher, film had a future in the classroom only if that control—in a literally hands-on manner—could be assured. Lemke, whose efforts represented the most serious attempt to accommodate films and schools, took up the difficult challenge to integrate motion pictures into the grade-school learning experience.[139] He not only had to take into account these complaints but also to assuage implicit anxieties about the role of the teacher in a technologically mediated classroom; admittedly, his and other utopian proclamations about the coming ascendancy of visual media and the subsequent decline of lectures did not help matters in that regard.[140]

Even if teachers such as Lemke assumed that film's fidelity to nature (its *Naturgetreu* qualities) allowed it to adequately substitute for the object, that was only half of *Anschauungsunterricht*, or the "object lesson"; film needed to be integrated into the lesson plan as well. Lemke attacked this problem by following the principles for guided apperception that developed in the late nineteenth century, especially after a new generation of pedagogues in the 1880s further refined Herbart's interpretation of Pestalozzi. As a concession to those against commercial screenings for children, Lemke argued for a distinction between special screenings for schoolchildren (*Schülervorstellungen*) and screenings that incorporated methods unique to the classroom, which he dubbed *Schulvorstellungen*; Lemke thereby put the film program in the teacher's control.[141] He advocated that teachers understand the available films, actively work with the exhibitor to curate the programs, and then incorporate the screenings into lesson plans according to accepted Herbartian principles. Lemke also emphasized discussion sessions before and after film screenings in the commercial theaters. This plan required much more preparation: preselection of the films, discussion among the faculty about modifying or accommodating the program to the existing curriculum, and teacher training in using the films and leading discussion. For a brief time, Lemke even held intensive teacher-training seminars on educational film issues and techniques.[142]

In fact, Lemke's suggestions for organizing film into a lesson plan followed Herbartian principles step by step. One of Herbart's enduring legacies in pedagogy is a five-step program for guiding apperception within a lesson:

1. preparation (*Vorbereitung*)
2. presentation (*Darbietung*)
3. association (*Verknüpfung*)
4. generalization (*Zusammenfassung*)
5. application (*Anwendung*)[143]

Through these reform-minded and theoretical steps, teachers would introduce new knowledge of an object to a student by first reminding the student of already known material (preparation); then presenting the new material, repeating as necessary (presentation); encouraging associations between what is known already and what is new speculating on abstract principles linking the two objects (generalization); and finally thinking about how to apply this knowledge to new objects. In his own plan for the educational use of film, which he outlined in detail in *Die kinematographische Unterrichtsstunde* (The Cinematic Lesson Plan, 1911), Lemke followed these steps closely by explaining how film could be deployed through each (see fig. 3.6).

He also recommended specific films and groupings of films that followed the idea of apperception: introducing new objects and concepts through abstract connections to the familiar. In this way, Lemke and others accommodated film to principles of *Anschauungsunterricht* that were already widely accepted by reform-minded teachers: first, by connecting the specific features of the cinematic image to the visual emphasis in that tradition (*Naturgetreu* and the *object* lesson), then by demonstrating that film could be incorporated into the curriculum in a familiar way (apperception, film, and the object *lesson*). With these efforts, and his own journal devoted to

2. Lektion.

Aus der Lebensgeschichte des Weins.
(Films: Weinlese oder Champagnerbereitung)

A. Lehrstunde.

1. **Vorbereitung.** Ihr sollt mich heut im Geiste nach dem schönen Rheinstrom begleiten, auf dessen sonnigen Bergen der Wein wächst. Schaut her!

2. **Darbietung.** Der Film wird zweimal vorgeführt. Daran schließt sich die Aufforderung des Lehrers: „Erzählt, was ihr gesehen habt!"

3. **Wiedergabe.** Um zu zeigen, in welcher Weise sich diese vollzieht, führe ich das an, was mir die Schüler nach der zweiten Darbietung sagten: „Wir sehen Weinstöcke; die Winzer kommen und schneiden die Weintrauben ab. Ein junges Mädchen lacht dabei, sie hat ein gebogenes Messer. Nun werden die Weintrauben abgeliefert; Männer tragen große Körbe, die auf einer Stange hängen, zum Wagen; dieser fährt die Weintrauben zur Kelterei. Dann versammeln sich die Arbeiter wieder und erhalten etwas in ihre Körbe. Der Wein ist nun auf Fässer gezogen, diese werden geschüttelt; dann sehen wir, wie der Wein in Flaschen verkorkt wird; diese werden geschüttelt, ein Mann spritzt etwas ab, dann werden die Flaschen verkorkt und verpackt, das Verpacken geht sehr schnell."

4. **Klärung der Vorstellungen.** Bei dieser Wiedergabe durch die Schüler war folgendes zu erklären: einmal hatten die Schüler falsch aufgefaßt, als sie sagten: Die Winzer empfingen etwas in ihre Körbe. Diese Szene gehörte nämlich nicht zur Weinbereitung, sondern stellte dar, wie die Winzer in der Nähe ihres Arbeitsplatzes von Händlern Essen kauften; die Winzer gehen nämlich während der Arbeitszeit nicht nach Hause. Darauf mußten die Schüler hingewiesen werden. Ferner mußte ihnen der Vorgang erklärt werden, „wie etwas abgespritzt wurde"; sie hatten dabei den typischen Unterschied zwischen Wein und Champagnerbereitung aufgefaßt, und ich brauchte nur zu erklären: Champagner wird aus Wein in der Weise bereitet, daß man eine Likör-

FIGURE 3.6. A page from Lemke's *Die kinematographische Unterrichtsstunde* (The Cinematic Lesson Plan, 1911), in which he provides a Herbartian lesson plan for a specific film

educational film and slide material (*Die Lichtbildkunst in Schule, Wissenschaft und Volksleben*, Storkow 1912–1914), Lemke was on the leading edge of the educational use of film in Europe and the United States. For an excellent example of the use of film for adult edification, however, we must turn to Hermann Häfker (fig. 3.7).

After the failure or, at best, limited success of the attempts to create an alternative production and distribution system, reformers realized that focusing on exhibition held the most promise for fulfillment of their program. So like Lemke, Häfker looked to existing commercial theaters to establish an alternative exhibition venue to create a suitable educational or edifying environment. Miriam Hansen has argued that the peculiarities of early cinema exhibition presented the structural possibility of an alternative public sphere. The variety format; the sense of theatrical space; the combination of lectures, live music, sound effects, and so on; and the uneven development of modes of production, distribution, and exhibition—all contributed to "overlapping types of public sphere, of 'nonsynchronous' layers of cultural organization."[144] Between the "fissures of institutional development," alternative modes of reception and experience could emerge. The reformers, of course, hoped to "synchronize" these layers, not only by coordinating the modes of production, distribution, and exhibition, but also by integrating the various cultural spheres that commercial cinema was already grafting upon itself: literature, science, the tradition of the lecture series, and art.

Hermann Häfker's "model presentations" (*Mustervorstellungen*) are the best example of the reformist exhibition program aimed at adults. Some have called him Germany's first film theorist—he was certainly one of the very first to write regularly about the cinema.[145] He began the century working as a writer, journalist, and translator for a number of periodicals, covering a range of topics, from Shakespeare's sonnets to his own bicycle tour of Finland. He was one of the first writers for *Der Kinematograph* and a spirited contributor to and editor of *Bild und Film*, eventually writing three books on film for the Volksvereins publishing company. His *Bild und Wort* (Image and Word) society film exhibitions were prototypes for many "model presentations" that reformers tried to implement on a regular basis around Germany. His 1913 book, *Kino und Kunst* (Cinema and Art), was an elaborate justification of the artistic potential of cinema and an extension of his earlier work in the reform journals. In this monograph, he describes his attempts to create aesthetically pleasing and educationally effective cinema

FIGURE 3.7. Hermann Häfker

programs. As we have seen, Häfker was not alone in these attempts, but he was unique in providing theoretical justifications of his programs.

Like Lange and the other reformers, Häfker was concerned with the aesthetic sensibility of the masses and the influence of bad taste. His comments about taste were directed particularly to the contemporary state of film exhibition. Of the nickelodeons of the 1910s, Häfker noted, "the educated circles have been repulsed by the tastelessness of the programs."[146] Further, "it's not the What of the program, but the How of the presentation that makes the impression."[147] Of course, he certainly did not withhold complaints about the "sensational" films the producers presented to the audience. But unlike many of his contemporaries, such as Willi Warstat, who felt that censorship was the proper solution, Häfker continued to express his concern for the "tasteless" exhibition. This tastelessness referred, most generally, to the intrusion of modern life's hectic pace into the auditorium, where spectators were assaulted with a "breathless chase of one number after another, accompanied by intertitles, the uninterrupted noise of the projector, the lights, etc." In this regard, Häfker's goals were consonant with those of *Anschauungsunterricht*. Häfker demanded an exhibition that avoided the exciting and the extraordinary and instead tried to establish "a quiet and natural mood." He advised exhibitors to program their films in accordance with classical aesthetic principles, building tension and then release by alternating comedies with dramas and "scenes from the life of nature and simple people." The exhibitor should also refrain from putting all the films on one reel, allowing instead a short pause between them so that "the spectator's eyes would receive their necessary recovery time and the nerves a moment to relax."[148]

This last bit of advice points to a range of literature dealing with visual fatigue and the motion pictures. In this discussion, the equation of cinema with modernity was most explicit. Häfker expressed the concerns of the day quite well when he complained, "image and form, word and sound, color and line . . . rain like a hailstorm on the nerves of modern man, especially in, but not limited to, the city."[149] Cinema came to epitomize this hailstorm. Some of the first articles written on cinema in Germany were medical papers on the harmful results of the "flicker" effect. Other medical investigations dealt with the threat of eyestrain in the film theaters. Nearly all reformers or opponents of cinema criticized its threat to public health and vision, as we saw in chapter 2.[150] And as we saw in chapter 1, this outcry represented the larger preoccupation with fatigue that characterized debates coming out of the late nineteenth century.

As Anson Rabinbach has shown, the trope of fatigue was more than a scientific mania of the age; it expressed a profound anxiety of decline and social disintegration. In the medical, scientific, and even literary study of fatigue, there was "a tendency to equate the psychological with the physical and to locate the body as the site where social deformations and dislocations can be most easily observed."[151] In other words, metaphors of health and sickness were used to express national anxiety. Fatigue, together with neurasthenia, was more than a physical ailment—it was also perceived as a *moral* disorder, a sign of weakness and the absence of will. Neurasthenia, mentioned before in connection to cinema's suggestive power, was the most typical metaphor for the delicate condition of the national psyche.

Häfker viewed modernity as a series of "shocks"; he sought a haven to which he could escape the hailstorm of modernity. He just wanted to rest for a while, to give his nerves time to recuperate, and he wanted to make cinema such a haven. But cinema would never be this sanctuary, he said in 1908, "so long as the corresponding sense of illusion is missing and the correct mood is lacking."[152] There is so much in the modern world to disturb this mood, but treating film as an art form, especially exhibiting films "tastefully," could slow this flood of "the much-too-much" (*eine Eindämmung des Vielzuvielen*).[153] He planned to do just this with his "model presentations." In 1910 he presented to the Image and Word association in Dresden a model program that was to be the prototype for other cities.[154] The selection consisted mostly of nature films, but plans for further exhibitions included travelogues, scientific films, and *actualités*. Originally, he intended to continue the exhibitions in coordination with local schools, but the project fizzled due to lack of readily available films for continuous programming.

The 1910 presentation, entitled "Spectacles of the Earth," highlighted Häfker's preferred form, the nature film: "The first part showed high mountains and deserts; the second part concerned ethnological subjects (Laplanders, Chinese, Arabs, Indians, cannibals, etc.). The third part dealt with 'The Thousand Games of Water' (Victoria Falls, Niagara Falls, storms on the coast, surfs, rapids, geysers, underwater volcanoes from New Zealand)."[155] The films were accompanied by lectures, slides, music, and nature sound effects, all of which Häfker tried to orchestrate into a *Gesamtkunstwerk* of Wagnerian proportions. The presentation began with a lecture of what to expect, what to look for, and "in which sense to take it." It would then alternate films with slides and lectures, carefully presenting each. Häfker provides a detailed—and obviously quite proud—description of the final section of the program:

> Then it became dark once again. You could hear the sound of water, and as the curtain parted, you could see an actual waterfall, etc. At the end of this section there was a beautiful image—one of the few that were also artistically impeccable. ["Trip on the Avon River in New Zealand"]. The spectators didn't know at first exactly where they were, when, as if by magic an invisible, delicate music sounded, completely in rhythm and harmony, as if made for the image (and, of course, purposefully arranged), accompanying the scene to its conclusion. As the lights came up in the auditorium in front of the closed curtain, the loud applause was not only for all that had been seen up to that point, but for the last scene and the genuine musical enjoyment that accompanied it. The proscenium seemed a magical realm, a mysterious land of light, life, and music.[156]

Häfker's further descriptions show the pains he took to assure a proper environment and mood. He reports that three men worked the slides to guarantee precise timing; curtains hung all around the auditorium to dampen the sound; colored stage lights splashed the proscenium as the audience seated themselves (59–60).

These preparations certainly have many precedents in traditions of theatrical and orchestral performance, and the format is adapted from the long tradition of lecturing in performance halls.[157] Like other reformers, Häfker insisted that focusing on the viewing environment was the first step toward cinema's eventual aesthetic respectability. But Häfker set himself apart from his contemporaries with his claim that the entire cinematic apparatus—image, light, music, sound effects, lectures—could be used in combination for the artistic presentation of film, calling this Wagnerian use of cinema *Kinetographie*. Häfker's efforts to guarantee the proper conditions show his concern was primarily with the spectator's relation to the film. The conditions of reception were vitally important to his program and his conception of the function of art. The full effect of the "total presentation" (*Gesamtvorführung*)—here illustrated by the audience's reported confusion/illusion that they were in New Zealand—required the spectators' complete and undistracted attention. It required, in short, their *contemplating* the film as they would an artwork. He hoped that he could educate audiences to this way of viewing films.

Häfker took his cue from Lichtwark's *Exercises in the Contemplation of Art Works*, which provided the foundation for the training of taste and vision, a way of viewing art that Häfker transferred to film. This way of viewing was certainly not unique, having immediate precedent in the German tradition of *Bildbetrachtung*, as exemplified by Lichtwark's exercises. His presentations did not simply provide an environment conducive to the passive reception of art; they set out to actually train the audience's vision. Through the lectures, Häfker guided the audience to what was important and "in which sense to take it"—that sense being, primarily, vision. But he did not want to stop there: "In order to draw attention to especially interesting images, perhaps one should *occasionally* employ little *signal lights*. They could be colored incandescent lamps placed around the screen that light up shortly before surprising scenes or scenes that are difficult to see" (57–58). These visual cues would reinforce his verbal guidance, perhaps eventually creating some sort of physiological response. (Apparently, Häfker did not consider that the lights might have been a distraction.) There was also a moral dimension to this way of looking. In Häfker's discussion of approaches to art, contemplation was exemplary of a certain economy of energy, in that focused attention on the artwork is a way of exercising the will against the excessive stimuli of modernity. If neurasthenia was a type of mental fatigue caused by the difficulties of dealing every day with modern life, art provided not only a haven of unity and harmony in a distracted and disorganized world, it also offered an opportunity to train the attention and exercise the taste.[158] Art and the artistic presentation of film were workouts for the mind; museums and film theaters could be mental health clubs.

◊ ◊ ◊

Lemke's and Häfker's education of vision and taste exemplifies the strikingly ambivalent tone of contemporary reform movements, in that they were both nationalistic and progressive, both protective of traditional values and open to modern innovations. Yet that combination of tradition and modernity was and is common for all early adopters of new (media) technology who try to incorporate their new tool into an established disciplinary method, or especially an established mode of expert viewing. In their case, their technophilia or excitement over the new medium prompted them to find within it some potential for compromise or appeal to their less excited colleagues. It is hardly coincidental that the problems most cited about film and its exhibition—its quick tempo, excessive detail, and jumbled programs—corresponded to some of the most likely complaints about modernity in general, namely, quick pace and disorder. Lemke's and Häfker's plans, anxiously aware of these charges, attempted to contain them by enveloping film within a protective casing of order, dwelling, and observational method, whether called *Anschauungsunterricht* or contemplation. Whether motion pictures, per se, could be incorporated into "traditional aesthetics" is the question of the next chapter. But before we move to that topic, let us briefly review the ideological connection between vision, taste, morality, and education.

Schiller, unlike most philosophers, was not suspicious of the senses, least of all of the sense of sight.[159] According to Schiller, knowledge of the physical world passes through the senses and is therefore contingent on them, but vision provides the opportunity to transcend the physical world and enter the aesthetic on the way to the moral realm. The key to this journey is "contemplation." Schiller declared, "As long as man, in that first physical state, is merely a passive recipient of the world of sense . . . he is still completely One with that world. . . . Only when, at the aesthetic stage, he puts it outside himself, or *contemplates* it, does his personality differentiate itself from it." Upon entering the aesthetic, the subject renounces his or her passions and creates the possibility of becoming a *moral* being. Contemplation is the exercise through which this process begins. The very act of perception, the very apparatus of vision is both inextricably implicated in the sensual world and ironically outside of it. "From the moment a man *sees* an object, he is no longer in a merely physical state," Schiller noted.[160] That is, while exercise of the other senses testifies to one's *proximity* to the natural world, vision offers the opportunity for *distance*. The aesthetic of contemplation, exemplified by what Benjamin called the "aura" of an artwork, is based on distance. The aesthetic of distraction, again illustrated by Benjamin's discussion of cinema and architecture's tactile qualities, is based on proximity.[161] Schiller again: "If desire seizes directly upon its object, contemplation removes its object to a distance, and makes it into a true and inalienable possession by putting it beyond the reach of passion."[162]

Hence the whole concept of subjectivity—becoming a knowing subject by objectifying and therefore "possessing" Nature—is dependent upon the refusal of passion and sensuality. Once "outside" this sphere, the moral becomes possible. For Schiller, the renunciation of Nature is not a goal in itself as much as a necessary step toward the fulfillment of humanity's moral potential. Like the act of vision, always in the physical world while simultaneously having the potential to transcend it, humanity balances on the fine line between the sensual and the moral. Schiller called this line "the aesthetic."

Taste, like vision, is both embedded in Nature and somehow removed from it. Even more than vision, taste implies participation in the social world. An artwork affects the individual, but the exercise of aesthetic judgment implies universality. When we find ourselves agreeing that something is beautiful or sublime, we are exercising a unique and precious form of intersubjectivity based on our recognition of shared capacities for aesthetic experience. This is what Kant meant when he, following Vico, called taste a *sensus communis*—a communal sense.[163]

The concept of taste provides the ideal illustration of the relation between aesthetics and ideology. While society could impose moral behavior on its subjects by appealing only to Reason, it is more efficient to employ the emotions in this task. As the medium between Nature and Reason, the aesthetic allows this operation. Schiller explained:

> The ethical State can merely make it (morally) necessary, by subjecting the individual will to the general; the aesthetic State alone can make it real, because it consummates the will of the whole through the nature of the individual. Though it may be his needs which drive man into society, and reason which implants within him the principles of social behavior, beauty alone can confer upon him a *social character*. Taste alone brings harmony into society, because it fosters harmony in the individual.[164]

Faced with a society that they felt was becoming more alienated and fractured, reformers latched onto the promise of harmony and unity offered by the aesthetic realm. Lichtwark and Häfker focused on vision to effect a renewal in taste. Their exercises in the contemplation of artworks were, like Lemke's Herbartian ordering of the "cinematic lesson plan," attempts to ward off the distractions of modernity, prophylactics against the "much-too-much." If spectatorship had been characterized as an addiction that lulled audiences into an impressionable somnambulism, Häfker, Lemke, and other reformers hoped to counteract this state by inscribing cinema into an aesthetic of contemplation and reflection. The audience's vision required *training* so that mental and physical fatigue would not set in; it was a way of "pumping up" moral weaklings. While Häfker's *Gesamtkunstwerkeffekt* would provide the illusion necessary for the aesthetic experience, it was not intended to lull the audience into distractedness. Rather, it provided access to the *sensus communis* through a *disinterested*, distanced aesthetic experience. Nature films were both safely asexual and reminders of potential harmony. Yet the use of nature films is ironic; the reformers' emphasis on vision and distance and disavowal of *Kinodrama* and "sensational" films amounted to a refusal of sensuality and corporeality—in short, a refusal of Nature. Training audiences to conform to certain rules of observation—an ascetic education of their vision—was part of an ideology that combined educational practices and Kantian aesthetics to establish some sense of social order.

This legitimation strategy—anesthetizing/aestheticizing cinema and its audiences—was a response to modernity's perceived assault on the body and the body politic, often exemplified by cinema's "flicker." Häfker and others felt that training the aesthetic sensibility could fend off the "shocks" of modern life. The combined concepts of "the child" and "taste" served as a fulcrum for the reformers, allowing them to "uplift" the motion pictures and incorporate cinema into their ideology. As Germany's *Kinoreformers* attempted to redeem and legitimate cinema as Art, they recognized within it the potential for recovering a lost utopia of unity and, ultimately, a means for social control. Yet even by 1912, presenting contemplation as a solution to modernity's ills seemed slightly old-fashioned, as we shall see.

4

THE PROBLEM WITH PASSIVITY

AESTHETIC CONTEMPLATION AND FILM SPECTATORSHIP

> *But the difficulties which photography caused for traditional aesthetics were child's play compared to those presented by film.*
> — WALTER BENJAMIN (1936)[1]

As cinema's bandwagon—already heavy with reformers and trade journal reporters—rolled toward World War I, literary intellectuals, pundits, and other belletrists climbed aboard (sometimes climbing down again after their thousand words) just to see what the ride was like. Judging from the sharp spike in the number of film essays written between 1912 and 1914, it was apparently de rigueur to offer a learned editorial on the way modern life and culture found expression through this new phenomenon. Just as a range of opinions comprised the reformist discourse explored in chapter 3, so the tone of this collection of articles, which Anton Kaes felicitously dubbed the *Kino-Debatte*, extended from peevish outrage and haughty condescension to diplomatic concession or even roguish delight in the new medium.[2] This expansion of the discourse in Germany corresponded to cinema's more visible public profile at this time, due to a stronger domestic film industry; the production of longer and more emotionally involving story films; the rise of film stars, such as Asta Nielsen and Henny Porten; the emergence of picture palaces and the successful "embourgeoisement" of the cinematic experience; and, most tellingly, the development of the *Autorenfilm*, a short-lived strategy that attached literary and theatrical luminaries to industry projects.[3] Perhaps because those weighing in—including Ernst Bloch, Max Brod, Alfred Döblin, Georg Lukács, Kurt Pinthus, Walter Serner, and others—were or were to become such prominent names in the German literary tradition, this part of the conversation about film has received the most attention in secondary surveys of the period. We should be quick to note, however, that reformers and trade journal writers did not disappear during this time—on the contrary, they exerted a clear influence on the direction of the discussion, even if negatively—but the sheer number of new voices in the mix has tended to shift our attention from the pedagogical and commercial sections of the debate.

This chapter will be no different in that respect. It will, however, back away from the scholarly emphasis on the relation between literature and film. While depictions of the *Kino-Debatte* have been as varied as the debate itself, scholars usually focus on the battle between image and word in the contemporary discussion of cinema's relationship to German culture. It is indeed hard to ignore the incessant complaints about the supposed decline in literacy attributed to the consumption of sensational and superficial images rather than great literature or the many declarations that cinema could never be considered genuine art as long as it lacked words to express the depths of the human soul. Anton Kaes was absolutely correct when he noted that the tension between old and new "found expression in a vigorous discourse about the relationship between literature and cinema," or, simply, that the debate about cinema was "a debate about the literature of the time."[4] While theories of film started to disengage themselves from literary or theatrical models by the 1920s, before World War I, cinema and its champions felt the need to justify themselves in terms of literature. For Sabine Hake, literature was "the primary reference point" for writers coming to terms with modernity through their essays on cinema.[5] Peter Jelavich emphasized the attempts to conform film to "traditional bourgeois aesthetics, which demanded clarity of authorial voice and rootedness in the written word."[6] Helmut Diederichs similarly charted the ways in which this group divided the ground between literature and film.[7] Stefanie Harris has demonstrated, on the other hand, how cinema's unique form shaped the literary work of such writers as Kurt Pinthus.[8] As Heinz-B. Heller usefully pointed out, these were *literary* intellectuals after all, so it comes as no surprise that their response to film would be from a position firmly grounded in their chosen medium.[9]

Focusing so closely on early cinema's relationship to literary form and turf, however, unintentionally narrows our understanding of cinematic experience to that particular relationship, leaving relatively unexplored the question of film and aesthetic reception in general. The above quotation from Benjamin—his point that the challenge that photography presented to traditional aesthetics was "child's play" compared with film—summarizes a fairly common conception: that film was emblematic of a change in aesthetic standards at the fin de siècle as mass reception and distraction replaced individual contemplation as the dominant or most appropriate mode of aesthetic reception. Indeed, in film and media studies this account is more or less taken on faith in Benjamin's word alone. I have no argument with the general outlines of this story, but even his well-known "Work of Art" essay compresses the events considerably. So we have a dual historiographical problem, as I see it: the emphasis on the literary misses a large swath of the cinematic experience, specifically the relationship between viewer and image, while the leap from contemplation to distraction in film history and theory is too often taken for granted without spelling out the character of "traditional aesthetics" and the transition to whatever replaced it. This chapter argues that a close, renewed examination of the *Kino-Debatte* is essential to solving both historiographical problems.

Previous chapters were concerned with the criteria for film's legitimacy within any given discipline and the adaptability of motion picture technology to an expert mode of viewing as a major factor in establishing that legitimacy; this chapter will explore film's legitimacy within the realm of aesthetics and its degree of adaptability to the expert mode of viewing known as aesthetic contemplation. It presumes that, following a trend in German aesthetics since Kant, the question of aesthetic value hinged on *reception* more than form. That is, the major statements on aesthetics in the long nineteenth century—especially those of Kant, Schiller, Schopenhauer, and others, but excluding those within the Hegelian tradition, which was concerned with how meaning inhered in form rather than how we experienced it—were concerned primarily with the role of aesthetics as a way of being in the world, more than whether any particular form was more artistic than another. They were concerned with the function of art within a moral, ethical, social, and philosophical system. The value and legitimacy of art in this system depended primarily on the singularity of the experience it aroused. The exact nature of that experience has been a topic of constant exploration since these statements, but the significance of that experience for the system in general—whether philosophical, moral, or political—cannot be underestimated. Form prompts experience, to be sure, but these and other statements emphasized the universal character of the aesthetic experience rather than the variety of experiences created by various forms. More often than not, evaluations of any given form, such as music, rested on the ability of that form to catapult the reader/listener/viewer into a particular kind of aesthetic experience. Better experience usually equaled more hallowed form.

So to understand fully the relationship between early cinema and aesthetics, we must focus on the relationship between the cinematic experience and aesthetic experience as it was understood, or between watching movies and the expert mode of viewing called aesthetic contemplation. This chapter asks the following questions: to what extent did the cinematic experience, as these authors described it, conform to their understanding of aesthetic experience? Specifically, how did the experience of film align with their understanding of aesthetic contemplation as an implicitly expert mode of viewing? To what extent did the descriptions of cinematic experience participate in changes to expert conceptions of this mode of viewing?

To answer these questions requires first understanding what the authors presumed about aesthetic experience. The trouble with aesthetic experience, of course, as countless analytic philosophers have complained, is its notoriously slippery surface. It is very difficult to define logically, or even on an individual basis. Fortunately, for the purposes of this project we need not come to a philosophically rigorous conclusion; instead, we need only outline what the authors *thought* aesthetic experience was. Even that is elusive, because any given writer borrowed ideas or presumptions, often haphazardly, from a long and varied aesthetic tradition. Indeed, we might find it more historiographically productive to think of descriptions of aesthetic experience (rather than aesthetic experience per se) as statements often expressing competing values. If one writer latched onto Kant's idea of disinterest and detachment as the point of aesthetic experience, another might champion, after Schopenhauer, the idea of losing oneself in the artwork. If one author saw repose as the fundamental criterion of any aesthetic encounter, another might have emphasized the value of the free play of associations while engaged with a work of art. My point is that these key ideas or terms—detachment, loss of self, repose, free play, and others—describing any aesthetic experience (not just cinematic) *functioned within a system of implicit dichotomies.* Let me explain.

If we gather common ideas about the nature of aesthetic experience in the major statements of Kant, Schiller, Schopenhauer, and others within the tradition of nineteenth-century German aesthetics, it appears that aesthetic contemplation entailed (1) *detachment or disinterest*, meaning that aesthetic judgment is without desire, passion, or self-interest; (2) *repose*, in that the object is lingered over without haste, giving enough time for the free play of the imagination;[10] (3) *activity*, meaning that the mind is alive with associations and correlations; and (4) *loss of self*, in that the contemplation of the object can lead to either a transcendental immersion into the object (Schopenhauer) or a moment of awareness of one's participation in a larger community (Kant). Under this category we can also subsume discussions about the role of the body in aesthetic experience; for some, aesthetic contemplation was a purely mental operation; others looked to the body as a model for understanding aesthetic pleasure, while avoiding the purely sensual. Some terms are mutually reinforcing (disinterest and participation in a community, for example), while others are contradictory (detachment and immersion seem to be opposite ideas).

But we can arrange these terms—detachment, repose, activity, and loss of self—into spheres or larger categories that also contain their opposites. These categories are: *space* (distance/proximity or detachment/immersion), *identity* (loss of self/self-awareness), *time* (repose/haste), and *agency* (activity/passivity or free will/determination) (see fig. 4.1). This arrangement shows especially clearly the *ideological* implications of these terms and of the descriptions of aesthetic experience they evoked. That is, to give aesthetic experience its moral or philosophical weight within a larger system—the pivotal role of Kant's *Critique of Judgment* with regard to his other *Critiques* comes to mind—the terms writers and philosophers used to describe aesthetic experience were rhetorically linked to fundamental ideological values (or ideologically inflected categories of experience) such as agency and identity. Understanding the rhetorical role these terms played allows us to glimpse the stakes of any historical debate regarding aesthetic experience. Pairing the terms with their opposites also underlines their mutual dependence in those historical debates.

This scheme also allows us to escape the analytic philosophers' futile attempts to logically reconcile descriptions, an evasion much more in keeping with the sharpest understandings of aesthetic experience as indeterminate. Schiller argued, for example, that the aesthetic state is a *mediating* moment "midway between matter and form, passivity and activity," wherein indeterminacy is the preferred state of being.[11] Accepting the indeterminacy of the aesthetic state should turn us away from pat definitions and toward an understanding of aesthetic contemplation as dichotomous,

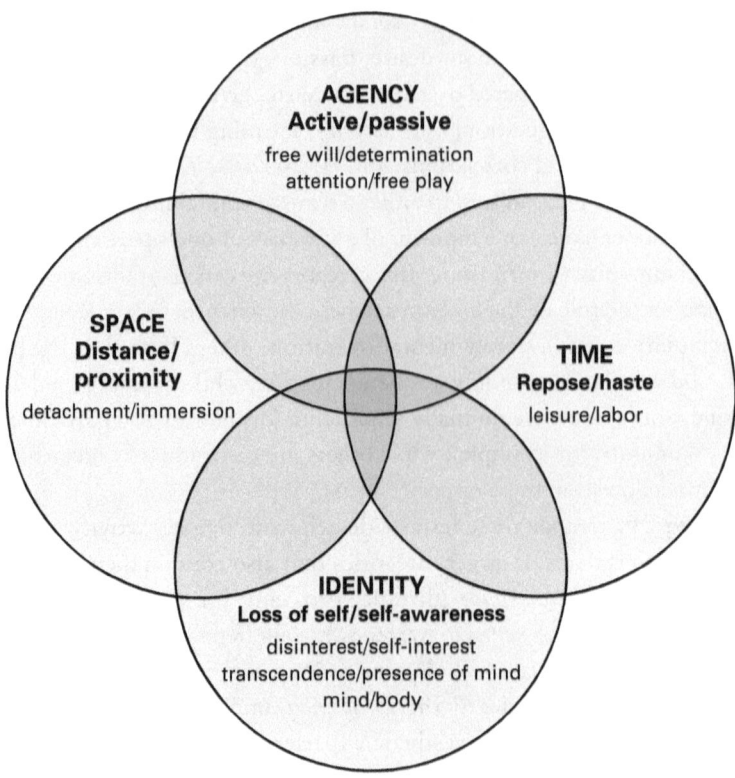

FIGURE 4.1. Aesthetic experience as a series of interlocking dichotomies

fluid, oscillating, or unfixed. (Again, I want to stress that questions about what the aesthetic state actually is and even whether it exists are irrelevant when considering historical descriptions of it.) For example, descriptions of contemplation almost always preferred repose over haste—thereby presuming that both the object and the subject were more or less in a state of stillness—but the quick, gestalt-like expert glance over the artwork as a whole was not ruled out as part of the process.[12] However, detachment was almost a universal feature of these descriptions, so its opposites, self-interest or passion, were hardly ever included as viable elements of the experience.[13] On the other hand, descriptions of the aesthetic experience often oscillated between sides of a dichotomy: claims that emotional detachment or psychic distance were crucial to aesthetic experience (Kant, Schiller) must be balanced or reconciled with descriptions that emphasized emotional projection or immersion into the artwork (Schopenhauer, Adolf Hildebrand). Likewise, the relationship between active and passive stances in aesthetic experience is often unclear from the descriptions; mastery alternates with surrender (as in many descriptions of the experience of the sublime), while focused attention alternates with the free play of associations or loss of self in the experience. This approach, then, matches a theoretical understanding of aesthetic experience as indeterminate to a historiographical scheme that does not try to resolve contradictions but allows them to stand to highlight their ideological function.

Obviously these categories overlap; questions of agency and identity are indeed hard to separate. But that is precisely the point: the overlapping conceptual categories strengthen the ideological weight of any given description. So this scheme might be useful for an investigation of *any* historical description of aesthetic experience; it is definitely useful to understanding the early debates about motion pictures. To summarize, viewing the historical claims for aesthetic experience in terms of a series of dichotomies, rather than as a series of firm characteristics, has three advantages. First, we avoid the futile philosophical debate about what the aesthetic experience actually is and replace it with the recognition that historical descriptions of aesthetic contemplation—what writers and aestheticians *thought* it was—have been varied and contradictory, but not infinitely so. Second, by subsuming the dichotomies under the general categories of identity and agency, we emphasize that most philosophers (in the neo-Kantian tradition) recognized aesthetics as a moral category that mediated between opposite poles of their choice. For example, the Kantian/Schillerian tradition of putting Art between Reason and Sensuality extends in spirit to at least Theodor Adorno, if not also Jacques Rancière and other contemporary philosophers who hope to find for Art a place apart and some emancipatory potential. Seeing the aesthetic experience as a series of unstable dichotomies recognizes not only Art's socially mediating function but also that historical descriptions of aesthetic experience emphasized its productive indeterminacy.

Finally, with regard to this specific project, these dichotomies allow us to see more clearly how the discussion of motion pictures was ultimately an *aesthetic* debate, in the sense that these essays on cinema may be about the status of literature, the content of the films, or whatever, but the core issue was the larger ideological problem of *the moral significance of the aesthetic experience*. These writers obviously cared about the decline in reading skills, about the loss of ability to concentrate, the superficiality of the mass audience, the novelty of the moving image, and many other things that may or may not have indicated the decline of civilization. But, at bottom, all these complaints can be traced to the *character* of the aesthetic experience, which apparently mattered most to these writers. "Film presentations are generally rejected from an artistic perspective," wrote Emilie Altenloh about the taste of the upper classes.[14] While Altenloh undoubtedly refers to judgments of artistic *form* ("von *künstlerischen* Gesichtspunkten"), this chapter agues that complaints or judgments of form were ultimately in the service of a better aesthetic experience. Modernity was therefore an affront to their aesthetic sensibilities and values, which were rooted in a philosophical tradition that privileged a particular, expert mode of viewing a stationary image or object. The extent to which cinema challenged or could be accommodated to these values is the question of this chapter.

To burrow further, we could argue that these dichotomies grew out a more fundamental, historical division between active and passive viewers. As we have seen, film spectators were often characterized as hypnotized by the cinematic image, even addicted to it, and captivated by the ease with which it entered the stream of consciousness, while cinema's fragmented temporality, quick pace, and motley form matched the habits and psyche of the modern city dweller, who was at once impatient and distracted, unable to concentrate on a single, main idea. Hence movies were perceived as effortless, whereas real art required work; movies were a jumble, whereas real art had a distinct form; movie spectators needed their next film like a drug, while real art encouraged a detached observer; movies moved by far too quickly, disallowing the continuous, leisurely attention required of real art. If these sets seem familiar, it is partly because they have been repeated over the centuries whenever audiences are described; theater audiences were probably the first victims of such characterizations.[15] Rancière's explanation of the paradox of spectatorship sheds some light on this issue:

> There is no theatre without a spectator. . . . But according to the accusers, being a spectator is a bad thing for two reasons. First, viewing is the opposite of knowing: the spectator is held before an appearance in a state of ignorance about the process of production of this appearance and about the reality it conceals. Second, it is the opposite of acting: the spectator remains in her seat, passive. To be a spectator is to be separated from both the capacity to know and the power to act.[16]

The idea of aesthetic contemplation from Kant and Schiller to early twentieth-century professional aestheticians looks like a deliberate attempt to counter the problem of passivity in spectatorship by endowing the viewer with both a protocol and a philosophical justification that connect aesthetic enjoyment to reasoning and knowledge production ("the capacity to know"), on one hand, and political maturity or citizenship ("the power to act") on the other. Endowing aesthetic contemplation with agency and linking it to individual or civic identity gave this mode of viewing the moral weight it needed to counter a mode of viewing perceived to be common to the passive, irresponsible crowd. The contemporary discussion of the cinematic experience could not avoid confronting this weighty connection between agency and contemplation—often to film's disadvantage.

Precisely because contemplation had long been endowed with high moral import—with values such as reason and citizenship—any changes to the conditions of contemplation put these values at risk, especially as mass reception became more common with new urban demographics and forms. If cinema were not at the heart of these transformations, for many essayists it was a worst-case scenario. At the same time, however, the discussion about cinema was not merely a reactionary defense of Enlightenment-era ideals; it was a collective attempt to find some common ground, to ask (often implicitly) whether contemplation was the proper way to engage with these new forms, and if not, what was? In fact, Benjamin's own position on the relative value of contemplation and distraction was not limited to his words in the "Work of Art" essay. As Carolin Duttlinger has shown, Benjamin's conception of attention (*Aufmerksamkeit*) played a central and complex role from his earliest essays, and if he advocated distraction (*Zerstreuung*) as the perceptual stance most appropriate toward modernity in the "Work of Art" essay, that was only one part of a long conversation he had with himself about the relationship between a mode of viewing and one's position toward and within modern life.[17] Ultimately, this conversation was about the relationship between "a way of being with images," in Dudley Andrew's ingenious phrase,[18] and a way of being in the world. For Benjamin as well as the writers of the *Kino-Debatte* decades earlier, the two stances were one and the same.

So this chapter will explore the status of aesthetic reception in the context of discussions about cinematic spectatorship. The first section will trace the broad themes in the *Kino-Debatte* about movies, modernity, and spectatorship in relation to the rhetorical or ideological categories of time and agency. Through an examination of Schiller and Schopenhauer, it will stress the philosophical and subsequently ideological connection between temporality, indeterminacy, and aesthetic experience, and how this constellation persisted in the descriptions of early cinema. The second section will inspect discussions of aesthetic experience in professional aesthetics of the time, especially around the concept of *Einfühlung*, or "feeling into" the artwork. These discussions about the relationship of emotions and the body to aesthetic experience were attempting to come to grips with, on one hand, unresolved ideas about reception in nineteenth-century aesthetics, and on the other, modern aesthetic reception characterized by turn of the century democratic access to art. This section will demonstrate that discussions of the cinematic experience also pressed on some of the same troubling issues—such as emotional projection, immersion and detachment, and visual pleasure—regarding the nature of aesthetic contemplation in the modern world that professional aestheticians discussed. This section explores the categories of space and identity. The third section focuses on Georg Lukács's famous early essay on film, "Thoughts Toward an Aesthetic of the Cinema" (1913), to demonstrate the nascent political critique of contemplation that was to be such a central feature of his and other theorists' stance after World War I. This final section will also return to Benjamin, Siegfried Kracauer, and the discussion of contemplation and distraction to realign our history of these theories in light of the *Kino-Debatte*'s discursive construction of spectatorship.

AGENCY AND TEMPORALITY IN THE AESTHETIC EXPERIENCE OF CINEMA

The scheme outlined above could apply to any historical description of aesthetic experience after Kant, but this section will recast some familiar themes in the *Kino-Debatte* in terms of the roles time and agency played in contemporary understandings of aesthetic experience. Let us begin with a relatively obvious case: motion pictures as an emblem of social acceleration. This was a bright, visible thread in early debates about cinema all over the world. Basically, commentators complained that the incessant tempo of motion pictures—its relentless push forward and its quick change from one view to the next—exemplified the problems of modern social acceleration.[19] Educator Georg Kleibömer discussed cinema's pace at length in his 1909 essay, "Cinema and Schoolchildren." He opened his attack by describing the

Great Lisbon Earthquake's profound effect on young Johann Wolfgang von Goethe, and then wondered whether the eruption of Mount Pelée in 1902 had the same effect:

> And let us search for just one child for whom this natural event had an indelible significance. How is it that we all well know that this search will be in vain? Because in our time all the world's grand strokes of fate are immediately known to us in every detail; because one harrowing incident displaces the other before it can be "worked over" by our mind. Now even the most tragic events only touch the surface of the soul.[20]

Superficiality abounds, according to Kleibömer, because we do not have enough time to absorb the emotional significance of events before we are compelled to move to the next one. "The moral education of a child," he instructed, "should include a *few* events that give him an understanding of certain emotional values," but "above all, such emotions must not confront other feelings in close temporal proximity." And in the cinema? "In rapid succession jest and seriousness alternate on the screen, since variety must reign. . . . So first *horror, fear* in the highest degree, then *compassion* increased to the utmost." He therefore criticized "the danger that this rapid traversal of the entire scale of feelings poses for the truth and depth of children's emotions." Kleibömer saw filmic form as a synecdoche for the rapidity with which modern information was conveyed; he assumed that because of that rapidity, children could not absorb the significance of events, and this was the root cause of their perceived superficiality (as opposed to, perhaps, the possibility that they were six-year-olds not named Goethe).

Kleibömer was not alone in his opinion; writers of all sorts, not just cranky reformers, connected movies and modernity in this way. Austrian author Karl Hans Strobl similarly argued, "The cinema is one of the most perfect expressions of our time. Its hasty, erratic tempo corresponds to the nervousness of our lives; the anxious flickering, the whisking away of its scenes is the extreme opposite of a measured, regular stride, of confident persistence."[21] In the pages of *Der Kinematograph* a commentator noted matter of factly, "And so the cabaret, with its indiscriminate, democratic disposition, appeals best to the nervous haste and variety of our time, which of course is not say that the theater is doomed. But now the cinema, the projection house, successfully competes with the cabaret for the affection of the people just as the latter surpassed the legitimate stage."[22]

Right-wing editor of *Der Kunstwart* and proto-Nazi Wilhelm Stapel was even more explicit:

> Because of the rush of images, you get used to absorbing only an approximation of the impression; you do not get a clear and conscious understanding of the image in its details. Therefore only the coarse, surprising, sensational impressions stick. The sense for the intimate, the precise, the delicate is lost. The patrons of the cinema "think" only in garish, vague ideas. Any image that lights up their mind's eye takes up all of their attention; they no longer mull and reconsider it, they no longer indulge in the particularities and the reasons.[23]

Devotees of the connection between early cinema and modernity will find this theme familiar. Noteworthy, however, is not simply this well-known correspondence but the strong implication that cinema's tempo interfered with the usual aesthetic experience of viewing images. Stapel was very clear in this regard: the spectator does not "get a clear and conscious understanding of the image in its details. . . . they no longer mull and reconsider it." If one is to understand the image, one must attend to the details slowly. Now this could just be a commonsensical caution that one cannot simultaneously do anything hastily and well, but it was not usually put in those terms. Instead, the complaint was specifically about the superficiality that results from processing (or being forced to process) visual imagery too rapidly. What was the connection between viewing tempo and shallowness?

To appreciate the moral implications of this equation, we must return to Schiller and Schopenhauer, who based their philosophies on a common understanding of time. The importance of temporality for Schiller's aesthetics is not only rarely discussed, it is key to understanding later ideological objections to film. So the following will explicate their aesthetic schemes while emphasizing temporality and indeterminacy, which will clarify certain aspects of the debates about cinema. Even if these early nineteenth-century philosophers were in many ways considered obsolete by the early twentieth century, their ideas about the moral and cultural importance of aesthetic experience—and the relationship between that significance and the passing of time—were widely and implicitly assumed. Schiller, especially, presumed that everyday, sensuous existence is necessarily tied to the inexorable succession of moments, while abstract thought and reason—or the Platonic idea—exist outside this temporal course in a timeless, static, changeless existence. Schiller called our animal nature the "sense drive" and the spiritual or intellectual side of our existence the "form drive." He described the sense drive:

> Since everything that exists in time exists as a succession, the very fact of something existing at all means that everything else is excluded.... when man is sensible of the present, the whole infinitude of his possible determinations is confined to this single mode of being. Wherever, therefore, this drive functions exclusively, we inevitably find the highest degree of limitation. Man in this state is nothing but a unit of quantity, an occupied moment of time—or rather, he is not at all, for his Personality is suspended as long as he is ruled by sensation, and swept along by the flux of time.[24]

Here and elsewhere, Schiller defined his understanding of the different facets of human existence in terms of their relationship to time. In the sense drive, one is nothing but "an occupied moment of time"; by being occupied by (in terms of both concerned with and controlled by) sensuous existence—ruled only by needs and desires of the moment—one is "swept along by the flux of time." This conception of temporality, for Schiller and others, implied a limit on human potential: "Wherever, therefore, this drive functions exclusively, we inevitably find the highest degree of limitation." The form drive, however, "embraces the whole sequence of time, which is as much as to say: it annuls time and annuls change. It wants the real to be necessary and eternal, and the eternal to be necessary and real" (81). Ideas aspire to timelessness, being outside of the succession of events and therefore outside of perception. Reason is timeless, in Schiller's scheme, but also ethereal. So Schiller invented another motive force to harmonize, balance, or counter the other two:

> For as long as he only feels, his Person, or his absolute existence, remains a mystery to him; and as long as he only thinks, his existence in time, or his Condition, does likewise. Should there, however, be cases in which he were to have this twofold experience *simultaneously*, in which he were to be at once conscious of his freedom and sensible of his existence, were, at one and the same time, to feel himself matter and to know himself as mind, then he would in such cases, and in such cases only, have a complete intuition of his human nature. (95, emphasis in original)

For Schiller, the desiring, physical side and the spiritual, intellectual side of our nature have their limitations, but when we are able to experience both at the same time we are without limitations. That is, if our lopsided, specialized existence is determined by our physical desires (that is, as matter without form), or by our intellectual aspirations (that is, form without matter), then the third way escapes determination entirely; it is indeterminate. Schiller called this third mode of existence "the play drive," which is exemplified by aesthetic experience. It is a moment of indeterminacy, of oscillation between Sense and Reason, or between physical necessity and law. That moment of indeterminacy holds the greatest human potential, when free will is exercised most productively in the service of "a complete intuition of [our] human nature." Schiller concluded:

> Our psyche passes, then, from sensation to thought via a middle disposition in which sense and reason are both active *at the same time*. Precisely for this reason, however, they cancel each other out as determining forces, and bring about a negation by means of an opposition. This middle disposition, in which the psyche is subject neither to physical nor to moral constraint, and yet is active in both these ways, pre-eminently deserves to be called a free disposition . . . we must call this condition of real and active determinability the *aesthetic*. (141, emphasis in original)

The exercise of free will, then, is the goal of aesthetic experience. But not entirely: the goal is to be "sensible of existence" without being determined by it; to be aware of all aspects of our nature while they are held in abeyance. Aesthetic experience is a balancing act between physical necessity and moral obligation, actively taking part in both sides of the human condition without letting either determine us. Schiller thereby gave the aesthetic experience a strong ethical charge; it serves as the moment when we are most fully human, overly determined by neither of our natures: "For, to mince matters no longer, man only plays when he is in the fullest sense of the word a human being, and *he is only fully a human being when he plays*" (107, emphasis in original). For Schiller, aesthetic experience was the grandest kind of play, which brings our fullest human potential into relief.

Especially interesting and rarely noted, however, is the emphasis Schiller placed on the temporal dimension of this experience. The sense drive is "swept along by the flux of time," the formal drive "annuls time," but the play drive is repeatedly described as an experience of simultaneity: "a middle disposition in which sense and reason are both active *at the same time*." The importance of simultaneity in his scheme marks Schiller's move as a very modern gesture; both simultaneous and indeterminate ("this condition of real and active determinability [*Bestimmbarkeit*]"), his category feels very modern indeed. It is not too great a leap to argue that simultaneity creates this indeterminacy; we are both in and outside of time, which puts us in a position of being in neither one drive nor the other, but both states at the same time. This clarifies his formulation of the play drive: "the play-drive, therefore, would be directed towards annulling time *within time*, reconciling with absolute being and change with identity" (97, emphasis in original). The annulment of time experienced in the formal drive is reconciled with the experience of time within the sense drive, thereby resulting in the "annulling time *within time.*" Whatever one makes of the validity of his interpretation, there can be no doubt that Schiller's conception of the aesthetic experience was ultimately temporal.

Schopenhauer also took pains to underline the temporality of the experience, even if it was different from Schiller's: "It is the state where, simultaneously and inseparably, the perceived individual thing is raised to the idea of its species, and the knowing individual to the pure subject of will-less knowing, and now the two, as such, no longer stand in the stream of time and of all other relations."[25] Both subject and object are raised to a transcendent state in Schopenhauer's conception of contemplation, during which the object becomes an idea and the subject loses itself and its desire in the process, thereby also finding temporary relief from the misery of longing and striving. The emancipatory potential of aesthetic experience derives from its ability to provide a momentary escape from time; this is different from Schiller's more moral and community-oriented aesthetics, but it affirms rather than cancels Schiller's conception of the aesthetic as a primarily temporal experience. For both philosophers, and in the most common conceptions of aesthetic experience in the long nineteenth century, human limitation was tied to "the flux of time," and its emancipatory potential was linked to the escape from that flux or being able to maintain a balance between competing temporalities.

This review of Schiller's and Schopenhauer's philosophies is necessary, because it reminds us of the *ethical or moral mission* they assigned to aesthetic experience, thereby giving it an ideological weight that it carried through the nineteenth and twentieth centuries. It also establishes *the importance of time* within traditional aesthetics, especially the notion of repose in the aesthetic stance. Finally, it emphasizes that Schiller and others calibrated their conception of contemplation to highlight its *indeterminacy*, or more precisely, the importance of the indeterminate quality of the aesthetic experience as an emblem of human liberty. All of these features of aesthetic experience—especially the focus on experience, rather than artistic form—spread and took root in everyday understandings of aesthetic issues that we see in the *Kino-Debatte* and other, early discussions about viewing motion pictures. Again, the goal is to demonstrate that these debates were largely discussions about aesthetic (as opposed to narrowly literary) experience, especially this model of contemplation, to which modern spectatorship was detrimentally compared. Indeed, the unfavorable comparisons functioned primarily as a defense of contemplation's ideological value. We could dismiss the comparisons out of hand as ideologically compromised, but we would lose the deeply ambivalent and complex character of aesthetic reception. Contemplation was not yet dead; whether the values associated with it actually ever died is still open to question.

Those values were primarily ethical in imperial Germany. Modernity, it appears, was too pushy; it shoved its spectators along, giving them no pause for reflection, which was an offense to values of repose. So when writers such as Gaupp, Sellmann, Stapel, Strobl, or others noted the potential effect of the fast pace of movies and modernity, they were not simply complaining about the vapid superficiality of "kids these days"; they were defending a very long tradition that insisted on leisurely approaches to the image. To allow oneself to succumb to the "flux of time" while attending to an image—or to let the image impose such limitations—was an abdication of the duty to self-cultivation, of the obligation to become fully human. Contemplation was depicted as a moment of escape, but also a moment of self-awareness; the aesthetic experience had a special place in state and cultural self-conceptions in Germany, which the righteous defended vigorously.[26] But because of its pivotal role between emancipation, civic duty, and self-awareness, contemplation represented a complex, ambivalent stance toward the world that was not so easily dismissed, as Benjamin recognized (and which I will discuss at greater length later in the chapter).

Another good example within the discussion concerns the fragmentary, disjointed quality authors attributed to modern life and to cinema. If movies and modernity were too pushy and quick, they were also disjunctive: the cinema "is short, rapid, even encrypted, and it stops for nothing. There is something compact, precise, even military about it. This fits very well with our age, an age of extracts."[27] These comments on form were often framed as an analogy between movies, modernity, and the mind of its spectators; modern life was disordered, as were movies and the thought processes of its audience. This homology—a discursive construction of film's spectator—was usually an implicit defense of aesthetic ideals of detachment and effort. Theater critic Hermann Kienzl's brief commentary is exemplary:

> The psychology behind the triumph of cinema is urban psychology. Not only because the metropolis is the natural focal point from which all social life radiates, but especially because the metropolitan soul—this inquisitive and inscrutable soul perpetually on the run, stumbling from one fleeting impression to the next—is really the soul of the cinema! Indeed, some city dwellers even lack the stamina and concentration for mental and emotional absorption—and probably also the time, especially in Berlin, a metropolis agitated by work-fever. The same trivial drive for relaxation that leads city dwellers to the operetta or farce after work to replenish their exhausted energies leads them likewise to seek the effortless pleasures offered by movie theaters.[28]

Here we have the usual stereotypes about the metropolis and its inhabitants: they are driven by passion, desire, necessity; their actions are hurried and jumbled. Needless to say, these were also transgressions against the aesthetic values of detachment and repose. Kienzl pretended to offer merely a description of the situation, but it was an implicit defense of the principles outlined earlier.

There is also the implication, well before Hugo Münsterberg, that movies were popular and effective because they somehow mimicked the structure of the modern mind, or at least the mind of its audience. For many commentators, the link between movies, modernity, and mind made film "the most psychological representation in our time."[29] The "effortless pleasures" of the cinema replenished the modern soul in a way—a trivial way, they would say—that other entertainments could not. Most writers disparaged movies and the mind of the masses in a dual dismissal, but others saw it differently:

> Only film technology permits the rapid sequence of images that roughly corresponds to our own imaginative faculty and in some measure imitates its jerky unpredictability [*Sprunghaftigkeit*]. Part of the fatigue to which we finally fall prey while watching theatrical works of art results not from the noble effort of aesthetic enjoyment, but rather from the exertion in adapting to the plodding, affected movement of life on stage. Spared this effort in the cinema, one is free to devote a considerably more uninhibited commitment to the illusion.[30]

Whereas Kienzl and others assumed that the mind of an intellectual is ordered and disciplined, Lou Andreas-Salomé concluded, after spending a year with Sigmund Freud and Viktor Tausk, that the mind is no such thing. When Kienzl remarked that "city dwellers lack the stamina and concentration for mental and emotional absorption," he compared the stalwart literary mind with the flighty, disordered, lazy mind of the moviegoer. But Andreas-Salomé acknowledged that movies moved quickly and disjunctively, which matched her understanding of the "imaginative faculty." This put both in a favorable, or at least neutral, light. Indeed, her sympathetic view of cinema's aesthetic potential depended on this structural similarity; the speed at which we think—slowed down considerably by theater—was matched only by cinema. Likewise, she saw in early cinema's unpredictability—we can imagine that to early, novice audiences the continual surprise at the next view might have given them the impression that film narrative or editing was capricious (which it sometimes was)—a disconnectedness that fit her understanding of human thought. Actually, the primary difference between her idea of mind and Kienzl's was that she refused to create two models divided by class, one for elite intellectuals and one for the mass audience. Her model assumed that, when it comes to thinking, we are all a hot mess, trained or not. Münsterberg also developed a single model, but presumed that all our imaginative faculties are well ordered. In any case, many authors assumed a structural similarity between movies and mind, but their aesthetic judgment depended on whose mind movies resembled.

But Kienzl and Andreas-Salomé shared one assumption: true art required effort and movies were effortless. The "noble effort of aesthetic enjoyment" was a persistent trope in these and other discussions about the difference between high and low cultural forms, in which high art was usually differentiated by its complexity or difficulty and low art by its simplicity. For Andreas-Salomé, cinema's effortlessness was a result of its structural similarity with the mind, especially with regard to its motion and pace. It had also something to do with the age-old comparison of images and ideas, a comparison common to the *Kino-Debatte*, too, as when Hermann Duenschmann, apparently a big fan of Gustave Le Bon, compared the film audience with a crowd: "The crowd thinks only in images and can only be influenced through images that act suggestively on their imagination."[31] This implied that the learned think in words, but the visual status of ideas has always been ambiguous. At least since Descartes, many philosophers have defined "ideas" as images, not words. Descartes consistently made the analogy that ideas are "like portraits drawn from Nature."[32] Or Locke, in his *Essay Concerning Human Understanding,* wrote that ideas are "Pictures drawn in our Minds,"[33] or that the "Idea is just like that Picture, which the Painter makes of the visible Appearances joyned together."[34] Now an important caveat here is that Descartes and Locke were thinking not of concepts, but of sensory ideas—the images that our brain creates from the sensory information gathered by normal perception.[35] But the leap to concepts is not too hard to make and has been made since Plato, perhaps. The point is that one support for this notion that films were "effortless" was the isomorphism between images and the way the mind is thought to work. (This was also a support for the assumption within discussion about educational media that visual instruction is more efficient than instruction with other kinds of materials.)

Yet images per se cannot be easier to comprehend; otherwise the aesthetic contemplation of paintings would have no pedestal upon which to place itself. So there must have been something other than the similarity between images and mind that supported the claim that movies were effortless. Ulrich Rauscher gives a clue: "I fear the cinema has one disadvantage for the audience: because it tells its story so comfortably, because it takes over the operations of sense-making itself, the cinema, which could foster the creativity of our literati, will bring about a general laziness of the public's imagination."[36] Film is effortless, because it "takes over the operations of sense-making itself" (*weil er die Versinnlichung der Vorgänge selber übernimmt*) for the audience, leaving nothing for the imagination to work on. The "noble effort" of aesthetic enjoyment was perceived as the free play of associations that accompanied a productively ambiguous artwork; the imagination filled in interpretive gaps with its own associations, thereby bringing subject and object together as the hermeneutic circle between part, whole, and subject tightened. Just as photography was not considered legitimate aesthetic material because of its mechanical nature and its excessive detail—which left nothing to the imagination, presumably—so, too, motion pictures were dismissed because their detail and their narrative patterns left the viewer nothing to do but watch. There was more to it than that, however, because these same movies were disdained for their chaotic, confusing stories, which doubtless would leave much to the imagination. The problem, then, was that the succession of images was not merely obvious, but *preordained.* Whether the images were confusing or crystal clear, they were always *someone else's* images, and if we combine this concept with the "images = ideas" formula, we have a series of views *imposed* upon the viewer in a preordained temporal sequence, which does not happen with paintings or literary images. Hence the loss of free play, or more importantly, *free will*. The cinematic experience, in this view, was not *indeterminate*, but rather *determined*. (Benjamin quotes Georges Duhamel: "I can no longer think what I want to think. My thoughts have been replaced by moving images."[37]) By showing us where to look, editing and cinematic narrative forced these writers to look there; this was for many patrons a step too far.

Those patrons would presumably never allow themselves to enjoy cinema's flow of images. But the inadequacy of this rigid decree—that the flux of time jeopardizes agency and thereby imposes limits on human potential—for modern life becomes apparent in the *Kino-Debatte* (and elsewhere) as well, especially in discussions that emphasized *alertness* and *somnambulism*. Essays often invoked these tropes when describing the tolls of modern life. Recalling Georg Simmel's vivid portrait of "The Metropolis and Mental Life" (1905) or Freud's idea of the "stimulus shield" (1920), these essays also assumed that life in the big city dulled the senses and made one blasé, requiring ever greater thrills to be satisfied.[38] For many writers, cinema was just the ticket: "But every day and every hour cinema restores to our pampered senses, which in the heyday of technology have forgotten how to be astonished, the feeling of Pygmalion's enchantment with Galatea."[39] Anticipating film theory of the 1920s, this author and others claimed that cinema renewed vision by offering a point of view we otherwise could not have. For writers such as Rauscher, cinema's preordained succession of images was oppressive and imprisoning, whereas for this writer, it was liberating to see with another's eyes. Max Brod, too, was rejuvenated by a visit to the cinema:

> The vividness with which so much happens [in the film] has finally shaken me out of my semi-somnolent state. Now on the way home I become an inventor, imagining new images for the Biograph: a pursuit in which, instead of automobiles, locomotives or trolleys, two ships, a cruiser and a pirate ship, race against each other over the wide surface of the sea, the gap between them narrowing amid the most furious shooting.[40]

Or this early editorial: "I believe that through the cinema we have only now learned to see. We have been awakened to the pleasure of watching [*Schauen*]. . . . Reality appears much more clearly before us, and the interest is twice as great. We can almost let our mind fall asleep and reap with our eyes what the soul desires."[41] If these writers were drifting through their modern lives, semi-awake to its challenges and blasé to its pleasures, cinema roused them from this aesthetic slumber to give them a new outlook. Here, then, a trip to the cinema did exactly what any aesthetic experience should do: renew our perception. This idea, however, was relatively new in aesthetics and already expressed the waning dominance of the Schillerian model.

Indeed, that aesthetic contemplation was undergoing a transformation is evident in essays that evoked alertness and somnambulism in terms of modern life. Strobl both admired and was uneasy with social acceleration and its effect on our perception:

> But wherever the streets of world commerce run, wherever the billions cross paths, wherever goods are converted into money and money into power—at these hubs of human and commercial relations a sustained readiness is necessary. Constant presence of mind has replaced the old contemplativeness that let one's mind wander, because we no longer need to have it at hand.[42]

According to Strobl, a leisurely approach to the traffic of images or to the city was simply not viable. One required instead a "presence of mind" that was very much embedded in the present but not necessarily "swept away by the flux of time." Alertness, as we will see in our discussion of Benjamin, was considered a more modern response to time and succession than repose, yet was no less complex in its relationship to time.

Aesthetic experience, as discussed by Kant, Schiller, and Schopenhauer in the early nineteenth century, was a complex and contradictory structure that was nonetheless required to bear a particularly heavy ethical load. If it collapses logically with the barest nudge, it was still home to a family of values related to agency and identity. As we have seen, early discussions about film were often couched in terms of agency, especially with regard to the spectator's relation to the passing of time. Complaints about cinema's incessant temporal push and its preordained succession of imagery continued at least one line of thinking about aesthetic experience, which insisted that the viewer's stance before the image was also a stance before the world, one that recognized our limits but also strove to glimpse human potential in an ephemeral moment of transcendence. Aesthetic contemplation was understood as an exercise of free will, so to surrender oneself completely to the image or to the flow of time was an abdication of duty; indeterminacy meant liberty, but the rush of preordained images meant constraint. The temporal complexity of Schiller's and Schopenhauer's visions of aesthetic experience was therefore at odds with conceptions of cinematic temporality as constraining or controlling. But with statements attesting to the importance of alertness in modern life or positing that seeing the world from another viewpoint could reawaken the senses, the discussion began to question the utility of a leisurely aesthetic stance in favor of an approach to images—and the world—that was not merely cognitively active and unhurried, but *reactive* and *quick*. So in the *Kino-Debatte* we can begin to see the "difficulties" new forms might have posed for what Benjamin called "traditional aesthetics."

EINFÜHLUNG, IDENTITY, AND EMBODIED VISION

But an understanding of aesthetic contemplation as inherently tied to leisure and free will—or, basically, the cultured gentleman's prerogative—is only partially correct. We must also consider the relationship between aesthetic experience and identity or subjectivity. One might think that the gentleman's prerogative also presumed a stable, unified self, which would be confirmed in aesthetic experience.[43] Kant's theory of the beautiful, for example, held that pleasure arose from the harmony between the artwork and our cognitive faculties, which assumed a resonance between them that was stable and stabilizing. His theory of the sublime, however, maintained that its pleasure derived from the oscillation between the physical threat of nature's awesome power or infinitude and our cognitive framing of that boundlessness. Faced with this power, the threat of physical annihilation presented a literal loss of self, while the mastery implied by cognitively framing such power was self-affirming. So the existence of that particular kind of aesthetic experience, especially given the importance of the sublime in aesthetic theory, meant that aesthetic experience per se was never inherently stabilizing.

This split between body and mind, or between loss of self (whether through physical danger or an experience of transcendence) and self-awareness (whether cognitive or corporeal), was exacerbated by the structure of Schopenhauer's system. On one hand, Schopenhauer insisted that a loss of self was absolutely necessary to any experience calling itself aesthetic, if it were to have any mollifying effect at all. As Schopenhauer described it, the viewer attends to the object and gradually loses self-awareness as the associations and Ideal become prominent. Likewise, his description of aesthetic contemplation emphasized the purely visual and cognitive—as opposed to corporeal—experience of viewing. For him, the aesthetic gaze divorced itself from the world, from the flux of time, and even from the artwork itself, the specifics of which did not figure largely in his understanding of aesthetic experience. Indeed, Kant, Schiller, Schelling, Schopenhauer, and Hegel all ignored the specific features of any given artwork in favor of using the object as a prompt for abstract thought and for mounting ever more complex philosophical systems. Johann Friedrich Herbart continued this approach in his formalist aesthetics, which "proposed a simplified theory of form, one that defined aesthetics essentially as the science of elementary relations to lines, tones, planes, colors, ideas, and so on," without reference to the specter of "content" or "meaning." Similarly, "the perfect aesthetic frame of mind for Herbart was a state of absolute indifference, that 'quiet seriousness' that lies 'between depression and excitation.'"[44] In this tradition, their understanding of the gaze matched their understanding of the object: neither emphasized detail or materiality; both could be characterized as disembodied or ephemeral.

On the other hand, Schopenhauer also simultaneously pursued another approach, which he did not see as contradictory, but complementary. Part of the second volume of *The World as Will and Representation* speculated on the *physiology* of the perceptual act, an approach that aesthetically minded scientists such as Hermann von Helmholtz, Gustav Fechner, and Wilhelm Wundt pursued further in their experiments on the nature of (aesthetic) perception.[45] Their findings, however, indicated that the human body was a transient, variable, and unstable field upon which any subjectivity rested at its own risk. That is, if a stable and unified subjectivity previously depended on a happy conception of the body as similarly stable and unified, experimental physiology and psychology of the mid-nineteenth century invalidated that guarantee. As Jonathan Crary has argued, the fleeting nature of the human body left barely any solid ground for subjectivity.[46]

Many aestheticians during the late nineteenth century looked for an approach to questions about aesthetic perception that would navigate between the Scylla of abstract, disembodied philosophizing about art and the Charybdis of fragmenting, "scientistic" approaches to the aesthetic experience. They were less interested in philosophical or physiological questions about how we perceive form and space than in the psychological problem of how we take delight in the specific characteristics of form and space.[47] Robert Vischer's seminal 1873 dissertation, *On the Optical Sense of Form*, coined the term *Einfühlung*, or "feeling into," as an answer to this question.[48] Vischer meant it to explain "the symbolism of form," or how spatial form has meaning for us: the subject "unconsciously projects its own bodily form—and with this also the soul—into the form of the object" (92). But *Einfühlung* eventually came to mean emotional projection in general, and then, with Edward B. Titchener's translation of the term as "empathy," it transformed into a broad psychological concept.[49] In aesthetics, however, *Einfühlung* was discussed in terms of several problems, including a renewed interest in analyzing the specific formal and material features of artworks and the experience they prompted[50]; describing the emotional content of the aesthetic experience in terms of projection of self in relation to distinct forms; and understanding the role of the body as a measure of both art and aesthetic experience, especially in terms of aesthetic pleasure.[51] That is, writers taking up the idea of *Einfühlung* explained aesthetic pleasure as a resonance between the structure of the body and the structure of the artwork, thereby explicitly acknowledging the embodied nature of perception.[52]

This brief history is necessary, because it clarifies "traditional aesthetics" and its "challenges," which were twofold: (1) new artistic *forms*, such as photography, fit uncomfortably in the canon of fine or even applied art; and (2) new aesthetic *experiences*, especially mass reception. The question of form is outside this chapter's purview, but we can address the nature of mass reception, which differs from individual contemplation in at least two pertinent ways: (1) the emotions of the audience amplify those of the individual, thereby making emotional projection into the work much more prominent; and (2) because of this amplification, and the awareness of being in an audience, the sense of embodiment is more pronounced as well. We could therefore view the discussion of *Einfühlung* in the late nineteenth and early twentieth centuries as an exploration of conflicting dichotomies nascent in "traditional aesthetics" (such as the relations between mind and body, or between free will and destabilizing aesthetic experience), while also attempting to reconcile the contradictions inherent in these grand aesthetic systems with changes in class demographics as more and more people gained access to and were interested in art.[53] Specifically, I would argue that *Einfühlung* aesthetics could be seen as an attempt to reconcile issues especially pertinent to mass reception (embodiment, emotional projection) with individual contemplation and its ideological implications. While mass reception was not explicitly acknowledged in discussions of *Einfühlung*, the theories pave the way for a deeper understanding of the role of emotion and corporeality in aesthetic experience. Indeed, many contemporary theorists have looked back to *Einfühlung* aesthetics as a possible framework for new theories of embodied spectatorship.[54]

The dichotomies circling around the category of identity were less fraught ideologically than those concerning agency. Whereas no one argued seriously for "determination" against "free will" (although "surrender" might have been an option), philosophers were able to describe aesthetic experience in terms of "loss of self" just as persuasively as "self-awareness." The issue of embodied perception, however, brushed against the almost universal requirement of disinterest, in that any hint of sensuality immediately lowered the status of aesthetic experience to mere pleasure. Like paper and scissors, physicality always lost against reason in this ideological game. But because the ideological stakes were generally not as high in this category, this section will not focus on those stakes; instead it will emphasize descriptions of aesthetic experience. Specifically, in an attempt to understand precisely the nature of film's "challenge" to "traditional aesthetics," this section will draw upon theories of *Einfühlung*, which were particularly adept at describing emotional projection and embodied perception in aesthetic experience, two issues also present in the debates about cinema. In fact, if the legitimacy of cinema as an aesthetic form in part hinged on the answers to these questions, it was also true that these two issues—which I want to stress were present in "traditional aesthetics" from the beginning, if we look closely—mounted the strongest theoretical and practical challenge to individual contemplation as it was generally understood. Emotional projection and embodied perception threatened to change the questions about aesthetic legitimacy itself. If *Einfühlung* aesthetics tried to reconcile individual contemplation with features of aesthetic experience that were also prominent in mass reception, then the presence of a form such as cinema pressed harder on those questions.

This is not to say, however, that early discussions of cinema participated directly in debates within professional aesthetics or that professional aestheticians participated in the *Kino-Debatte*; to my knowledge no writings on film during this period in Germany ever mentioned the term *Einfühlung*.[55] But the early discussions of film brought up the same issues. To be sure, salons crowded with paintings and people had already prompted much complaint as mass reception elbowed its way into the art world and demanded a seat at the table. But cinema was not merely one of many emblems of modernity; the nature of the experience itself—the feeling of movement, synesthesia, emotional projection, and spatial depth—coincided precisely with issues in aesthetic reception that professional aesthetics hoped to pin down. We can therefore read the *Kino-Debatte* as part of an extended conversation in German culture about the nature of aesthetic experience. The debates were not just a series of complaints about film encroaching on literary form and territory; they were also a very sophisticated debate about the nature of aesthetic reception, about what it meant, in a broad sense, to contemplate an image.

As an opening example, consider this quotation from a 1912 article describing an *actualité* that featured Wilhelm II on one of his hunting trips:

> The Kaiser sits motionlessly—only the image twitches slightly here and there, flickering and spotting up as if something were boiling under the projected surface. Suddenly, he raises his rifle and my ear hears the gunshot without its being fired. The audience silently rejoices over their Kaiser's fine shot; a wave-like movement goes through the room. The mountain goat rolls.[56]

This otherwise incidental passage is exemplary of the early debates about film in its focus on audience reaction. This writer described, for example, a process of emotional projection: the audience understood the representation by projecting itself into the scene of the hunt. They rejoiced *with* the kaiser as well as *over* his excellent aim. The writer also emphasized the *physical* response of the audience, demonstrating the importance of the body for this process of projection. The "wave-like movement" that ripples through the room was an effect of this projection. But here we should note what is different about the cinematic experience over the solitary enjoyment of a painting. This writer described the mutual amplification of audience emotion (which Benjamin later theorized in his "Work of Art" essay) that is inherent to the cinematic experience and that manifested itself here as a physical reaction. There is a reciprocity between spectator and screen, just as *Einfühlung* aesthetics described the mutual relationship between viewer and artwork, but here that feeling of resonance was amplified by the number of people in the room all feeling it at once (fig. 4.2).

Note, too, that if *Einfühlung* aestheticians thought about that resonance between painting or sculpture and viewer in terms of "movement," here that term is no longer metaphorical but strikingly literal: the movement of the image, of the kaiser, and of the audience were all working in concert to provide the basis for emotional projection. If before—as I will explain later—the idea of "movement" in aesthetics was limited to what we might term "inner movement," or emotional resonance, now that same reciprocity depended on a *physical* sensation of movement. Furthermore, just as writers on *Einfühlung* argued that our senses were not strictly delimited and that stimuli for one sense could also stimulate another, so we see here a

FIGURE 4.2. A prewar audience enjoying a night at the Union-Theater in Berlin, 1913

similar recognition of the confusion of the senses. In noting that he "hears the gunshot," the writer emphasized a *synesthetic* response; he recognized that the film experience is not merely visual, but that vision itself is embodied and connected to the other senses. (This "gunshot" could have been a sound effect, of course, but we should also consider that it could have been psychosomatic, an effect of deep immersion.)

This speaks to another theme in *Einfühlung* aesthetics: the idea of contemplation as the ground for aesthetic experience. Our usual assumption that the spectator is held in continuous rapt attention is troubled by this passage, for it also emphasizes the *materiality* of the experience: "the image twitches slightly here and there, flickering and spotting up as if something were boiling under the projected surface." The film experience seemed to require the audience, like the motionless kaiser and the rolling mountain goat, to hold opposite attitudes—stasis and movement, contemplation and distraction—at the same time. At once immersed in the illusion while blithely noting the twitching image, the audience alternated between attention and detachment. Already, the cinematic experience highlighted contradictions at the heart of contemplation. The rest of this section, then, will explore the issues of emotional projection, synesthesia, and movement as they relate to identity and embodied perception in both *Einfühlung* aesthetics and early discussions of cinema in the *Kino-Debatte*.

Emotional projection was a key issue in aesthetics during this period. An essay from the debates on film will help explicate this topic. Alfred Polgar felt that attending the cinema was a sensual experience. In his 1911 essay, "Drama in the Film Theaters," he described seeing pretty girls on the screen; they smiled at him and he felt that he should smile back, even wait for them outside the theater once the show was over. He knows it is silly to think of such things, but there is something strangely compelling in the way the film takes on a life of its own, the way it "smiles back at me."[57] This could be the sigh of another lonely film guy, or the first glimmer of a phenomenology of film, in which the film itself has a subjectivity that engages with the spectator, such as that suggested by Vivian Sobchack.[58] But it also hints at the connection between emotional projection, *Einfühlung*, and identity.

Polgar's emotional investment in the girls on the screen recalls one of the central concepts of Vischer's dissertation and of this trend in German aesthetics: the imputation of one's inner life to an inanimate object, which Vischer termed *Einfühlung*. Vischer wrote:

> We have seen how the perception of a pleasing form evokes a pleasurable sensation and how such an image symbolically relates to the idea of our own bodies—or conversely, how the imagination seeks to experience itself through the image. We thus have a wonderful ability to project and incorporate our own physical form into an objective form. . . . What can that form be other than the form of a content identical with it? It is therefore our own personality that we project into it. (104)

According to Vischer, we project our emotions and sense of our body (such as our orientation in space) onto an inanimate object, just as we project these aspects of ourselves onto other people to understand their expressions. To us, the work of art itself is permeated by human sensations. In fact, subjectivity is present not only in our attitude toward the work but also within the work itself. "There is also a will *within* the image" (114), Vischer wrote, meaning that we project our experience on to the relation *between* the parts of the work as well. Vischer's concept of *Einfühlung* described how the projection of human feelings onto inanimate objects plays a role in the creation, shaping, and reception of artworks.

Likewise, Adolf Hildebrand's *The Problem of Form in the Fine Arts* (1893)—which had already gone through nine editions by 1914—was an incredibly popular exploration of the problems posed by sensual perception. He was especially interested in legibility—how we can understand the relation between changing appearances in people, nature, and art. Hildebrand explained:

> Nature in its movements and transformations produces changes in its appearance, which we interpret as the visible signs of those processes. We perceive the signs and imagine the process; we participate in it, so to speak, perform along with it, and accept the internal activity as the cause of the external appearance. Just as the child learns to understand laughter and tears by joining the process and is able to feel, through muscular activity that he himself calls forth, the inner cause of the pleasure or pain, so does all gestural expression and all movement on the part of others become for us a comprehensible expression of internal processes, a comprehensible language.[59]

Hildebrand extended this analogy to inanimate objects and representations as well. We understand images by imputing them with a "story" of their processes, similar to our own; our own bodily sensations and experiences enliven the images and make them comprehensible to us. We understand art through a process of empathy. Taken together, the theories of Vischer and Hildebrand helped to establish *Einfühlung* as a popular explanation of the aesthetic experience.

Polgar also made a comparison between film and music to further his thesis about the power of the cinematic image. Music, he wrote, has the power to "move" us; it moves our "inner humanity" in a way quite mysterious. It could also provoke a synesthetic response: "a powerful suite of images, color, feeling, and idea." Film, Polgar argued, functions in the same way: it is "a similar fertilization of all the other senses through the stimulation of an optical sense."[60] This idea of synesthesia was important to *Einfühlung* theorists, because it helped to describe what actually happens during aesthetic reception, but also because it indicated the centrality of the body as a measure of aesthetic response. *Einfühlung* theory postulated that the harmonious correlation of form to the physical or sensory structure of the viewer was the key to aesthetic pleasure. That is, according to the theories, if the object has a structure similar to our body, it is pleasing to us;

if not, it strikes us as unpleasant. By extension, the symmetry of the body accounts for the tendency toward symmetry in art. A sympathetic response to this harmony could take place on a number of levels. There is, for example, in terms of symmetry, the physical correspondence of the structure of sense organs and various objects (a horizontal line matches the horizontality of our eyes, for instance). In terms of regularity, the formal arrangement in series matches the rhythmic function of our organs. In other words, "the body, in effect, imposes an 'organic norm' in viewing the world, according to which regularity, symmetry, and proportion induce pleasing sensations by emulating the normal human body."[61]

But this "viewing" is not merely visual. Vischer noted that sometimes "a visual stimulus is experienced not so much with our eyes as with a different sense in another part of our body" (98). Sunglasses could have the effect of "cooling" the skin, while "loud" colors might offend our auditory nerves. "For in the body there is, strictly speaking, no such process as localization." The embodiment of vision meant that "the whole body is involved; the entire physical being is moved" (99). On one hand, this embodiment implied the possibility—even necessity—of synesthesia, at least at some level. The senses resonate with one another; the boundaries between them are not simply blurred, but porous, causing a stimulus to one sense to spark a response in the others. On the other hand, Vischer also argued that this resonance between the senses is the model for the relation between the arts: "These reflexes or reciprocal vibrations of the senses are the physical cause of the unity of the arts" (99). While this parallel between the structure of the body and the relation between the arts was just a footnote to his argument, it points to an ambiguity in Vischer's book and in subsequent discussions of *Einfühlung*, as we will see: the question of embodied spectatorship did not always imply physical participation in the act of viewing; instead, it often implied shared forms and orientations between the artwork and the human body per se. That is, the body was more often a *model for* art than a *participant in* art. Vischer's evocation of synesthesia functioned ambiguously by pointing both to embodied spectatorship and to the body as an idealized model.

However, Ernst Bloch's 1914 essay, "Melody in the Cinema, or Immanent and Transcendental Music," pressed harder on this issue via the example of silent film accompanied by music. A musician by training, Bloch placed greater emphasis on the redemptive and sensual nature of sound. But he argued that, while it was exclusively visual, the film image gained another dimension when combined with the proper music. Indeed, his article was a plea for music that was appropriate for the film and not simply random notes played to cover the sound of the projector. He found synesthetic possibilities in film music, which could present a complete sensual experience initiated by the ear, not the eye.

> As visitors to the cinema we must initially rely exclusively upon our eyes. Now, the sense of touch conveys the impression of reality most immediately; in front of the film screen, however, we must renounce everything—pressure, warmth, scent, sound, and the feeling of being sensually encompassed [*sinnliches Mittendarinsein*]—that gives the appearance of things its fully "real" character. The skin, the nose, ears, and all other senses are switched off, while the eyes are overwhelmed. Only an optical impression of black and white is excerpted from the real world, and since this impression is presented in the most confusing momentary motion and without any stylization, it produces the uncanny impression of a solar eclipse, a silent and sensuously deprived reality that is heightened only in its speed and its concentration, but without departing from our world aesthetically and ideally. But now the ear takes on an unusual function: it fills in as the replacement for all the other senses. Because the rustling, rubbing, and noisy collision of objects, because above all, human voices (which are themselves ringing with emotion) can blend seamlessly into [musical] tones—indeed, precisely because there is nowhere a natural or manufactured sound in the world that competes with music—this art form [film music] is able to reflect the colorfulness of lived reality and achieve at one stroke a sensuous totality that never makes us aware of our individual senses.[62]

Like many writers during this period, Bloch presented film as a lack: in front of the screen we must "renounce everything" that gives the world its "fully 'real' character," such as "pressure, warmth, scent, sound." We may disagree with his assessment of the cinematic experience—we might rightly wonder where he found such a cool, odorless, silent, and spacious film theater—but we can also see this as a condemnation of the poverty of any form that relies on vision alone. In this case, Bloch argued that film's silent, black-and-white world presented only an attenuated version of the world, a meager representation that could not satisfy the need for sensuality until it was matched with music. Then, as if music were color itself, this world came alive, and the filmic representation was able to regain the "colorfulness of lived reality." It seems at first glance that music was the sole hero here, rescuing bored spectators everywhere from the paucity of filmed reality. But it was precisely the *combination* of film and music that created this embodied experience. Bloch called film "a silent and sensuously deprived reality that is heightened only in its speed and its concentration," but it was exactly that speed and concentration that allowed a "sensuous totality," in that film and music are both temporal forms. It was their combination (as in opera, for example) that swept the spectator into a powerful representation of "the colorfulness of lived reality."

But film is not like opera; "its speed and its concentration" are like no other forms. What did he mean by "concentration" (*Konzentration*)? On one hand, he referred to the temporal density of filmic representation: its ability to create temporal ellipses through editing and to pack more events into a limited amount of time—film as a precipitate of time. On the other hand, it recalled the concentration of the film viewer, the spectator's absorption and immersion in the film, the kind of contemplation that almost allowed the viewer to forget "pressure, warmth, scent, sound," and so on. Both were required for this new form of embodied spectatorship. Almost despite himself, Bloch argued for a synesthetic experience unique to cinema. In so doing, he extended further the issue that Vischer and others explored: the role of the body in the aesthetic experience. Bloch's description of film and music's "sensuous totality" (*Gesamtsinnlichkeit*) certainly pushed synesthetic principles of artistic formation and reception much further than what we find in most theories of *Einfühlung*.

Bloch's description of film and music and Polgar's case of the pretty girls bring up an interesting question in relation to this theory of aesthetic response. Generally speaking, Vischer, Hildebrand, and other theorists of *Einfühlung* limited their discussion to *static* forms, such as painting, sculpture, and architecture. We cannot blame them for not including cinema in their investigations, but even theater and dance or other temporal forms were rarely, if ever, evoked. Music featured prominently, but only because it is abstract; these writers were working on the problem of legibility—how is it that we understand something that is not like us?—so music's abstractness, not its temporality, was intriguing to them. Polgar's pretty girls were just a representation, certainly—black and white, silent, two-dimensional—but they *moved*, and the moving image's precise role in *Einfühlung* theory was still indeterminate. To what extent could *Einfühlung* aesthetics accommodate a temporal form such as moving pictures? Vischer, Hildebrand, and others frequently discussed "movement," but we should first understand what they meant by this term.

Movement is implied by the very term *Einfühlung*, which translates literally as "feeling into." Yet it also goes without saying that the combination of "feeling" and "into"—that is, the comparison of emotion and space—must be metaphorical. Nevertheless, the emphasis in *Einfühlung* aesthetics on physiology and on spectator response left open the possibility that "movement" during the aesthetic experience might be more than merely metaphorical. This was indeed the case for both Vischer and Hildebrand, who discussed physical movement of the spectator in terms of the movement of the eye over the object before it. But this activity, as Vischer explained, is largely confined to the eye: "We achieve this muscular activity, by moving the eye while looking at the object: that is, by scanning [*Schauen*]" (94). Hildebrand agreed; both theorists distinguished seeing (*Sehen*) from scanning (*Schauen*) and likened the latter to touch as a more sensual, tactile appropriation of the image.[63] Vischer allowed that there might be a more visceral or holistic physical response to art when he ventured that "mental stimuli can bring about motor stimuli in the lower organs, and vice versa. The whole body is involved; the entire physical being is moved" (99). Ultimately, however, this response was not what he had in mind when discussing the aesthetic experience. Instead, both he and Hildebrand stressed the importance of the imagination as a mediator between raw stimuli and a purely aesthetic response. "I might imagine myself," Vischer wrote, "moving along the line of a range of hills guided by kinesthetic imagination (be it direct or mediated by the reflex stimuli of sensitized nerves). In the same way, fleeting clouds might carry me far away. This is no longer seeing [*Sehen*] but a *watching* [*Zusehen*]: the forms appear to move, but only *we* move in the imagination" (101, emphasis in original). In other words, our projection into the work—and our physical response to it—depends on an act of imagination, ensuring the aesthetic experience remains within the domain of the "higher" functions. The movement of the eye initiates the process, but in our response any "movement" is to be construed as virtual, not real.[64]

If *Einfühlung* was a combination of emotional projection and (a limited form of) embodied spectatorship, the work of architectural historian August Schmarsow is especially appropriate. His inaugural lecture from 1893, "The Essence of Architectural Creation," outlined a theory of architecture as "the creatress of space": "Our sense of space [*Raumgefühl*] and spatial imagination [*Raumphantasie*] press toward spatial creation [*Raumgestaltung*]; they seek their satisfaction in art. We call this art architecture; in plain words, it is the *creatress of space* [*Raumgestalterin*]."[65] Schmarsow argued that we understand architectural form by projecting ourselves into it—by correlating our bodily axis, for example, to the vertical lines of a building. For Schmarsow, this was a process of spatial creation (*Raumgestaltung*), closely linked to proprioception, our sense of spatial orientation, or what he called "the intuited form of space": "The intuited form [*Anschauungsform*] of space . . . consists of the residues of sensory experience to which the muscular sensations of our body, the sensitivity of our skin, and the structure of our body all contribute" (286). Our bodily activity creates or contributes to a physical "memory" of sensation, which serves as the basis for a physiological understanding of architectural form. Like Vischer and Hildebrand, Schmarsow found the imagination central to this process: "It is an act of free aesthetic contemplation when, with the aid of our imagination, we transport ourselves from the exterior that we see before us into the center of the interior space; when, by inquiring into its axial system, we strive to open up a remote organism to the analogous feeling within ourselves" (293). This "striving" to open up the interior of a building by means of our already existing sense of space is to Schmarsow a creative process, and "movement" was vital to it.

As in the work of Vischer and Hildebrand, this "movement" seemed virtual or imagined, detached from real muscular tension, but in Schmarsow's essays something more seems to be at play: "We cannot express its relation to ourselves in any way other than by imagining that we are in motion, measuring the length, width, and depth, or by attributing to the static lines, surfaces, and volumes the movement that our eyes and our kinesthetic sensations suggest to us, even though we survey the dimensions while standing still" (291). Here Schmarsow hinted that our "kinesthetic sensations" play a part, that in the appropriation of architectural form there is a productive tension between stillness and movement in the body at the moment of apperception. Indeed, architecture may be the "creatress of space," but Schmarsow emphasized the creative aspect of spectatorship as well:

> Just as, in response to external events, shared emotional feelings intensify in their rise and fall into moods or press in their growth toward blissful delight or convulsive pain in order to move farther out and fill the immediate surroundings with the vibrations of the inner life and to influence those surroundings, if only through the fleeting sound of the human voice—so, too, the purely imagined impressions and their integration or combination into three-dimensional visual forms involuntarily project themselves into the world outside and develop further into a sensorially perceptible reality. (292)

Schmarsow argued that *emotion* was key to understanding space and that this emotion projected outward to create a larger reality. Furthermore, when he wrote that "the spatial construct is, so to speak, an emanation of the human being present, a projection from within the subject" (289), Schmarsow suggested not simply that architectural space was a mutual construction of architect and viewer (à la Kant), but that the apperception of architecture involved a simultaneous construction of lived space, that the architectural representation of space was not merely imagined but felt emotionally and physically—and that this emotional and physical "movement," in a real way, *creates* the space represented. (In this respect, Schmarsow's ideas might be useful for thinking about our bodily and emotional engagement with cinematic space.)

On the other end of the spectrum, however, we find Munich psychologist Theodor Lipps, regarded as the chief proponent and theorist of *Einfühlung* aesthetics. Between 1890 and 1914, he was a prolific—not to say incontinent—explicator of *Einfühlung*, both in aesthetic terms (emotional projection of ourselves into art) and psychological terms (as the fundamental way we understand others).[66] On the question of movement, Lipps was contradictory. In a typically confusing passage, Lipps defined his terms: "Finally, one could figuratively say that 'activity' is the inner breath or heartbeat, or more generally, it is the inner motion. But 'motion' here is not meant as a simple event within me, but rather the fact that I move. Of course, this 'motion' has nothing to do with space."[67] In postulating that "I move" but that "this 'motion' has nothing to do with space," Lipps hoped to have it both ways, succinctly expressing the tantalizing tension in the definition of movement in *Einfühlung* aesthetics. Yet, in the end, Lipps was fairly clear that this movement was limited to "some inner, striving motion" (*der inneren strebenden Bewegung*).[68] Indeed, we might assume (or hope) that kinesthetic affect is an important aspect of the aesthetic experience, but Lipps was adamant that if we *feel* any bodily response, if we are aware of our bodies in any way, then we are no longer in the realm of the aesthetic proper. Real bodily movement or physical response may be mimicry or erotic pleasure, according to Lipps, but at the point that it is physical, it ceases to be aesthetic. He stated it flatly:

> But the pleasure which I feel while I am paying attention to my physical states or the processes in my physical organs cannot be identical, neither wholly nor in part, with the joy which I feel when I do *not* pay attention to the processes in my physical organs, but devote my whole attention to the aesthetic

object. In short, *A* cannot equal non-*A*. . . . *Einfühlung* means, not a sensation in one's *body*, but feeling something, namely, *oneself*, into the aesthetic object. . . . Actually, the sensations of my own bodily state are only present in aesthetic contemplation in order to be entirely absent to me.[69]

This, of course, meant that any sexual feeling or arousal could not be part of the aesthetic experience:

> The sexual has nothing, absolutely nothing, to do with the aesthetic. Those who employ it to explain aesthetic feeling know as little of the meaning of beauty and aesthetic contemplation as those who warn against "nudity" in art because they fear that even chaste nudity will threaten morality: first their own morality, then that of others to whom they ascribe their own crudity.[70]

In sum, Lipps insisted that the movement implied by *Einfühlung* was metaphorical and must remain so if it were to be aesthetic; he firmly stood by a strict, conservative delimitation of art and body that required the terms of aesthetic pleasure to be divorced from bodily response.[71]

The cinematic experience, however, brings us back to the question. As we become immersed in the cinematic image, we might, as Lipps demanded, forget about our bodies, but this very forgetfulness might result in involuntary movement, such as gripping the arm of the chair (or the person next to you) during a car chase or moving slightly side to side during a well-choreographed fight sequence. It might even result in erotic pleasure. During the time that Lipps wrote about *Einfühlung*, these pleasures in the cinema would not have even been included in the category of "the aesthetic," but the character of the moving image—among other things—may have forced a reconsideration of the definition of aesthetic pleasure. We can see this renegotiation start to take place with Walter Serner's 1913 essay, "Kino und Schaulust," which is one of the first to suggest the sexual dimension of visual pleasure.[72] Serner's discussion of the link between cinematic movement and bodily reaction implicitly challenged the division between visual stimuli and visceral response that we have seen in traditional and *Einfühlung* aesthetics. I am not suggesting that Serner was actively intervening in this discussion; indeed, his attitude toward cinema was ambivalent at best. But his essay indicates that what remains virtual and latent in *Einfühlung* aesthetics—movement in the image and in the spectator—was even at the time visceral and literal in the experience of cinema.

Serner was explicit that the *Schaulust* he described was not the polite, distanced gaze of the aesthete: "Not the harmless kind, for which everything is only movement or only color or both, but a frightening desire that is no less powerful than the very deepest kind. It is a desire that boils the blood and makes it rage, until an unfathomably powerful excitement common to all desire races through the flesh."[73] This "unfathomably powerful excitement" was, of course, sexual desire. Serner was at pains to point out that this brand of sexual desire was not merely erotic but was tinged as well with the thirst for violence. Cinema appealed to our basest instincts, our darkest needs. But it was not simply the content of the films that made this appeal; it was cinematic movement itself:

> It was neither the quick victory over something that was dying nor an astonishing profitability that with one stroke put this [cinema] into place from the outset. It was the exciting adventurousness of a tiger hunt, of a daring mountain run, of a death-defying automobile ride; it was the breathtaking pursuit of a wounded, bleeding thug over the dizzyingly high rooftops of New York; it was the eerie district filled with misery, sickness, and crime, and the entire horrific detective genre with murder and violence, Browning and Navajo; and it was all the bloody, blazing images of fire and death, of horror and terror, on which all eyes sate themselves after long privation. A gaze, it was, that had rhythm and life and was desire.[74]

The tiger hunt, the mountain ride, the chase—these were the components of cinema as much as the bloody pictures of fire and death. Serner named the *thrill* of cinema as an explicitly visceral reaction in which the spectator was swept away by the movement of the image. The eye was no longer detached from the body but connected "through the sheer endless rush, which is all the more welcome, since it knows that movement is the greatest component of the desire to look and its innermost essence."[75] *Einfühlung* aesthetics saw the body as a model for art, as a category for understanding, but mostly not as an active participant in the aesthetic experience, except for the initial roaming gaze. And there is no indication that Serner or anyone else at this moment viewed what he described as aesthetic. But as the filmic image and cinematic experience *became* associated with Art through other means (for example, the immersive form of film narrative, movie palaces, and so on), this residual experience of movement and visceral response could not be logically left behind, even if it were attenuated in practice or, as Tom Gunning argues, took its place in lower forms of cinema.[76] In any case, after cinema, aesthetics could no longer logically justify the well-mannered separation between contemplation and physical response.

THE POLITICS OF CONTEMPLATION

Whatever their differences regarding the role of the entire body in aesthetic response, all writers on *Einfühlung* (indeed, most writers on aesthetics in general) assumed that contemplation was absolutely central to the aesthetic experience. Recall that Schmarsow declared, "It is an act of *free aesthetic contemplation* when, with the aid of our imagination, we transport ourselves from the exterior that we see before us into the center of the interior space" (293, emphasis added). Likewise, Lipps consistently invoked contemplation as the prerequisite to, even the ground for, *Einfühlung*: "The contemplation of the observed movement awakens the tendency to a corresponding inner response [*Selbstbetätigung*]; and by corresponding we mean that which would be connected with the execution of such a movement if I were to perform it. And this tendency is at the same time *actualized* in the act of contemplation."[77] Lipps argued that whatever movement we might feel when encountering a work (be it a painting or dance or another form), we do not actually move, nor do we need to move. Instead, the impulse toward imitation of that movement in the work is resolved or dissipated by the act of contemplation itself. In Lipps's version, the contemplative act transmutes *motion* into *emotion*. Contemplation thereby allows emotional projection to take place, undisturbed by the intrusion of physical sensation. It is a surrender to the object and to the process: "the aesthetic experience is the way that I feel moved when I engage in aesthetic contemplation, when I surrender myself completely to the representation," wrote Lipps.[78] In this respect, *Einfühlung* aesthetics grafted common notions of aesthetic contemplation onto the process of *Einfühlung*. We could even say that the theory of *Einfühlung* was constructed to explain what happens during the contemplation of artworks, but without disturbing the moral and ideological significance of the operation. That is, the meaning of "contemplation" in *Einfühlung* aesthetics was not markedly different from what we might find in aesthetics generally, even if elements of *Einfühlung* theory tread close to an expansion of that meaning.

Could this model of contemplation accommodate the cinematic experience? An initial answer might be no, it could not, especially given the standard history, which dictates that contemplation faded and distraction took its place. The evidence in this chapter elaborates that history. The first section demonstrated that objections to cinema's rush of images were grounded not simply in contemplation's leisurely approach but also in the historically strong tension between the flux of time and free will. That is, writers were not just *unaccustomed to* the rush of imagery but felt *determined by* the choice and flow of imagery that seemed to push them along ever more impatiently through the present, giving them a strong, lived sense of time's passage. Meanwhile, other writers reveled in the sense of presence and the present that motion pictures gave them, which sharpened their senses and put them on alert, as if caught on foot in midday traffic. Likewise, the second section sketched the trouble with contemplation from another angle as aesthetics tried to accommodate aspects of the aesthetic experience that were nascent in early theories but never fully accounted for, especially emotional projection and embodied perception. As these features became more prominent in the theories, it was more difficult to graft a common conception of contemplation (as largely inattentive to the body) onto this model. The cinematic experience, however, was varied. As we saw in chapter 3, reformers such as Häfker worked hard to fit film presentations into the ideologically dominant model of contemplative experience. We could also argue, after Charles Musser, that some film genres, such as landscape films, were designed from the start to provide for a contemplative gaze.[79]

We could further contend, after Miriam Hansen and Heide Schlüpmann, that cinema's public sphere had already created an alternative to detached aesthetic contemplation. They argue persuasively that the usual interpretation of these early discussions of film audiences in terms of class misreads to a certain extent the masculine anxieties of the moment, which should more properly be classified in terms of gender and sexuality. The presence of women in the public sphere in general and in the movie theater specifically, according to Hansen and Schlüpmann, indicates not only a change in gender dynamics in pre–World War I Germany, but also a change in aesthetic modes of reception—specifically, a shift from detached aesthetic contemplation to a more immersive, emotionally involved mode of viewing that has often been characterized as female. Their point is that, if we are going to look to the debates as evidence of a shift in aesthetic reception, we would do well to consider that most interlocutors were male and that many in the audience were female; contemplation was an ideological tool in service of not just class-based but also gender-based goals.[80]

Hansen and Schlüpmann provide an important corrective and an avenue of research that has been enormously productive. I would only indicate that, as chapter 3 demonstrated, children also presented an important model of spectatorship. In fact, this book argues that gender, class, and age were only some of the categories through which spectatorship was discursively constructed, and that our understanding of this process is best grasped via the larger division between *expert* and *lay* viewers. An expert could very well have an immersive, emotionally involved reaction to a film or a painting, as in Diderot's famous description of one of his encounters with a painting: "I was motionless, my eyes wandered without fixing themselves on any object, my arms fell to my sides, my mouth opened.... I shall not tell you how long my enchantment lasted. The immobility of being, the solitude of a place, its profound silence, all suspend time; time no longer exists, nothing measures it, man becomes as if eternal."[81] Apart from the final venture into metaphysics ("man becomes as if eternal"), this description recalls numerous reports of ordinary moviegoers, female and otherwise. But because it was written by an expert, it is free of the damning taint of passivity, which characterized all spectators—young and old, female and male. Diderot was absorbed by the image, but he wandered through it of his own accord. Passivity, not necessarily immersion or emotion, was the feminized trait that characterized lay spectators of all sorts.

Yet class does matter; if contemplation held most aesthetic theories in a tight ideological grip, it did so in service of implicit class privilege and agency. After World War I, however, when class barriers were subject to attack, the reaction to contemplation took on a more overtly political tone. While nineteenth-century philosophers might have viewed contemplation (and themselves) as active, revolutionaries saw contemplation (and the philosophers) as bourgeois and passive in favor of the status quo. This critique of contemplation was voiced most vociferously by the Dadaists, who vowed to no longer assume the role of mere "passive spectator [*passiver Beschauer*] of this comical world, of an aesthete like Oscar Wilde," for, as Richard Huelsenbeck warned at the time, "to sit in a chair for a single moment is to risk one's life."[82] For the Dadaists, contemplation was irredeemably tied to passivity. They condemned the class dimensions of contemplation as a bourgeois appropriation of art for art's sake that ignored the revolutionary potential of the aesthetic experience. Early discussions of cinema reversed this complaint. Various *Kinogegner* rejected motion pictures for their perceived association with the working classes and further argued that the nature of the filmic medium precluded the possibility of contemplation, which they defended as if protecting class boundaries. Either way, it seemed, motion pictures were not usually connected to contemplation. If they were, as in Häfker's case, it was in defense of film, contemplation, and middle-class values. In fact, the extent to which film was considered suitable for either contemplation or a distracted mode of viewing depended precisely on the class alignments of any given writer.

After World War I, then, the passive and active connotations of contemplation and distraction traded places. The rehabilitation of distraction began, at least in film theory, with Siegfried Kracauer's 1926 essay, "The Cult of Distraction," in which he described film presentations in Berlin's lush movie palaces (fig 4.3).[83] For Kracauer, these inherently fragmented, varied, and artistically incoherent presentations (which included live revues, advertisements, films, lighting schemes, music, and other spectacular elements) structurally matched "the disorder of society" and, hence, had the potential to present the working classes with a true picture of "the uncontrolled anarchy of our world" (327). That picture could also function as an image of the mass audience or masses (*die Masse*) of and for itself; that is, for Kracauer, if the mass audience could recognize itself in an image of itself, it would be a step closer to political consciousness: "Here, in pure externality, the audience encounters itself; its own reality is revealed in the fragmented sequence of splendid sense impressions. Were this reality to remain hidden from the viewers, they could neither attack nor change it; its disclosure in distraction is therefore of *moral* significance" (326, emphasis in original). But Kracauer complained that picture palaces in 1926 papered over this fragmentation with "the glue of sentimentality" or an "artistic" unity, thereby hiding the true picture of alienation behind a surface sheen. They functioned in this case, for Kracauer, as distraction *from*, whereas he advocated distraction *for* political awakening. Although never named in this essay, contemplation was implicitly no more than a phantom appendage of an "idealist culture that haunts us today only as a specter" (327).

Benjamin also gave distraction an explicitly political valence. In his 1929 "Surrealism" essay, for example, he discussed the "transformation of a highly contemplative attitude into revolutionary opposition."[84] But Benjamin's understanding of the nature of that political charge differed from Kracauer's. His "Work of Art" essay (1935–1939) remains the definitive sign of the rising political stock of distraction and the devalued currency of contemplation between the wars.[85] This essay has received more than its share of commentary, so I only want to underline certain keys differences between Benjamin and Kracauer. For Kracauer, distraction functioned as

FIGURE 4.3. Kammer-Lichtspiele Theater in Berlin, 1912

an image of the masses for the masses; the homology between leisure and labor, between the disjunctive form of entertainment and the alienated character of work and modern, urban life was something that the audience *could see* or picture through their experience in the picture palace. Indeed, Kracauer assumed that, once they grasped this picture, the mass audience would *reflect* on it and come to an understanding of its political situation.

So the image that distraction presented would be structurally no different than any image the audience would contemplate and reflect on.

Benjamin, on the other hand, argued that modernity and technological reproducibility had destroyed the cult value of images, so he searched for an alternative mode of thinking that would not rely on images per se, or even vision. Distraction, for him, was this revolutionary mode that struck at the heart of knowledge as usual. For Benjamin, contemplation comes at knowledge through the encounter with—or reliance on—a transcendental Ideal (recall Schopenhauer's understanding of aesthetic experience as loss of oneself and a nirvana-like communion with the Ideal). It is a top-down approach that already presumes form, like a Kantian a priori. That "already-given" rigs the game of knowledge production, stacking the deck against experience. Contemplation, for Benjamin, succumbed to the tyranny of forms of knowledge already given. Distraction, however, represented another way of thinking that worked from the bottom up; it takes the world and experience for what they are, not what they are presumed to be: "A person who concentrates before a work of art is absorbed by it. . . . By contrast, the distracted masses absorb the work of art into themselves."[86] The tables have turned: contemplation is now passive, in that the viewer is ruled by the rules and form of the work of art. Distraction, on the other hand, is now active, in that the masses control the work and care not for its form or the usual ways of appropriating it. Benjamin evoked touch and habit—senses other than vision—to describe this mode; distraction was a famously tactile form of knowledge: "The values of distraction should be defined with regard to film, just as the values of catharsis are defined with regard to tragedy . . . Distraction, like catharsis, should be conceived as a *physiological* phenomenon."[87] Like Bergson, he hoped to turn us away from our common, ordinary perception that actually keeps us chained to the status quo.[88]

Georg Lukács also offered a thorough and trenchant analysis of the class dimensions of contemplation. From his early *Theory of the Novel* (1916) to *History and Class Consciousness* (1923), Lukács presented a strong critique of the *vita contemplativa*.[89] For Lukács, contemplation was not merely a mode of aesthetic viewing, but a general reaction to the alienating character of modern life that must be surmounted for any transformation of the world to take place. It was a "pattern of consciousness in which man contemplates from a position of formal freedom his own integration in a system of alien compulsions and confuses this formal 'freedom' of his contemplation with an authentic freedom."[90] Contemplation, for Lukács, was simultaneously tied to both transcendence and complacency. In the preface to the 1967 edition of *History and Class Consciousness*, Lukács wrote, "Above all, I was absolutely convinced of one thing: that the purely contemplative nature of bourgeois thought had to be radically overcome."[91] Even in *Theory of the Novel*, Lukács organized his argument around two modes: contemplation and action. Specifically, he found two broad trends in the history of the novel, "abstract idealism, which concentrates on pure action, and Romanticism, which interiorizes action and reduces it to contemplation."[92] Novels such as *Don Quixote* (1605) emphasized action without reflection, while novels such as Novalis's *Heinrich von Ofterdingen* (1802) expressed the inward turn of romanticism, away from the disappointment and failures of modern alienation. Lukács explained:

> In Romanticism, the literary nature of the *a priori* status of the soul vis-à-vis reality becomes conscious: the self, cut off from transcendence, recognizes itself as the source of the ideal reality, and, as a necessary consequence, as the only material worthy of self-realization. Life becomes a work of literature; but, as a result, man becomes the author of his own life and at the same time the observer of that life as a created work of art.[93]

Lukács set up agency/identity dichotomies similar to Benjamin's: the paths to identity or self-realization were twofold, either through transcendence or self-awareness. If one denied the possibility of transcendence, then only a materialist, humanist approach would be possible, which was what Lukács advocated. Likewise, once this path was chosen, agency returned and one could be an active "author" of one's life rather than a passive "observer," which Lukács condemned. In *Theory of the Novel*, he advocated humanism (exemplified by Goethe's *Wilhelm Meister*) as a middle ground between action and contemplation, but in his later, more explicitly Marxist works, Lukács focused primarily on a radical critique of contemplation.

But the first rumblings of this dissatisfaction could be felt before World War I, especially in Lukács's 1913 essay, "Thoughts Toward an Aesthetic of the Cinema" ("Gedanken zu einer Aesthetik des 'Kino'").[94] This brief but provocative article for the *Frankfurter Zeitung*—it was chosen for publication over Bloch's piece on film and music—has been given insightful consideration by a number of scholars over the years,[95] but it is worth another look, because it both addresses the most insistent themes of the *Kino-Debatte* (the distinction between theater and cinema, word and image) and looks past them to anticipate themes in film theory of the 1920s. While it does not speak to most issues common in *Einfühlung* aesthetics (for example, emotional projection, synesthesia, or embodied spectatorship), it offers in nascent form a critique of contemplation that Lukács developed in his later work; in this way, it articulates cinema's implicit challenge to "traditional aesthetics" that will come out more forcefully in Lukács's later work and the writings of Benjamin, Kracauer, and others.

Lukács did this through what at first appears to be a standard comparison of theater and cinema, a strategy common to much early writing on film. The difference between theater and film, according to Lukács, is the difference between presence and absence, specifically, the presence of the actor on stage. But it is not the gestures or words of the actor that give theater its effect, nor the action of the drama. Instead, it is "the power with which a person, the living will of a person, without mediation and without restraining direction, streams forth onto an equally living mass or multitude" (13). The very *presence* of the actor—the force of his or her personal will, his or her drive toward the fulfillment of his or her being, his or her destiny—is the essence of theater. This living, breathing essence of another person is inherently transitory—tied to the now of the moment, to the *present*—but its very ephemerality is "not a deplorable weakness, but rather a productive limit, the necessary correlate and most observable expression of destiny in drama" (13–14). That is, the presence of the actor in the transitory moment shared by everyone in the theater is the most visible expression of the enactment of becoming that is drama itself. This sense of becoming, of the lived moment, is available to us all merely through the act of living intensely, but "this so-called 'life' never attains an intensity that could raise everything up to the sphere of destiny" (14). Drama, in other words, frames and concentrates this sense of destiny and fate for us: "Throughout the presentation of the drama this metaphysical feeling grows in immediacy and perceptibility: out of the deepest truth of people and their position in the universe grows a self-evident reality" (14). The stage, by virtue of its emphasis on *presence* and *the moment*, becomes a vehicle for the profound truths of human existence. Actors are an instantiation of these metaphysical truths in that they focus the force of living, or the direction of fate, into an ephemeral here-and-now shared by the actor and the audience; in fact, the mere presence of a great actor is "already and without any great drama destiny, divine service, tragedy, mystery" because of his or her unique ability to be "absolutely present" (14).

Cinema, on the other hand, does not share this emphasis on "presence" or "the present." Motion pictures lack this "presence" not because of any inherent defect, but simply because in films "there are only movements and actions of people—but *no people*" (14, emphasis in original). The people on the screen are not present—nor do they exist in the present—so they have no presence. Even so, the representations in cinema are "no less organic and alive than those images of the stage" (14). They are simply different: "they maintain a life of a completely different kind. In a word, they become *fantastic*" (14, emphasis in original). They present "a new aspect" of life: "one without the present, a life without fate, without reasons, without motives, a life without measure or order, without essence or value, a life without soul, of pure surface, a life with which the innermost of our soul does not want to coincide. . . . The world of the 'cinema' is thus a world without background or perspective, without any difference in weight or quality, as only the present gives things fate and weight, light and lightness" (14). At first glance, this judgment sounds very much like it might have come from a *Kinoreformer*, like a condemnation of the cinema as "soulless" and suitable for only the basest amusements that require no thought or reflection. But Lukács was after something else here. He saw in cinema the same utopian potential that one could find in fantasy: "'*Everything is possible*': this is the worldview of the 'cinema'" (15, emphasis in original). Cinema was not just the poor cousin of the stage; it offered a completely different set of possibilities.

The basis for this difference lies in their formal properties as well as their dissimilar relationship to time and the word. The stage is populated with living, breathing people; drama focuses on exactly that moment of their shared presence with the audience. Motion pictures, however, do not picture people, but representations of people. Nevertheless, "not only in their technique, but also in their effect, cinematic images" are "equal in their essence to nature" (14). The "technique" Lukács speaks of here is the photographic quality of the cinematic image, while its "effect" is the illusion of movement that the apparatus creates. This movement, this temporal flow, also distinguishes cinema from theater. The stage has a paradoxical relation to time: "it is the flow of grand moments, something internally deep at rest, almost 'arrested,' something become eternal, as a direct result of the painfully strong 'present'" (14). Theater's emphasis on presence and the present gives it, oddly, a sense of stillness. The essence of cinema, on the other hand, "is movement in itself, an eternal variability, the never-resting change of things" (15).[96] These different concepts of time, according to Lukács, correspond to the fundamental difference between the stage and cinema: "The one is purely metaphysical, distancing itself from all that is empirically alive;

the other is so strongly unmetaphysical, so exclusively empirically alive, that through this sheer extremity of its nature another entirely different metaphysics arises" (15). One is fate, thought, stillness, presence, metaphysics. The other is fantasy, surface, movement, absence, action. But, as Tom Levin has noted in his reading of this essay, the intriguing aspect of cinema is that it is both "empirical" and "fantastic."[97] That is, its photographic image grounds it in reality, in the everyday, while its temporality—its rush of images, as well as their arbitrary relation to laws of physics and causality—frees it of necessity and allows it flights of fantasy. As Lukács declares, "The fundamental law of connection for the stage and the theater is inexorable necessity; for the 'cinema' it is unlimited possibility" (15). The uniqueness of cinema lies not simply in its ability to portray fantastic worlds but, as Levin argued, in its dialectical relation between the real and the possible. "Everything is true and real, is equally true and equally real" in the cinema, according to Lukács. The combination of reality and possibility and the *equality* of the two even gives cinema a utopian dimension.[98]

But I would also suggest that, in Lukács's essay, cinema provided a critical dimension alongside its utopian potential. In his comparison of stage and film in terms of "metaphysics" and "soullessness," there was, I think, a critique of contemplation that emerged more forcefully in his later work, especially *History and Class Consciousness* (1923). Consistently throughout the essay, Lukács distinguished the "soul" of the theater (its stillness, its metaphysical depth, its emphasis on destiny and the human condition, its presence) with the "soullessness" of the cinema (its constant movement, its emphasis on surface, its metaphysical "lightness," its fantastic quality, its absence). In this distinction, Lukács was not alone. Many other writers of this period also called cinema "soulless." Franz Pfemfert, editor of *Die Aktion*, bemoaned in 1911 cinema's role in the "soullessness" of modern society. For Pfemfert, the naked reality presented by the cinematic image—which he blamed for the *lack* of fantasy in the filmic world—could best be utilized for educational purposes.[99] But Kurt Pinthus, introducing his 1913 *Kinobuch*, a collection of scenarios written specifically for cinema, warned his readers, "Always remember: this is not high art, not 'soulful' art, not theatrical art. We write only pieces for cinema, not for theater."[100]

Here we find what Lukács meant by "soul." Pinthus wrote, "not 'soulful' art, not theatrical art," as if they are one and the same. Pinthus and Lukács referred, of course, to German inwardness (*Innerlichkeit*), that national tendency toward contemplation and reflection (and the demand for it in art). In his consistent opposition of "action" (cinema) to "soul" (theater), Lukács indicated that cinema offered an alternative to this cul-de-sac of reflection. His distinction between cinema and theater corresponds to the dichotomy between abstract idealism and romantic inwardness found in his *Theory of the Novel*, which in turn was the foundation for his critique of contemplation in *History and Class Consciousness*. The latter book, especially, condemned as bankrupt the alienated, contemplative character, in favor of political or radical action. His discussion of cinema's "soullessness," then, was not a condemnation, not even a slight in favor of the theater, but the articulation of a new possibility, a new form that more closely corresponded to the practical needs of the people. Cinema's emphasis on surface and lightness was an opportunity to escape from the suffocating and futile tradition of *Innerlichkeit*, a formulation remarkably similar to those of Kracauer and Benjamin. Lukács thereby subtly pointed to the emergence of a new aesthetic, which Kracauer and Benjamin later articulated as "distraction." So when Lukács declared that, with cinema, "Man has lost his *soul*; in return, however, he gains his *body*" (16), he celebrated the potential for a new aesthetic standard that incorporated physiological, even tactile, responses to cinema, thereby pointing the way for film theorists of the 1920s and beyond.

◊ ◊ ◊

"What does all this mean? It means that real presentation banishes contemplation," Benjamin announced, only slightly ironically, in 1930.[101] It certainly seems that way, with Lukács proclaiming that our soul has been replaced by our body and Benjamin theorizing a physiological and tactile model of knowledge. Yet as Carolin Duttlinger has shown, Benjamin's own conception of contemplation and distraction was much more complex and dialectical than the version even the "Work of Art" essay provides. In his essays from the early 1920s, Benjamin saw in religious contemplation a presence of mind that allowed for both absorption and alertness, which was different from what he saw in the secular, solipsistic immersion of the modern age. Indeed, he and Bertolt Brecht agreed that modern contemplation and distraction were two sides of the same coin: both precluded critical engagement and response and therefore impeded political action. If his discussion of distraction in the "Work of Art" essay reversed this view, it is only because he dialectically leavened distraction with both habit (*Gewohnheit*) and alertness or presence of mind (*Geistesgegenwart*). Yet later, in some of his historical works, Benjamin's conception of modern contemplation again adds *Geistesgegenwart* as the essential ingredient. So Duttlinger concludes that Benjamin's writings consistently reject "solipsistic contemplation in favor of a more flexible, perpetually alert presence of mind."[102]

The balance between immersion and detachment has been historically difficult to find, not only because it is challenged with every new art form, but because it is, as Benjamin's work so clearly shows, a *moral and political* stance as well as an aesthetic stance, so finding the balance also implies finding a proper stance toward the historical moment. This is why the task continues to fascinate, and why contemplation, no matter what opprobrium it receives, will never die. For example, as Jocelyn Szczepaniak-Gillece has shown, contemplation and immersion made a comeback in film theory in the 1930s through the 1950s, as modernist architects attempted to design movie theaters that would create an ideal, immersive viewing experience and hence an ideal spectator, who was contemplative yet alert and responsive.[103] Indeed, the change in ideas about contemplation seems to respond to Benjamin and Brecht's complaints that the inherently individualist nature of contemplation separates the viewer from the collective, thereby inhibiting communication, solidarity, and political action. These architect-theorists, such as Benjamin Schlanger and Frederick Kiesler, hoped to induce an almost Schopenhauerian disembodied viewing experience, while also making that experience resolutely communal. Like Benjamin, they accepted and worked with cinema's collectivity. This is the primary difference between the writers of the *Kino-Debatte* and later theorists: for the most part, the earlier writers were unwilling to relinquish the historically strong connection between contemplation and individual viewing. Mass reception, for them, held little aesthetic potential (and the fear of social democracy often shut down any political potential). Yet we have seen glimmers of this possibility in various individual essays across the debates; if they often argued individually against film's aesthetic potential, as a group the essays helped clear the way. If *single* writers clung to models of individual, expert viewing, the *collection* of essays put mass reception on the film-theoretical agenda. Later shifts in film theory were therefore anticipated collectively by the film essays that appeared in the years before World War I.

CONCLUSION

TOWARD A TACTILE HISTORIOGRAPHY

Of course, after World War I everything changed. On 1 July 1918, the newly established state film production company, UFA, created a *Kulturabteilung*, or cultural division, to make and distribute a regular series of educational films. The leader of this division, Ernst Krieger, had been a major in the imperial army and had served alongside Alexander Grau, the wartime minister of culture. The war had offered a number of lessons, but in this case it had persuaded Krieger, Grau, and others beyond a doubt of the value of motion pictures as a means of mass education and indoctrination—which after the war could be directed toward peacetime social issues. Krieger hired a number of doctors, scientists, and educators fresh from the front to make educational films encompassing titles from *The Alps* (1918) to *Marital Hygiene* (1922). The division also funded scientific research films, thereby giving the state's seal of approval and support to scientific and medical cinematography.[1] The number of universities and institutes using motion pictures to document and to explore increased dramatically after the war, as film found a secure place in laboratories and lecture halls. Hermann Häfker's dream of a state-run, educational cinema had come true, even if—in UFA's case, at least—it had to make room for fictional films as well. Indeed, cinema's aesthetic pretensions received a boost after 26 February 1920, when *The Cabinet of Dr. Caligari* opened in Berlin to much hype and wide acclaim. Its remarkable mise-en-scène, acting, and narrative framework struck a chord with all audiences, even those beyond Germany's borders; its self-consciously artistic presentation drew from a distinctly German style, marking it immediately as "Art." *Caligari* became a lightning rod of debate about the nature of film, even while people lined up at the box office. Ushering in German cinema's golden era, *Caligari* gave cinematic form some measure of aesthetic legitimacy. Meanwhile, the credibility of film in terms of aesthetic reception received more attention and validation with new theories (Rudolf Arnheim, Balàzs, Benjamin, Kracauer) that acknowledged its significance for the transformation of aesthetic standards.

So in terms of disciplinary legitimacy, cinema had much less to worry about after the war. Debates continued about its proper function in society or how its artistic potential could be best realized, but they were not nearly as heated as in the prewar years. German society seemed to have adjusted itself to cinema, and cinema to it. In addition to prompting UFA's role as state-sanctioned protector of educational (and propaganda) film, the war had also cut off the supply of imported films, thereby allowing domestic production companies to flourish; the complaint about foreign films could no longer fuel debates, and the supply of "quality" German films also rose. The international success of *Caligari* awakened the establishment to the artistic and economic potential of film, giving it a stamp of legitimacy that before had been only grudgingly offered at home. Firmly entrenched in German culture by the 1920s, cinema spent considerably less time defending its right to exist.

On the other hand, little has changed. We are still as preoccupied as ever—perhaps even more—with social acceleration and the dangers of distraction. If motion pictures exemplified the quick, relentless pace of modernity and the social consequences of inattentiveness, their secure place in mass culture did nothing to calm nervousness about their perceived effects. In fact, those concerns have doubled or tripled in our new century, predictably, as the emergence of new media forms seems to underline the lack of control we have over the pace of change and the technologies that come with it. The perceived effects have not changed, only the media to which they are attached. The rhetorical patterns we have seen over the course of this investigation have remained remarkably steady over the century; with each new media form we have a new round of experts decrying the infantilization of its audience while appropriating it for their own disciplinary projects. The difference between expert observers and lay spectators, as we have seen, resides in the amount of control one can wield over the technology and its product. The audience is infantilized precisely because it is perceived to surrender to the technology without maintaining a proper distance or detachment from it; the expert is able to use the same technology because he or she is able to master the medium—which testifies to his or her own self-mastery as well. This dichotomy between resistance and surrender seems at first indistinguishable from middle-class ideals of self-mastery or restraint. These class ideals definitely reinforce the disciplinary distinctions, but because experts have the technology *in their hands*, often literally exercising their control in this way, their disciplinary alignment is a more historiographically proximate explanation of their ideological stance. Hence the importance of examining these rhetorical patterns, which still persist in our media landscape today, in terms of *disciplines* as well as class.

Expert/lay distinctions are perhaps just as intractable as class distinctions, and they can be just as blunt. Yet by implicating education, training, and disciplinary communities, they offer an obvious and clear point of entry for historical investigation. Indeed, this project has argued that a full description of the rhetorical dichotomies common to debates about media—which would allow us to see their porousness and mutual dependence—requires an acknowledgment of the heterogeneity of film technologies, products, and experiences. Taken as a whole, the expressions of film's potential or dangers are dynamic and conflicted—a result of the many-sided quality of the products or technologies the historical agents encounter on the ground. Any given agent might be intransigent in this or that essay or speech, but the collection of voices within a discipline or community are, like the object itself, heterogeneous. The heterogeneity of object and audience is, in my opinion, best (but not exclusively) expressed through the specific institutional appropriations of film in the nontheatrical context, and if we want to understand these applications, it seems apparent that we need to understand the disciplinary agendas at play and their relationship to the application. I am convinced that the historical interplay between these heterogeneous agendas and objects holds more insight into "what cinema is" than our own top-down theoretical proclamations.

So the way that experts *handled* the technology, both literally and figuratively, fills out the history of nontheatrical uses of media such as film. Production, distribution, and exhibition histories are traditionally important paths of inquiry for film and media studies, but we can also add descriptions of how experts have studied, manipulated, and adapted the image to their own ends, which reveals presumptions and expectations about their audience, their own expertise, and the role of film in their discipline. Experts have emphasized different aspects of film form in their use or study of the image: the still image is good for some tasks, the moving image is appropriate for other approaches; the detail of an image is an intriguing landscape open to discovery for some, while others (or even the same) gravitated toward the usefully abstract animated image. We are certainly aware of all these various facets of film, but we could focus more on patterns of use and their relationship to historically persistent rhetorical patterns. Investigating those varied uses also allows us to see the precise relationship between film form and disciplinary agendas. Moreover, the variety of approaches or entry points for expert analysis of the image also testifies to the ambidexterity of film, or its usefulness as both a still and moving image, among its other, varied manifestations. By emphasizing film's variety of forms and ways of "handling" it, film's malleability prompts what we might call a *tactile historiography*.

Indeed, it was precisely film's adaptability that appealed to experts, that brought them to the idea that they could appropriate it, transform it, and make it their own. These experts sculpted cinema's *dispositif*—the technology, the image, the audience, and the relationship between all three—in markedly different ways.[2] The scientific disciplines had the most rigid framework within which motion pictures could fit; the technology was used primarily to control duration and to aid correlation. From the aesthetic debates, we can see the negotiation between the cinematic experience—as varied as it was—and ideals of individual contemplation. In any case, the extent to which film could be shaped according to preexisting standards and practices determined its acceptability. What was this shaping, exactly? For researchers, it involved a very literal hands-on tinkering with the technology to adapt it to standards of evidence and imaging. It also involved, as we have seen, tinkering with the object of study to adapt it to the representational technology. For someone like Lemke or Häfker, on the other hand, it sometimes meant tinkering with the technology or the screening venue, but more often corralling funders, exhibitors, town councils, and other members of the community and driving them toward a common goal. Persuading these groups required a similar negotiation between the film experience or *dispositif* and established, often disciplinary agendas or ideals ("modernizing education," for example). For essayists and aestheticians, it meant thinking through how some aspect of film form could fit established or emerging ideals. To press film into service, then, meant shaping it to an existing framework of institutional resources, policies, practices, and ideals.

So tactile historiography is sensitive to the historical impressions left on cinema (the filmic "material" that includes the technology, the image, and the audience). These impressions are of at least two types. On one side, we have the institutional or disciplinary framework to which film is made to fit; this is a historically specific but more or less stable configuration of disciplinary ideals, established practices, rules, policies, norms, and conventions. In education, for example, limited funding meant that elementary schools would not include film projection equipment in the classroom until the 1920s or so. So in the early 1910s educational screenings were held in commercial film theaters, which of course shaped the educational experience, sometimes objectionably. Yet teachers undoubtedly managed these educational screenings in such a way that the norms of the classroom were imported to the theater, so the use of film in this context expressed a combination of two institutional frameworks or two sets of institutional or disciplinary norms: commercial and educational.

On the other side, historical agents such as Lemke, who worked to make film fit into educational norms, left on these experiences the impression of their specific sculpting, tinkering, or negotiating. In this case, Lemke's specific adjustments in part consisted of his arrangement of the films and discussion into a program designed around Herbartian principles. A discursive capitulation to disciplinary norms in education, this move shaped the experience in a local, perhaps fleeting way, in that the audience might have been only vaguely aware of this justification, so its impression was likely not durable. Still, the effort itself led to the acceptance of film as a tool that *could* be managed, as opposed to the earlier conception of film as totally unsuitable. This shift did indeed leave a lasting impression on the practices of educational film in Germany. Norms change, as do specific, local adjustments to them, which then incrementally extend or change the norms (or not). This historiographic approach sees film as an adaptable material, the form of which at any given historical moment expresses these norms and adjustments.

Yet if form matters, *how* does it matter? How can an understanding of film form contribute to an understanding of broader historical trends, or vice versa? Usually film historians illustrate this relationship through close analyses of individual films, the best of which present analogies between textual and historical structures, as in Tom Gunning's essay on

D. W. Griffith's *The Lonedale Operator* (1909), which demonstrates the homology between the mise-en-scène, the editing patterns of the film, and the emergence of new gender roles and modern forms of transportation and communication.[3] In this approach, the historian illuminates film style and historical moment simultaneously—each expresses the other. I advocate, in addition to this method, an approach that similarly demonstrates the mutually expressive relationship between disciplinary practice, ideals, and agendas, on one hand, and film form broadly speaking, on the other, which could include the specific quality of the image, the structure of the technology at the time, or the experience itself. The particular style of, say, a research film could be pertinent, but the way the technology is used in a research setting might reveal more substantial and specific patterns that help to explain the medium's use in that particular laboratory at that particular time. The style of the research film could also express these patterns of use, so that is an area worthy of further investigation, but this project has stressed the experts' own understanding of film form expressed through their patterns of use.

We must acknowledge, however, that all of these aspects of film form—style, technology, image, experience, and so on—are unstable and subjective; they change shape with the historical moment, of course, but what counts most as "form" depends on what the experts take it to be. This historiographic approach acknowledges that disciplines may see film differently than we do. Ludwig Braun, for example, saw motion pictures as an extension of serial photography; he therefore understood the technology as an image generator that could be best used to examine minute differences in the rhythmic function of rapidly moving objects such as the heart. In this case, Braun took film form to be a series of slightly different exposures, and the specific photographic quality of the image, for example, counted very little for his task. (If the technology could have generated successive line drawings of the heart just as easily, he would have been thrilled.) He took what he needed and left the rest. This unspoken selection or emphasis usually follows disciplinary logic, so this understanding of form—of what film is—also expresses that logic; form and discipline illuminate each other. Braun understood film to be "incremental exposures," which expressed the serial logic in scientific and especially medical thinking at the time. The analogy between "incremental exposure" and "series" has explanatory power, because it demonstrates—more powerfully than, for example, just listing advantages and disadvantages—*why a tool would be useful*. If the form of a wrench is crucial to understanding why it works on a bolt, the relationship between film and any given discipline is not so straightforward; analogies and correspondences can help us see how film fit an agenda.

So this historiographic strategy differs from the usual, first, in its emphasis on patterns of use rather than style. Style is important and can be helpful, but patterns of use apply to adjustments to the technology, the circulation of images, the multiple function of any given film, or the role of moving images in a selection of representational options—a partial list at best. Patterns of use provide many more points of contact with the organization's goals or the discipline's logic than style alone. An analysis of style helps us understand, for example, the conventions within a given genre, which is good, but it limits us to the genre, when patterns of use take us into the heart of the discipline itself. Second, this approach emphasizes the notion of film form that arises from patterns of use. These patterns of use make sense only against a background of disciplinary problems and solutions. So the third way that this approach differs from the norm is in its emphasis on the disciplinary (as opposed to economic, technological, biographical, or aesthetic) context. This requires, as I have noted, some significant immersion into those contexts, but we would be rewarded with a deeper understanding of why this or that media technology mattered to the discipline. Having this insight into a discipline also helps us, as film historians, better argue the significance of film for that discipline. If we are determined to venture into the realm of the nontheatrical film, and we are convinced of its importance for this or that task, then we cannot back away from the possibility that film had some impact on the way an organization, community, or even discipline understood its agenda, its object, and itself. Having expertise in the discipline helps make the case for the significance of moving images on collective "thought styles," in Ludwik Fleck's phrase, of which disciplinary logic is one expression.[4]

This project has emphasized the correspondence between material and discipline. If we are interested in the mechanism of cultural legitimacy—the process by which a new technology or form comes to be accepted by a group or discipline—then this approach makes sense: any new form is not only understood in terms of what came before, but it must play by the rules set out by the discipline. The rules of evidence or legibility in the physical sciences, for example, are to a certain extent malleable or accommodating to new technologies, but if the technology does not offer representational solutions at least somewhat familiar with more or less agreed-upon conventions, it will not be considered useful.[5] But if we are to explain not just legitimacy and acceptance but also the extent to which the technology *transforms* the discipline, then we must also consider the *strangeness* or difference the new form brings to the endeavor. This strangeness evokes wonder, as we have seen in the revelationist film theory of Balázs, Epstein, or Benjamin, or in the descriptions of researchers as they gaze into the microscopic sublime.[6] This wonder may propel agents to replicate the experiments or experiences; it may reveal new vistas that shift the discipline's horizon of experience. In other words, the strangeness of the experience of film might have acted as an engine of change in a given discipline. Cell biology provides a good example; descriptions of cell "behavior"—and the conclusions about cell life that rested upon such observations—could not have existed without the temporal manipulations afforded by time-lapse cinematography.[7] That is, the descriptions and therefore understanding of cell behavior depended on the technology's ability to match one timescale—that of the cells—to another, our human perception. Cell biologists really saw life through film's eye, and it changed what they thought life was; it changed their "thought style." The strangeness of film transformed biological conceptions of life, even if only slightly.

Yet these and other transformations are incremental, not revolutionary. Hans-Jörg Rheinberger's idea of "differential reproduction" in experimental method is pertinent here; experiments are designed to be reproducible, to be sure, but that is not so important as the slight difference that each experiment affords. The difference between what is known and what is unexpected is usually only slight, but the gap between them is the real goal and product of experimental practice, because that is the space where new knowledge is produced.[8] Likewise, the use of a new media technology as a tool is almost always experimental, and the difference between current understanding and the new view it provides is sometimes a surprisingly productive opening. The use of this tool is an attempt to solve a problem, and we can see over the course of the disciplinary appropriations of this tool a series of linked solutions to a common problem—like a series of experiments. Looking at the history of the disciplinary uses of film helps us to understand its role in a larger series of linked solutions and to see film's incremental effect on the problem itself. It is as if Braun's method of comparing minute differences in his incremental exposures and then projecting them to get a sense of the whole in its duration could be applied to historiography: we note the slight differences in applications of film technology in a discipline, but we are able to see the cumulative effect only by running the series through the flip book of history, so to speak. Or, to apply a biological metaphor, film technology could be the mutant gene in the disciplinary organism, the true effects of which we notice only after a generation or two.[9]

So this project has envisioned film's *dispositif*—its technology, its image, its audience—as material in a grand, cross-disciplinary series of experiments. Experts shaped this "material," adjusted or trimmed it to fit a set of needs, but the crucial difference that film provided to these disciplines came from its resistance to these efforts. The material was not infinitely malleable. The technology could be made to do only so much, the image was only so informative, the audience only so agreeable. We see this resistance in the strangeness of the view to the experts, but also in its inappropriate fit; some disciplines, such as geology, found little use for it. We see it also in the excess or residue, such as pleasure or thrill, which spills sloppily over the disciplinary framework; pleasure continues to be difficult to fit even into aesthetics. Expert vision wrestles with this remainder, even as it depends on it. Aesthetic contemplation, for example, is at once surrender and mastery; the oscillation between them relies on the irreducibly material, sensual, and singular character of the artwork, which is the basis for immersion and pleasure, but also the ground of its resistance to rationalization and abstraction—both of which are also part of the aesthetic experience.[10] Similarly, the filmic *dispositif* as "material" partly resists expert efforts to sculpt it; this is especially true when it comes to the audience, which has been subject to much rationalization and abstraction. The "shape of spectatorship" in this regard was not just the negative outline or boundary with expert vision but the product of expert modeling. The form spectatorship has taken depended on the discursive and practical molding of experts but also on the resistance of the audience and cinema itself to this kneading and on the spectators' willingness to take the experiment to places beyond which they were prodded.

NOTES

INTRODUCTION

1. Adolf Sellmann, "Das Geheimnis des Kinos," *Bild und Film* 1, nos. 3–4 (1912): 65–67, here 65. All translations are my own unless otherwise noted.
2. The scholarship on "useful film" and educational film is growing. See, e.g., the special issues on "Gebrauchsfilm" in *montage/AV: Zeitschrift für Theorie & Geschichte audiovisueller Kommunikation* 14, no. 2 (2005), and no. 3 (2006); Vinzenz Hediger and Patrick Vonderau, eds., *Films That Work: Industrial Film and the Productivity of Media* (Amsterdam: Amsterdam University Press, 2009); Charles R. Acland and Haidee Wasson, eds., *Useful Cinema* (Durham, N.C.: Duke University Press, 2011); and Devin Orgeron, Marsha Orgeron, and Dan Streible, eds., *Learning with the Lights Out* (New York: Oxford University Press, 2012).
3. Max Weber, "Science as a Vocation," in *The Vocation Lectures*, ed. and with an introduction by David Owen and Tracy B. Strong, trans. Rodney Livingstone (Indianapolis: Hackett, 2004), 1–31, here 7 (emphasis in original). This and the following paragraph have been adapted from my essay, "Science Lessons," *Film History* 25, nos. 1–2 (2013): 45–54.

4. See Norbert Bolz, *Am Ende der Gutenberg-Galaxis. Die neuen Kommunikationsverhältnisse* (Munich: Fink, 1993), 115, or chap. 3; or Thomas Elsaesser, "Die Stadt von Morgen: Filme zum Bauen und Wohnen in der Weimarer Republik," in *Geschichte des dokumentarischen Film in Deutschland, Band 2: Weimarer Republik (1918–1933)*, ed. Klaus Kreimeier, Antje Ehmann, and Jeanpaul Goergen (Stuttgart: Reclam-Verlag, 2005), 381–410; as well as Vinzenz Hediger and Patrick Vonderau, "Record, Rhetoric, Rationalization: Industrial Organization and Film," in Hediger and Vonderau, *Films That Work*, 35–49; and Petr Szczepanik, "Modernism, Industry, Film: A Network of Media in the Baťa Corporation and the Town of Zlín in the 1930s," in Hediger and Vonderau, *Films That Work*, 349–376.
5. On experimental systems, see Hans-Jörg Rheinberger, *Toward a History of Epistemic Things: Synthesizing Proteins in the Test Tube* (Stanford, Calif.: Stanford University Press, 1997).
6. Ludwik Fleck, "Some Specific Features of the Medical Way of Thinking [1927]," in *Cognition and Fact: Materials on Ludwik Fleck*, ed. Robert S. Cohen and Thomas Schnelle (Dordrecht, Netherlands, and Boston: Reidel, 1986), 39–46.
7. Ludwik Fleck, "Scientific Observation and Perception in General [1935]," in Cohen and Schnelle, *Cognition and Fact*, 59–78, here 60.
8. Fleck, "Scientific Observation and Perception in General [1935]," 61.
9. Sir James Paget, "An Address on the Utility of Scientific Work in Practice," *British Medical Journal* (15 October 1887): 811–814, here 811.
10. Johann Heinrich Pestalozzi, *How Gertrude Teaches Her Children* [1801], ed. Ebenezer Cooke, trans. Lucy E. Holland and Frances C. Turner, 2d ed. (Syracuse, N.Y.: Bardeen, 1898), tenth letter, 220 (emphasis in original).
11. Ludwik Fleck, "To Look, To See, To Know [1947]," in Cohen and Schnelle, *Cognition and Fact*, 129–151, here 137.
12. Robert Vischer, *On the Optical Sense of Form* [1873], in *Empathy, Form, and Space: Problems in German Aesthetics, 1873–1893*, ed. and trans. Harry Francis Mallgrave and Eleftherios Ikonomou (Santa Monica, Calif.: Getty Center for the History of Art and the Humanities, 1994), 89–124, esp. 93–95. In art history, see Norman Bryson's discussion of the terms in *Vision and Painting: The Logic of the Gaze* (New Haven: Yale University Press, 1983). For the use of "gaze" and "glance" in media studies, see John Ellis, *Visible Fictions: Cinema, Television, Video* (London: Routledge, 1982); and Timothy Corrigan, *A Cinema Without Walls: Movies and Culture After Vietnam* (New Brunswick, N.J.: Rutgers University Press, 1991).
13. Michael Hau, "The Holistic Gaze in German Medicine, 1890–1930," *Bulletin of the History of Medicine* 74, no. 3 (Fall 2000): 495–524.
14. Michel Foucault, *The Birth of the Clinic: An Archaeology of Medical Perception*, trans. A. M. Sheridan Smith (New York: Vintage, 1973), 109.
15. See William Egginton, "Intimacy and Anonymity, or How the Audience Became a Crowd," in *Crowds*, ed. Jeffrey T. Schnapp and Matthew Tiews (Stanford, Calif.: Stanford University Press, 2006), 97–110.
16. "The Work of Art in the Age of Its Technological Reproducibility (Second Version)," in *Walter Benjamin: Selected Writings*, vol. 3, *1935–1938*, ed. Howard

Eiland and Michael W. Jennings, trans. Edmund Jephcott, Howard Eiland, and others (Cambridge, Mass.: Belknap, 2002), 101–136, here 116.
17. Benjamin, "The Work of Art in the Age of Its Technological Reproducibility (Second Version)," 104.
18. Representative interventions in the debate about the role of modernity in a history of film style include the collection *Cinema and the Invention of Modern Life*, ed. Leo Charney and Vanessa Schwartz (Berkeley: University of California Press, 1995); David Bordwell, *On the History of Film Style* (Cambridge, Mass.: Harvard University Press, 1997), 141–46; Noel Carroll, "Modernity and the Plasticity of Perception," *Journal of Aesthetics and Art Criticism* 59, no. 1 (Winter 2001): 11–18; Ben Singer, *Melodrama and Modernity: Early Sensational Cinema and Its Contexts* (New York: Columbia University Press, 2001); Charlie Keil, "'To Here from Modernity': Style, Historiography, and Transitional Cinema," in *American Cinema's Transitional Era: Audiences, Institutions, Practices*, ed. Charlie Keil and Shelley Stamp (Berkeley: University of California Press, 2004), 51–65; Tom Gunning, "Modernity and Cinema: A Culture of Shocks and Flows," in *Cinema and Modernity*, ed. Murray Pomerance (New Brunswick, N.J.: Rutgers University Press, 2006), 297–315; and Frank Kessler, "Viewing Change, Changing Views: The 'History of Vision' Debate," in *Film 1900: Technology, Perception, Culture*, ed. Annemone Ligensa and Klaus Kreimeier (New Barnet, U.K.: Libbey, 2009), 23–35.
19. Among many other surveys discussed in chap. 4, see esp. Sabine Hake, *The Cinema's Third Machine: Writing on Film in Germany, 1907–1933* (Lincoln: University of Nebraska Press, 1993).
20. André Gaudreault, *Film and Attraction: From Kinematography to Cinema*, trans. Timothy Barnard (Urbana: University of Illinois Press, 2011).
21. See, e.g., Kevin Repp, *Reformers, Critics, and the Paths of German Modernity: Anti-Politics and the Search for Alternatives, 1890–1914* (Cambridge, Mass.: Harvard University Press, 2000); Andrew Lees, *Cities, Sin, and Social Reform in Imperial Germany* (Ann Arbor: University of Michigan Press, 2002); and Dennis Sweeney, "Reconsidering the Modernity Paradigm: Reform Movements, the Social and the State in Wilhelmine Germany," *Social History* 31, no. 4 (November 2006): 405–434.
22. Ben Singer names this ambivalence "ambimodernity" and brings it to the attention of the film studies community in "The Ambimodernity of Early Cinema: Problems and Paradoxes in the Film-and-Modernity Discourse," in Ligensa and Kreimeier, *Film 1900*, 37–51.

1. SCIENCE'S CINEMATIC METHOD

1. Charles Cros, "Inscription," in *Oeuvres completes*, ed. Louis Forestier and Pascal Pia (Paris: Pauvert, 1964), 135–136.
2. On the graphic method, see Merriley Borell, "Extending the Senses: The Graphic Method," *Medical Heritage* 2, no. 2 (March/April 1986): 114–121; Robert G. Frank Jr., "The Telltale Heart: Physiological Instruments, Graphic Methods,

and Clinical Hopes, 1854–1914," in *The Investigative Enterprise: Experimental Physiology in Nineteenth-Century Medicine*, ed. William Coleman and Frederic L. Holmes (Berkeley: University of California Press, 1988), 211–290; Soraya de Chadarevian, "Graphical Method and Discipline: Self-Recording in Nineteenth-Century Physiology," *Studies in History and Philosophy of Science* 24, no. 2 (June 1993): 267–291; and Robert M. Brain, "Representation on the Line: Graphic Recording Instruments and Scientific Modernism," in *From Energy to Information: Representation in Science and Technology, Art, and Literature*, ed. Bruce Clarke and Linda Dalrymple Henderson (Stanford, Calif.: Stanford University Press, 2002), 155–177.

3. Wilhelm Braune and Otto Fischer, *The Human Gait*, trans. Paul Maquet and Ronald Furlong (Berlin and New York: Springer, 1987). In addition to *The Human Gait*, Braune and Fischer's human motion studies included *Das Gesetz der Bewegungen an der Basis der mittleren Finger und im Handgelenk des Menschen* (Leipzig, 1887); *Über den Schwerpunkt des menschlichen Körpers mit Rücksicht auf die Ausrüstung des deutschen Infanteristen* (Leipzig, 1889), translated as *On the Centre of Gravity of the Human Body as Related to the Equipment of the German Infantry Soldier* by Paul Maquet and Ronald Furlong (Berlin and New York: Springer, 1985); *Bestimmung der Trägheitsmomente des menschlichen Koerpers und seiner Glieder* (Leipzig, 1892), translated as *Determination of the Moments of Inertia of the Human Body and Its Limbs* by Paul Maquet and Ronald Furlong (Berlin and New York: Springer, 1988); and Fischer's *Theoretische Grundlagen für eine Mechanik der lebenden Körper* (Leipzig, 1906).

4. Max Seddig, "Ueber Abhängigkeit der Brown'schen Molekularbewegung von der Temperatur," *Sitzungsberichte der Gesellschaft zur Beförderung der gesammten Naturwissenschaften zu Marburg* 18 (1907): 182–188; Seddig, "Über die Messung der Temperaturabhängigkeit der Brownschen Molekularbewegung," *Physikalische Zeitschrift* 9, no. 14 (15 July 1908): 465–468; Seddig, "Messung der Temperatur-Abhängigkeit der Brown'schen Molekularbewegung," Habilitationsschrift, Akademie in Frankfurt a. M., 1909; Seddig, "Exacte Messung des Zeitintervalles bei kinematographischen Aufnahmen," *Jahrbuch für Photographie und Reproduktionstechnik* 26 (1912): 654–657; and Seddig, "Messung der Temperatur-Abhängigkeit der Brown-Zsigmondyschen Bewegung," *Zeitschrift für Anorganische Chemie* 73–74 (1912): 360–384.

5. Hermann Braus, "Mikro-Kino-Projektionen von in vitro gezüchteten Organanlagen," *Verhandlungen der Gesellschaft deutscher Naturforscher und Ärzte* 83, part 2 (1911): 472–475.

6. For a fascinating account of cinema's relationship to time and science, see Mary Ann Doane, *The Emergence of Cinematic Time: Modernity, Contingency, the Archive* (Cambridge, Mass.: Harvard University Press, 2002).

7. Examinations of the relationship between illustrative materials and the agendas of scientists are increasingly popular in the history of science; a good, representative example is Martin Kemp, "Temples of the Body and Temples of the Cosmos: Vision and Visualization in the Vesalian and Copernican Revolutions," in *Picturing*

Knowledge: Historical and Philosophical Problems Concerning the Use of Art in Science, ed. Brian S. Baigrie (Toronto and Buffalo, N.Y.: University of Toronto Press, 1996), 40–84. Two especially influential collections are Michael Lynch and Steve Woolgar, eds., *Representation in Scientific Practice* (Cambridge, Mass.: MIT Press, 1990); and Caroline A. Jones and Peter Galison, eds., *Picturing Science, Producing Art* (New York: Routledge, 1998). Catelijne Coopmans, Janet Vertesi, Michael E. Lynch, and Steve Woolgar, eds., *Representation in Scientific Practice Revisited* (Cambridge, Mass.: MIT Press, 2014) is an excellent recent collection.

8. For more on the relationship between motion pictures and scientific experiment, see the section on "The Multiple Functions of the Medical Film" in chap. 2 in this volume.

9. On experimental systems and "dislocation" or "differential reproduction," see Hans-Jörg Rheinberger, *Toward a History of Epistemic Things: Synthesizing Proteins in the Test Tube* (Stanford, Calif.: Stanford University Press, 1997), esp. chap. 5. Generally speaking, experimental systems are designed to produce incremental differences, which ultimately produce new inquiries and systems. But the balance between old and new is not symmetrical; my theory of disciplinary appropriation accommodates both the *correspondence* between disciplinary ideals and film and the *difference* between them, but the history I tell in this book favors the former. For more on the latter, see the conclusion in this volume.

10. Gaston Bachelard, *The New Scientific Spirit*, trans. Arthur Goldhammer (Boston: Beacon, 1984), 13. See also Davis Baird, *Thing Knowledge: A Philosophy of Scientific Instruments* (Berkeley: University of California Press, 2004).

11. Rheinberger, *Toward a History of Epistemic Things*, 21.

12. Ian Hacking, *Representing and Intervening: Introductory Topics in the Philosophy of Natural Science* (Cambridge: Cambridge University Press, 1983).

13. For surveys of scientific uses of photography and film, see Karl Wilhelm Wolf-Czapek, ed., *Angewandte Photographie in Wissenschaft und Technik* (Berlin: Union Deutsche Verlagsgesellschaft, 1911); Martin Weiser, *Medizinische Kinematographie* (Dresden and Leipzig: Steinkopff, 1919); F. Paul Liesegang, *Wissenschaftliche Kinematographie* (Düsseldorf: Liesegang, 1920); Anthony R. Michaelis, *Research Films in Biology, Anthropology, Psychology, and Medicine* (New York: Academic, 1955); Virgilio Tosi, *Cinema Before Cinema: The Origins of Scientific Cinematography*, trans. Sergio Angelini (London: British Universities Film & Video Council, 2005); Timothy Boon, *Films of Fact: A History of Science in Documentary Films and Television* (New York: Wallflower, 2008); and Kelly Wilder, *Photography and Science* (London: Reaktion, 2009).

14. The literature on this transition between chronophotography and film is vast, but one place to start is Deac Rossell, *Living Pictures: The Origins of the Movies* (Albany: State University of New York Press, 1998). On Janssen, see esp. the work of Jimena Canales, who describes the history of Janssen's photographic revolver in the context of emerging cinematographic forms of representation in the following: "Photogenic Venus: The 'Cinematographic Turn' in Science and Its Alternatives," *Isis* 93 (2002): 585–613; "Sensational Differences: The Case of the Transit of Venus," *Cahiers François Viète* 1, nos. 11/12 (September 2007): 15–40; *A Tenth of*

a Second: A History (Chicago: University of Chicago Press, 2009), 87–115; and "Desired Machines: Cinema and the World in Its Own Image," *Science in Context* 24, no. 3 (September 2011): 329–359.

15. Jennifer Tucker, *Nature Exposed: Photography as Eyewitness in Victorian Science* (Baltimore, Md.: Johns Hopkins University Press, 2005); or Tucker, "Photography as Witness, Detective, and Imposter: Visual Representation in Victorian Science," in *Victorian Science in Context*, ed. Bernard Lightman (Chicago: University of Chicago Press, 1997), 378–408; see also Joel Snyder, "Res Ipsa Loquitur," in *Things That Talk: Object Lessons from Art and Science*, ed. Lorraine Daston (New York: Zone, 2004), 195–221; as well as Wilder, *Photography and Science*.

16. An excellent discussion of one scientist's dissatisfaction with photography can be found in Sarah de Rijcke, "Drawing Into Abstraction. Practices of Observation and Visualisation in the Work of Santiago Ramón y Cajal." *Interdisciplinary Science Reviews* 33, no. 4 (2008): 287–311. Ramón y Cajal found that photography could not capture the three-dimensionality of nerve cells as well as drawings.

17. Thomas Schlich, "'Wichtiger als der Gegenstand selbst': Die Bedeutung des fotografischen Bildes in der Begründung der bakteriologischen Krankheitsauffassung durch Robert Koch," in *Neue Wege in der Seuchengeschichte*, ed. Martin Dinges and Thomas Schlich (Stuttgart: Steiner, 1995), 143–174. See also Olaf Breidbach, "Representation of the Microcosm: The Claim for Objectivity in 19th Century Scientific Microphotography," *Journal of the History of Biology* 35 (2002): 221–250.

18. On the similarities between the graphic method and cinema, see Lisa Cartwright, "'Experiments of Destruction': Cinematic Inscriptions of Physiology," *Representations* 40 (Fall 1992): 129–152; and Cartwright, *Screening the Body: Tracing Medicine's Visual Culture* (Minneapolis: University of Minnesota Press, 1995).

19. On Germany's research infrastructure, see, e.g., Margit Szöllösi-Janze, "Science and Social Space: Transformations in the Institutions of *Wissenschaft* from the Wilhelmine Empire to the Weimar Republic," *Minerva* 43 (2005): 339–360.

20. Carl Cranz, "Über einen ballistischen Kinematographen," *Deutscher Mechaniker-Zeitung* 18 (15 September 1909): 173–177. See also P. W. W. Fuller, "Carl Cranz, His Contemporaries, and High-Speed Photography," *Proceedings of SPIE*, no. 5580, 26th International Congress on High-Speed Photography and Photonics (25 March 2005): 250–260; and Wilhelm Pfeffer, "Die Anwendung des Projectionsapparates zur Demonstration von Lebensvorgängen," *Jahrbücher wissenschaftliche Botanik* 35 (1900): 711–745. On Pfeffer, see esp. Oliver Gaycken, "'The Swarming of Life': Moving Images, Education, and Views Through the Microscope," *Science in Context* 24, no. 3 (September 2011): 361–380; and Gaycken, "The Secret Life of Plants: Visualizing Vegetative Movement, 1880–1903," *Early Popular Visual Culture* 10, no. 1 (2012): 51–69.

21. The best overview of Comandon's life and work is Béatrice de Pastre and Thierry Lefebvre, eds., *Filmer la science, comprendre la vie: Le cinema de Jean Comandon* (Paris: Centre national de la cinématographie, 2012).

22. On Messter, see Christian Ilgner and Dietmar Linke, "Filmtechnik: Vom Malteserkreuz zum Panzerkino," in *Oskar Messter: Filmpioneer der Kaiserzeit*, ed. Martin Loiperdinger (Basel and Frankfurt: Stroemfeld/Roter Stern, 1994), 93–134, esp. 128–134; as well as Frank Kessler, Sabine Lenk, and Martin Loiperdinger, eds., *Oskar Messter—Erfinder und Geschäftsmann*, KINtop Schriften 3 (Basel and Frankfurt: Stoemfeld/Roter Stern, 1994). On Ernemann, see, e.g., the nod to the manufacturer in Hans Hennes,"Die Kinematographie der Bewegungsstörungen," *Die Umschau* 15, no. 29 (1911): 605–606; as well as Hanns Günther, "Mikrokinematographische Aufnahmeapparate," *Film and Lichtbild* 1, no. 1 (1912): 4–6; 1, no. 2 (1912): 13–14.

23. On popular scientific cinema, see Thierry Lefebvre, "The Scientia Production (1911–1914): Scientific Popularization Through Pictures," *Griffithiana* no. 47 (May 1993): 137–153; Oliver Gaycken, "'A Drama Unites Them in a Fight to the Death': Some Remarks on the Flourishing of a Cinema of Scientific Vernacularization in France, 1909–1914," *Historical Journal of Film, Radio and Television* 22, no. 3 (2002): 353–374; and Gaycken, *Devices of Curiosity: Early Cinema and Popular Science* (Oxford and New York: Oxford University Press, 2015). For an excellent account of popular science films in the United Kingdom, see Boon, *Films of Fact*. On the entwinement of scientific experiment, projection, popular science, and Victorian physics, see Simon Schaffer, "Transport Phenomena: Space and Visibility in Victorian Physics," *Early Popular Visual Culture* 10, no. 1 (February 2012): 71–91. On popularization in science in general, see Roger Cooter and Stephen Pumfrey, "Separate Spheres and Public Places: Reflections on the History of Science Popularization and Science in Popular Culture," *History of Science* 32, no. 97 (1994): 237–267.

24. Readers were asked to write the editor for details; see "Verzeichnis wissenschaftlich und technisch wertvoller Films," *Film und Lichtbild* 1, no. 2 (1912): 16; or "An unsere Abonnenten!" *Film und Lichtbild* 2, no. 5 (1913): 84.

25. For reports of screenings, see "Wissenschaft und Lichtspiele," *Bild und Film* 1, no. 2 (1912): 49–50; "Kino und Wissenschaft," *Bild und Film* 1, no. 2 (1912): 55; W. Thielemann, "Kinematographie und biologische Forschung," *Bild und Film* 3, no. 7 (1913/1914): 171–172; and "Wissenschaftliche Abende," *Film und Lichtbild* 1, no. 3 (1912): 30–31.

26. "Ein neues wissenschaftliches Kino," *Film und Lichtbild* 1, no. 5 (1912): 62. It is possible that the Fata Morgana was the only one of its kind.

27. See Thierry Lefebvre, *La chair et le celluloid: Le cinéma chirurgical du Docteur Doyen* (Brionne: Jean Doyen éditeur, 2004), for a discussion of the controversy surrounding the theft and possible unauthorized exhibition of Parisian doctor Eugène-Louis Doyen's surgical films in the early 1900s. See also chap. 2 in this volume.

28. For example, Wilhelm Richter, a Berlin teacher and school reformer, often wrote on scientific cinema in the popular press, cheering all efforts to bring new views to public perception. See "Der Kinematograph als naturwissenschaftliches Anschauungsmittel," *Naturwissenschaftliche Wochenschrift* 12, no. 52 (28 December 1913): 817–820.

29. On the use of film in university teaching, see Franz Bergmann, "Der Kinematograph im Hochschulunterricht," *Bild und Film* 2, no. 2 (1912/1913): 48; Wilhelm Richter, "Hochschulkinematographie," *Bild und Film* 2, nos. 11/12 (1912/1913): 253–257; and L. Segmiller, "Das Skizzieren nach Lichtbildern bei Tageslicht und künstlicher Beleuchtung," *Film und Lichtbild* 1, no. 4 (1912): 35–39.
30. One survey of the contemporary use of microcinematography notes, "Films of the latter [Ernst Sommerfeldt] are available commercially from Ernemann and depict crystallographic phenomenon," so apparently whether a research film made it into theaters depended on both the researcher's willingness and the manufacturer's evaluation of its popular appeal. On this example, see Ernst Sommerfeldt, "Über flüssige und scheinbar lebende Kristalle; mit kinematographischen Projektionen," *Verhandlungen der Gesellschaft deutscher Naturforscher und Ärzte* 79, part 2 (1907): 202. The survey of microcinematography is Engelhard Wychgram, "Aus optischen und mechanischen Werkstätten IV," *Zeitschrift für wissenschaftliche Mikroskopie und für mikroskopishe Technik* 28 (1911): 337–361, esp. 351–361.
31. On "mechanical objectivity," see Lorraine Daston and Peter Galison, "The Image of Objectivity," *Representations* 40 (Fall 1992): 81–128; and Daston and Galison, *Objectivity* (New York: Zone, 2007).
32. On the difference between instruments for experimentation and for demonstration, see Thomas L. Hankins and Robert J. Silverman, "The Magic Lantern and the Art of Demonstration," in *Instruments and the Imagination* (Princeton, N.J.: Princeton University Press, 1995), 37–71.
33. Bruno Latour, "Visualization and Cognition: Thinking with Eyes and Hands," *Knowledge and Society: Studies in the Sociology of Culture Past and Present* 6 (1986): 1–40, here 5. Brian Winston has also considered Latour's work in relation to film: "The Documentary Film as Scientific Inscription," in *Theorizing Documentary*, ed. Michael Renov (London and New York: Routledge, 1993), 37–57.
34. Latour, "Visualization and Cognition," 7 (emphasis in original).
35. See esp. Nicolas Rasmussen, *Picture Control: The Electron Microscope and the Transformation of Biology in America, 1940–1960* (Stanford, Calif.: Stanford University Press, 1997); Bernike Pasveer, "Knowledge of Shadows: The Introduction of X-Ray Images in Medicine," *Sociology of Health and Illness* 11, no. 4 (December 1989): 361–381; and Edward Yoxen, "Seeing with Sound: A Study of the Development of Medical Images," in *The Social Construction of Technological Systems: New Directions in the Sociology and History of Technology*, ed. Wiebe E. Bijker, Thomas P. Hughes, and Trevor J. Pinch (Cambridge, Mass.: MIT Press, 1987), 281–303.
36. Anne Harrington, *Reenchanted Science: Holism in German Culture from Wilhelm II to Hitler* (Princeton, N.J.: Princeton University Press, 1996). For an account of earlier debates about mechanistic approaches in science, see Peter Hanns Reill, *Vitalizing Nature in the Enlightenment* (Berkeley: University of California Press, 2005).
37. Henri Bergson, *Creative Evolution*, trans. Arthur Mitchell (New York: Random House, 1944). Hereafter cited parenthetically. Bergson indicates in the opening footnote of chap. 4 that he began thinking about science and film during his 1902–1903 course on the "History of the Idea of Time" at the Collège de France.

38. The mechanistic approach to biology was represented in Germany by such "biophysicists" as Hermann von Helmholtz and Emil Du Bois-Reymond, who hoped to demonstrate that life operated under the same physical and chemical laws as other phenomena. See Helmholtz's *Über die Erhaltung der Kraft* (Berlin: Reimer, 1847) for a comparison of muscles and mechanics, or Du Bois-Reymond's *Untersuchungen über thierische Elektricität* (Berlin: Reimer, 1848–1884) for an exploration of life's basic physical forces.
39. Henri Bergson, *An Introduction to Metaphysics*, trans. T. E. Hulme (New York: Macmillan, 1955), 52.
40. Bergson, *Metaphysics*, 51.
41. At this point we should note the influence of Bergson's philosophy on film theory from the work of Jean Epstein to Gilles Deleuze. For Epstein, see representative selections in *French Film Theory and Criticism*, vols. 1 and 2, ed. Richard Abel (Princeton, N.J.: Princeton University Press, 1988); for Deleuze, see *Cinema 1: The Movement-Image* and *Cinema 2: The Time-Image*, trans. Hugh Tomlinson and Barbara Habberjam (Minneapolis: University of Minnesota, 1986 and 1989).
42. Michel Georges-Michel, "Henri Bergson nous parle au cinéma," *Le Journal* (20 February 1914): 7; translated by Louis-Georges Schwartz as "Henri Bergson Talks to Us About Cinema," *Cinema Journal* 50, no. 3 (Spring 2011): 79–82, here 81. See also Frank Kessler's German translation and discussion of this article: "Henri Bergson und die Kinematographie," *KINtop* 12 (2003): 12–16.
43. I want to thank Paula Amad, whose doctoral dissertation "Archiving the Everyday: A Topos in French Film History, 1895–1931" (University of Chicago, 2002) led me in this direction. See also Amad's *Counter-Archive: Film, The Everyday, and Albert Kahn's* Archives de la Planète (New York: Columbia University Press, 2010). For more on Bergson's understanding of movement, see Jimena Canales, "Movement Before Cinematography: The High-Speed Qualities of Sentiment," *Journal of Visual Culture* 5, no. 3 (December 2006): 275–294.
44. Walter Benjamin, "On Some Motifs in Baudelaire," in *Walter Benjamin: Selected Writings*, vol. 4, *1938–1940*, ed. Howard Eiland and Michael W. Jennings, trans. Edmund Jephcott and others (Cambridge, Mass.: Belknap, 2003), 313–355, here 314.
45. This is not to say that Bergson had no effect in Germany. Georg Simmel, e.g., was impressed and influenced by Bergson. See Gregor Fitzi, *Soziale Erfahrung und Lebensphilosophie. Georg Simmels Beziehung zu Henri Bergson* (Konstanz, Germany: UVK Verlagsgesellschaft, 2002). For Bergson's reception by such thinkers as Max Scheler, Roman Ingarden, Hans Driesch, Max Horkeimer, and others, see Rudolf W. Meyer, "Bergson in Deutschland. Unter besonderer Berücksichtigung seiner Zeitauffassung," in *Studien zum Zeitproblem in der Philosophie des 20. Jahrhunderts*, ed. Rudolf W. Meyer et al. (Munich: Alber, 1982), 10–64; and Günther Pflug, "Die Bergson-Rezeption in Deutschland," *Zeitschrift für philosophische Forschung* 45, no. 2 (April–June 1991): 257–266.
46. Cited in Michael Ermarth, *Wilhelm Dilthey: The Critique of Historical Reason* (Chicago: University of Chicago Press, 1978), 24.

47. Harrington, *Reenchanted Science*, 27.
48. Wilhelm Dilthey, "The Construction of the Historical World in the Human Studies," in *Selected Writings*, ed., trans., and with an introduction by H. P. Rickman (Cambridge: Cambridge University Press, 1976), 168–245, here 181.
49. My discussion of the science of work is indebted to Anson Rabinbach, *The Human Motor: Energy, Fatigue, and the Origins of Modernity* (Berkeley and Los Angeles: University of California Press, 1990). Hereafter cited parenthetically. Closely related to this idea of fatigue was the modern notion of "nervousness": see Andreas Killen, *Berlin Electropolis: Shock, Nerves, and German Modernity* (Berkeley: University of California Press, 2006).
50. In addition to Harrington's *Reenchanted Science*, a good overview of the debate, from the vitalist's point of view, is Frederick Burwick and Paul Douglass, eds., *The Crisis in Modernism: Bergson and the Vitalist Controversy* (Cambridge: Cambridge University Press, 1992). With regard to cinema and vitalism, see also Inga Pollmann, "Cinematic Vitalism: Theories of Life and the Moving Image" (PhD diss., University of Chicago, 2011).
51. Rabinbach, *Human Motor*, 181; see also Ernest Solvay, *Notes sur le productivisme et le comptabilisme* (Brussels: Misch and Thron, 1900) 2: 323. On the institute, see Daniel Warnotte, *Ernest Solvay et l'Institut de Sociologie: Contribution à l'histoire de l'énergétique sociale* (Brussels: Bruylant, 1946).
52. Rabinbach, *Human Motor*, 134; see also Angelo Mosso, *Fatigue*, trans. Margaret Drummond and W. B. Drummond (New York: Putnam, 1904); and Karen J. Fleckenstein, "The Mosso Plethysmograph in 19th-Century Physiology," *Medical Instrumentation* 18, no. 6 (November–December 1984): 330–331.
53. See Leo Koenigsberger, *Hermann von Helmholtz* (Braunschweig, Germany: Viewig, 1911); as well as David Cahan, ed., *Hermann von Helmholtz and the Foundations of Nineteenth-Century Science* (Berkeley and Los Angeles: University of California Press, 1993).
54. See especially Greg Myers, "Nineteenth-Century Popularizations of Thermodynamics and the Rhetoric of Social Prophecy," in *Energy & Entropy: Science and Culture in Victorian Britain*, ed. Patrick Brantlinger (Bloomington: Indiana University Press, 1988), 307–338; Ed Block Jr., "T. H. Huxley's Rhetoric and the Popularization of Victorian Scientific Ideas, 1854–1874," in Brantlinger, *Energy & Entropy*, 205–228; and *Degeneration: The Dark Side of Progress*, ed. J. Edward Chamberlin and Sander L. Gilman (New York: Columbia University Press, 1985). The concept of degeneration will receive much more attention in chap. 2 of this volume.
55. Recent studies of Marey include Marta Braun, *Picturing Time: The Work of Étienne-Jules Marey* (Chicago: University of Chicago Press, 1992); and Francois Dagognet, *Étienne-Jules Marey: A Passion for the Trace*, trans. Robert Galeta with Jeanine Herman (New York: Zone, 1992). See also Marey's *La methode graphique dans les sciences expérimentales et principlement en physiologie et en médecine* (Paris: Masson, 1885); *Movement*, trans. Eric Pritchard (London: Heinemann, 1895); and *La chronophotographie* (Paris: Gauthier-Villars, 1899).

56. Étienne-Jules Marey, "Études sur la marche de l'homme," *Revue Militaire de Médecine et de Chirurgie* 1 (1880): 244–46, cited in Rabinbach, *Human Motor*, 116.
57. For the culmination of this approach, see Frederick Winslow Taylor, *The Principles of Scientific Management* (1911; New York: Norton, 1947); and Frank B. Gilbreth, *Motion Study: A Method for Increasing the Efficiency of the Workman* (New York: Van Nostrand, 1911). On the use of motion pictures in this program, see Richard Lindstrom, "'They All Believe They Are Undiscovered Mary Pickfords': Workers, Photography, and Scientific Management," *Technology and Culture* 41, no. 4 (2000): 725–751; Ramón Reichert, "Der Arbeitstudienfilm: Eine verborgene Geschichte des Stummfilms," *medien & zeit: Kommunikation in Vergangenheit und Gegenwart* 5 (2002): 46–57; Philipp Sarasin, "Die Rationalisierung des Körpers: Über 'Scientific Management' und 'biologische Rationalisierung'" in *Geschichtswissenschaft und Diskursanalyse* (Frankfurt: Suhrkamp, 2003), 61–99; Elspeth H. Brown, *The Corporate Eye: Photography and the Rationalization of American Commercial Culture, 1884–1929* (Baltimore: Johns Hopkins University Press, 2005); Florian Hoof, "'The One Best Way': Bildgebende Verfahren der Ökonomie und die Innovation der Managementtheorie Nach 1860," *montage/AV: Zeitschrift für Theorie und Geschichte audiovisueller Kommunikation* 15, no. 1 (2006): 123–138; and Scott Curtis, "Images of Efficiency: The Films of Frank B. Gilbreth," in *Films That Work: Industrial Film and the Productivity of Media*, ed. Vinzenz Hediger and Patrick Vonderau (Amsterdam: Amsterdam University Press, 2009), 85–99.
58. This tradition goes back as far as 1836, with the publication of the work of Wilhelm Weber, who with his brothers Ernst and Eduard investigated human motion using a variety of innovative visual and graphic technologies. See *Mechanik der menschlichen Gehwerkzeuge: Eine anatomisch-physiologische Untersuchung*, in *Wilhelm Weber's Werke*, vol. 6, ed. Der königlichen Gesellschaft der Wissenschaften zu Göttingen (Berlin: Springer, 1894), 1–305. (Wilhelm Braune and Otto Fischer were closely involved in the selection and editing of Weber's works.) For a comprehensive history of human motion studies, see Andreas Mayer, *Wissenschaft vom Gehen. Die Erforschung der Bewegung im 19. Jahrhundert* (Frankfurt: Fischer, 2013).
59. Braune and Fischer, *The Human Gait*, 18.
60. Braune and Fischer, *The Human Gait*, 4.
61. Rabinbach, *Human Motor*, 189; see also Nathan Zuntz and Wilhelm Schumberg, *Studien zu einer Physiologie des Marsches* (Berlin: Hirschwald, 1901).
62. Rabinbach, *Human Motor*, 104; see also Étienne-Jules Marey, *Animal Mechanism: A Treatise on Terrestrial and Aerial Locomotion* (New York: Appleton, 1874).
63. Rabinbach, *Human Motor*, 144; see also Wilhelm Weichardt, "Ermüdungsbekämpfung durch Antikenotoxin," *Deutsche militärärztliche Zeitschrift* 42, no. 1 (5 January 1913): 12–13.
64. Rabinbach, *Human Motor*, 135; see also Mosso, *Fatigue*, 121.

65. "Novel Uses for Moving Pictures," *Moving Picture World* 1, no. 3 (23 March 1907): 39–40.
66. Michel Foucault, *Discipline and Punish*, trans. Alan Sheridan (New York: Vintage, 1979), 136.
67. Foucault, *Discipline and Punish*, 137.
68. Ian Hacking, *Representing and Intervening: Introductory Topics in the Philosophy of Natural Science* (Cambridge: Cambridge University Press, 1983).
69. Michael Lynch, "Discipline and the Material Form of Images: An Analysis of Scientific Visuality," *Social Studies of Science* 15, no. 1 (February 1985): 43–44 (emphasis in original). Daston and Galison, in their work on objectivity (cited in note 31), call it a "working object," which has become the more common term in the history and sociology of science.
70. Some of the most notable early sociological studies of the scientific process include Harold Garfinkel, Michael Lynch, and Eric Livingston, "The Work of a Discovering Science Construed with Materials from the Optically Discovered Pulsar," *Philosophy of the Social Sciences* 11, no. 2 (June 1981): 131–158; Karin Knorr-Cetina, *The Manufacture of Knowledge: An Essay on the Constructivist and Contextual Nature of Science* (Oxford: Pergamon, 1981); Bruno Latour and Steve Woolgar, *Laboratory Life: The Social Construction of Scientific Facts* (London and Beverly Hills, Calif.: Sage, 1979); and Latour, *Science in Action* (Cambridge, Mass.: Harvard University Press, 1987).
71. Chapter 1 appeared as "Versuche am unbelasteten und belasteten Menschen," *Abhandlungen der Mathematisch-Physischen Klasse der Königlich Sächsischen Gesellschaft der Wissenschaften* 21, no. 4 (1895): 151–322; chap. 2 as "Die Bewegung des Gesamtschwerpunktes und die äußeren Kräfte" in 25, no. 1 (1899): 1–130; chap. 3 as "Betrachtungen über die weiteren Ziele der Untersuchung und Überblick über die Bewegungen der unteren Extremitäten" in 26, no. 3 (1900): 85–170; chap. 4 as "Über die Bewegung des Fußes und die auf denselben einwirkenden Kräfte" in 26, no. 7 (1901): 467–556; chap. 5 as "Die Kinematik des Beinschwingens" in 28, no. 5 (1903): 319–418; and chap. 6 as "Über den Einfluß der Schwere und der Muskeln auf die Schwingungsbewegung des Beins" in 28, no. 7 (1904): 531–617.
72. Herman J. Wohlring, review of *The Human Gait*, *Human Movement Science* 8, no. 1 (February 1989): 79–83. Aerial photography, the counterpart to close-range photogrammetry, also developed from military applications. See Teodor J. Blachut and Rudolf Burkhardt, *Historical Development of Photogrammetric Methods and Instruments* (Falls Church, Va.: American Society for Photogrammetry and Remote Sensing, 1989); and Paul Virilio, *War and Cinema: The Logistics of Perception*, trans. Patrick Camiller (London: Verso, 1989).
73. Eadweard Muybridge, *Animal Locomotion: An Electro-Photographic Investigation of Consecutive Phases of Animal Movement* (Philadelphia: University of Pennsylvania, 1887); or *Muybridge's Complete Human and Animal Locomotion* (New York: Dover, 1979); and Albert Londe, *Photographie médicale* (Paris: Gauthiers-Villars, 1893). There also seems to have been surprisingly little crossover between Braune and Fischer's work and that of other German

chronophotographers, such as Ottomar Anschütz and Ernst Kohlrausch, but that is primarily a disciplinary issue; as the quotation above indicates, Braune and Fischer saw Anschütz's decision to pursue the popular potential of his devices, rather than any research applications, as a markedly different path. A potentially interesting overlap might be the German Ministry of War, with which Anschütz worked in 1884, but more research needs to be done here. On Anschütz, see Deac Rossell, *Faszination der Bewegung: Ottomar Anschütz zwischen Photographie und Kino*, KINtop Schriften 6 (Frankfurt: Stroemfeld/Roter Stern, 2001).

74. Braune and Fischer, *The Human Gait*, 6. Hereafter cited parenthetically.
75. Lynch, "Discipline and the Material Form of Images," 43.
76. Lynch, "Discipline and the Material Form of Images," 42.
77. Edmund Husserl, *The Crisis of European Sciences and Transcendental Phenomenology*, trans. David Carr (Evanston, Ill.: Northwestern University Press, 1970), 376.
78. Michael Lynch, "The Externalized Retina: Selection and Mathematization in the Visual Documentation of Objects in the Life Sciences," in *Representation in Scientific Practice*, ed. Michael Lynch and Steve Woolgar (Cambridge, Mass.: MIT Press, 1990), 153–186, here 170.
79. Bergson, *Creative Evolution*, 342 (emphasis in original).
80. Michaelis, *Research Films in Biology, Anthropology, Psychology, and Medicine*, 371.
81. Bergson, *Creative Evolution*, 238.
82. Jill Vance Buroker, "Descartes on Sensible Qualities," *Journal of the History of Philosophy* 29, no. 4 (October 1991): 585–611.
83. Bergson, *Creative Evolution*, 240 (emphasis in original).
84. An earlier, shorter version of this section appeared as "Die kinematographische Methode. Das 'Bewegte Bild' und die Brownsche Bewegung," *montage/AV: Zeitschrift für Theorie & Geschichte audiovisueller Kommunikation* 14, no. 2 (2005): 23–43.
85. Roberto Maiocchi, "The Case of Brownian Motion," *British Journal for the History of Science* 23 (September 1990): 257–283, here 257. See also Stephen G. Brush, "A History of Random Processes: I. Brownian Movement from Brown to Perrin," *Archive for History of Exact Sciences* 5, no. 1 (1968): 1–36; Mary Jo Nye, *Molecular Reality: A Perspective on the Scientific Work of Jean Perrin* (New York: American Elsevier, 1972); Milton Kerker, "Brownian Movement and Molecular Reality Prior to 1900," *Journal of Chemical Education* 51, no. 12 (December 1974): 764–768; Peter Clark, "Atomism Versus Thermodynamics," in *Method and Appraisal in the Physical Sciences*, ed. Colin Howson (Cambridge: Cambridge University Press, 1976), 41–106; Brush, *Statistical Physics and the Atomic Theory of Matter from Boyle and Newton to Landau and Onsager* (Princeton, N.J.: Princeton University Press, 1983), 79–104; and John Stachel, "Einstein on Brownian Motion," in *The Collected Papers of Albert Einstein*, vol. 2, *The Swiss Years: Writings, 1900–1909*, ed. John Stachel (Princeton, N.J.: Princeton University Press, 1989), 206–222. On the modern implications and theoretical offspring of Brownian motion, see

Erwin Frey and Klaus Kroy, "Brownian Motion: A Paradigm of Soft Matter and Biological Physics," *Annalen der Physik* 14, nos. 1–3 (February 2005): 20–50.

86. Clark, "Atomism Versus Thermodynamics," 42.
87. For Mach's objections to the kinetic theory, see Ernst Mach, *Die Principien der Wärmlehre: Historisch-kritisch Entwickelt* (Leipzig: Barth, 1896), 428–429; for a discussion of Mach and the historical context of the debate, see Stephen G. Brush, *The Kind of Motion We Call Heat: A History of the Kinetic Theory of Gases in the 19th Century* (Amsterdam and New York: North-Holland and American Elsevier, 1976), 274–299; for more on Mach, see John T. Blackmore, *Ernst Mach: His Work, Life, and Influence* (Berkeley: University of California Press, 1972). For Ostwald and other anti-atomists, see Mary Jo Nye, *Molecular Reality*; and Nye, "The Nineteenth-Century Atomic Debates and the Dilemma of an 'Indifferent Hypothesis,'" *Studies in History and Philosophy of Science* 7, no. 3 (1976): 245–268.
88. Brush, *Statistical Physics*, 97.
89. Louis-Georges Gouy, "Le mouvement brownien et les mouvements moléculaires," *Revue générale des sciences pures et appliquées* 6, no. 1 (15 January 1895): 1–7.
90. Maiocchi, "The Case of Brownian Motion," 260.
91. Felix M. Exner, "Notiz zu Brown's Molecularbewegung," *Annalen der Physik* 2, no. 8 (1900): 843–847.
92. Exner, "Notiz," 844–845.
93. Albert Einstein, "On the Movement of Small Particles Suspended in a Stationary Liquid Demanded by the Molecular-Kinetic Theory of Heat," in *Investigations on the Theory of the Brownian Movement*, ed. R. Fürth, trans. A. D. Cowper (New York: Dover, 1956), 1–18, here 1–2.
94. The best explanation of Einstein's use of Brownian motion as a statistical system is Martin J. Klein, "Fluctuations and Statistical Physics in Einstein's Early Work," in *Albert Einstein: Historical and Cultural Perspectives*, ed. Gerald Holton and Yehuda Elkana (Princeton, N.J.: Princeton University Press, 1982), 39–58. Klein remarks, "Einstein had *invented* the Brownian motion. To say anything less, to describe this paper in the usual way, that is, as his *explanation* of the Brownian motion, is to undervalue it" (47, emphasis in original). See also Jürgen Renn, "Einstein's Invention of Brownian Motion," *Annalen der Physik* 14, supplement (2005): 23–37.
95. Maiocchi, "The Case of Brownian Motion," 260 (emphasis in original).
96. Nye, *Molecular Reality*, 111.
97. Significantly, Brownian motion described an observable system in liquid, whereas up to this point discussions of random systems and fluctuations had focused only on gases. That is, the kinetic theory to this point applied only to gases, not liquids or solids. By removing that particular restriction in his focus on Brownian motion, Einstein also raised the stakes of the debate over the existence of atoms.
98. Einstein further simplified the system by limiting his mathematical derivations to two dimensions, thereby only taking into account the horizontal displacement of particles. In this way, Einstein's theory corresponds to experimental practice in

that the field of observation corresponds to the flat, two-dimensional field of the microscope.

99. Maiocchi, "The Case of Brownian Motion," 263–264 (emphasis in original).
100. Brush, "A History of Random Processes," 22–23.
101. On the invention and use of the ultramicroscope, see David Cahan, "The Zeiss Werke and the Ultramicroscope: The Creation of a Scientific Instrument in Context," in *Scientific Credibility and Technical Standards in 19th and Early 20th Century German and Britain*, ed. Jed Z. Buchwald (Dordrecht, Netherlands: Kluwer Academic, 1996), 67–115.
102. Victor Henri, "Études cinématographique des mouvements browniens," *Comptes rendus hebdomadaires des séances de l'Academie des Sciences* 146 (18 May 1908): 1024–1026; and Henri, "Influence du milieu sur les mouvements browniens," *Comptes rendus hebdomadaires des séances de l'Academie des Sciences* 147 (6 July 1908): 62–65. For a later use of motion pictures to record the distribution of particles in a gaseous system, which builds upon Seddig's and Henri's work, see Richard Lorenz and W. Eitel, "Über die örtliche Verteilung von Rauchteilchen," *Zeitschrift für anorganische Chemie* 87, no. 1 (12 May 1914): 357–374.
103. For a critique of Henri's results, see Aimé Cotton, "Recherches récentes sur les mouvements browniens," *La Revue du Mois* 5 (10 June 1908): 737–741.
104. The most influential paper of many is Jean Perrin, "Mouvement brownien et molécules," *Annales de chimie et de physique* 18 (September 1909): 1–114, translated by F. Soddy as "Brownian Movement and Molecular Reality," in *The Question of the Atom*, ed. Mary Jo Nye (Los Angeles, Calif.: Tomash, 1984), 507–601. The best commentary on Perrin is Mary Jo Nye, *Molecular Reality*. For Perrin's own discussion of Henri, Seddig, and others, see *Atoms*, trans. D. L. Hammick, 2d English ed. rev. (London: Constable, 1923), 109–133. On Perrin's visualization techniques, see Charlotte Bigg, "Evident Atoms: Visuality in Jean Perrin's Brownian Motion Research," *Studies in History and Philosophy of Science* 39 (2008): 312–322. See also Bigg, "A Visual History of Jean Perrin's Brownian Motion Curves," in *Histories of Scientific Observation*, ed. Lorraine Daston and Elizabeth Lunbeck (Chicago: University of Chicago Press, 2011), 156–180.
105. Nye, *Molecular Reality*, 97.
106. Seddig, "Messung der Temperatur-Abhängigkeit der Brown'schen Molekularbewegung," 12 (emphasis in original).
107. Seddig, "Ueber Abhängigkeit," 185.
108. Marey came upon this solution as early as 1891. See Braun, *Picturing Time*, 166.
109. Einstein to Jakob Laub, 30 July 1908, quoted in Stachel, "Einstein on Brownian Motion," 220. We should also note that in 1907 there was another attempt to verify Einstein's theories: Theodor Svedberg, *Studien zur Lehre von der Kolloiden Lösungen* (Uppsala: Akademische Buchdruckerei Edv. Berling, 1907). Einstein was not so kind: "The errors in Svedberg's method of observation and also in his theoretical treatment became clear to me at once. I wrote a minor correction at the time, which only addressed the worst, as I couldn't bring myself to detract from

Mr. S's great pleasure in his work." (Einstein to Jean Perrin, 11 November 1909, quoted in Stachel, "Einstein on Brownian Motion," 220).

110. Maryan Smoluchowski, "Essai d'une thèorie cinètique du movement brownien et des milieux troubles," *Krakau Anzeiger* 7 (1906): 585–586, quoted in Maiocchi, "The Case of Brownian Motion," 264.

111. Jean Perrin, *Les atoms*, 4th ed. rev. (Paris: Librarie Félix Alcan, 1914), 157; *Atoms*, 2d English ed. rev., 109–110 (emphasis in original). On Jean Comandon's Brownian motion films, see Jean Comandon with Albert Dastre, "Cinématographie, à l'ultra-microscope, de microbes vivants et des particules mobiles," *Comptes rendus hebdomadaires des séances de l'Académie des Sciences* 149 (22 November 1909): 938–941. Perrin showed the films of Henri and Comandon to the Société des Amis de l'Université de Paris in 1911: Jean Perrin, "La realité des molecules," *Revue Scientifique* 49, no. 2 (1911): 774–784, quoted in Nye, *Molecular Reality*, 153. See also Hannah Landecker, "Cellular Features: Microcinematography and Film Theory," *Critical Inquiry* 31, no. 4 (2005): 903–937.

112. On the debate in the 1920s between Bergson and Einstein regarding this category mistake, see Jimena Canales, "Einstein, Bergson, and the Experiment That Failed: Intellectual Cooperation at the League of Nations," *Modern Language Notes* 120, no. 5 (2006): 1168–1191.

113. Louis de Broglie, "The Concepts of Contemporary Physics and Bergson's Ideas on Time and Motion," in *Bergson and the Evolution of Physics*, ed. and trans. P. A. Y. Gunter (Knoxville: University of Tennessee Press, 1969), 45–62, here 49.

114. Suzanne Guerlac, *Thinking in Time: An Introduction to Henri Bergson* (Ithaca, N.Y.: Cornell University Press, 2006), 19.

115. Bergson was not alone in this stance; even Nobel Prize–winning Belgian physicist and chemist Ilya Prigogine argued that physics should account for the irreversibility of time. See, e.g., *From Being to Becoming: Time and Complexity in the Physical Sciences* (San Francisco: Freeman, 1980); and with Isabelle Stengers, *Order Out of Chaos: Man's New Dialogue with Nature* (Boulder, Colo.: New Science Library, 1984); and *The End of Certainty: Time, Chaos, and the New Laws of Nature* (New York: Free Press, 1997). For an introduction to this contradiction between time in physics and time as it is experienced, see David Z. Albert, *Time and Chance* (Cambridge, Mass.: Harvard University Press, 2000).

116. For different interpretations of the instant and the point in photography and cinema, see Thierry de Duve, "Time Exposure and Snapshot: The Photograph as Paradox," *October* 5 (Summer 1978): 113–125; and Doane, *The Emergence of Cinematic Time*, 208–230.

117. Henri Bergson, *Matter and Memory*, trans. Nancy Margaret Paul and W. Scott Palmer (Cambridge, Mass.: Zone, 1988), 247 (emphasis in original).

118. Bergson, *Matter and Memory*, 250 (emphasis in original).

119. Deleuze, *Cinema 1*.

120. Seddig, "Messung der Temperatur-Abhängigkeit der Brown'schen Molekularbewegung," 36–37.

121. Marey, *Movement*.

122. *Lucien Bull* (Brussels: Hayez, 1967), 11.
123. For a representative sampling, see Charles François-Franck, "La chronophotographie simultanée du coeur et des courbes cardiographiques chez les mammifères," *Comptes rendus hebdomadaires des séances et mémoires de la Société de Biologie* 54 (8 November 1902): 1193–1197; "Note sur quelques points de technique relatifs à la photographie et à la chronophotographie avec le magnésium à deflagration lente," *Comptes rendus hebdomadaires des séances et mémoires de la Société de Biologie* 55 (5 December 1903): 1538–1540; "Études graphiques et photographiques de mécanique respiratoire comparée," *Comptes rendus hebdomadaires des séances et mémoires de la Société de Biologie* 61 (28 July 1906): 174–176; and "Démonstrations de microphotographie instantanée et de chronomicrophotographie," *Comptes rendus hebdomadaires des séances et mémoires de la Société de Biologie* 62 (25 May 1907): 964–967. Although many of his citations are incorrect, Thierry Lefebvre, "Contribution à l'histoire de la microcinématographie: De François-Franck à Comandon," *1895* 14 (June 1993): 35–43, is an essential introduction.
124. L. Chevroton and F. Vlès, "La cinématique de la segmentation de l'oeuf et la chronophotographie du développement de l'Oursin," *Comptes rendus hebdomadaires des séances de l'Academie des Sciences* 149 (8 November 1909): 806–809.
125. Henri, "Études cinématographique des mouvements browniens."
126. Isabelle do O'Gomes, "L'oeuvre de Jean Comandon," in *Le cinéma et la science*, ed. Alexis Martinet (Paris: CNRS Éditions, 1994), 78–85.
127. Julius Ries, "Kinematographie der Befruchtung und Zellteilung," *Archiv für Mikroskopische Anatomie und Entwicklungsgeschichte* 74 (1909): 1–31.
128. Henri Bénard, "Les tourbillons cellulaires dans une nappe liquide. II. Procédés mécaniques et optiques d'examens, lois numériques des phénomènes," *Revue générale des sciences pures et appliquées* 11 (1900): 1309–1328; and Bénard, "Formation de centres de giration à l'arrière d'un obstacle en movement," *Comptes rendus de l'Académie des Sciences* 147 (1908): 839–842. See also José Eduardo Wesfreid, "Scientific Biography of Henri Bénard (1874–1939)," in *Dynamics of Spatio-Temporal Cellular Structures: Henri Bénard Centenary Review*, ed. Innocent Mutabazi, José Eduardo Wesfreid, and Étienne Guyon (New York: Springer, 2006), 9–40; and David Aubin, "'The Memory of Life Itself': Bénard's Cells and the Cinematography of Self-Organization," *Studies in History and Philosophy of Science* 39, no. 3 (2008): 359–369.
129. H. Siedentopf and E. Sommerfeldt, "Über die Anfertigung kinematographischer Mikrophotographien der Kristallisationserscheinungen," *Zeitschrift für Elektrochemie* 13, no. 24 (14 June 1907): 325–326; and Siedentopf, "Über ultramikroskopische Abbildung," *Zeitschrift für wissenschaftliche Mikroskopie und mikroskopische Technik* 26 (1909): 391–410.
130. A contemporary overview of Braus's career can be found in Walther Vogt's eulogy, "Hermann Braus," *Münchener medizinische Wochenschrift* 72, no. 8 (20 February 1925): 304–305. An assessment of his legacy in morphology is in Lynn K. Nyhart, "Learning from History: Morphology's Challenges in Germany ca. 1900," *Journal of Morphology* 252 (April 2002): 2–14.

131. A shorter version of this section appeared as "Science Lessons," *Film History* 25, nos. 1–2 (2013): 45–54.
132. My presentation of these theories and the history of this debate is indebted to Susan M. Billings's excellent survey, "Concepts of Nerve Fiber Development, 1839–1930," *Journal of the History of Biology* 4, no. 2 (Fall 1971): 275–305.
133. Ross Granville Harrison, "Further Experiments on the Development of Peripheral Nerves," *American Journal of Anatomy* 5, no. 2 (31 May 1906): 121–131. A brief overview of Harrison's career can be found in J. S. Nicholas, "Ross Granville Harrison, 1870–1959," *Yale Journal of Biology and Medicine* 32, no. 6 (June 1960): 407–412.
134. Peter J. Taylor and Ann S. Blum, "Pictorial Representation in Biology," *Biology and Philosophy* 6, no. 2 (April 1991): 125–134; and Nick Hopwood, "Producing Development: The Anatomy of Human Embryos and the Norms of Wilhelm His," *Bulletin of the History of Medicine* 74 (2000): 29–79.
135. Billings, "Concepts," 293–294.
136. Santiago Ramón y Cajal, "New Observations on the Development of Neuroblasts, with Comments on the Neurogenetic Hypothesis of Hensen-Held (1908)," in *Studies on Vertebrate Neurogenesis*, trans. Lloyd Guth (Springfield, Ill.: Thomas, 1960), 71–76, quoted in Billings, "Concepts," 294. See also De Rijcke, "Drawing Into Abstraction."
137. Hannah Landecker, "New Times for Biology: Nerve Cultures and the Advent of Cellular Life in Vitro," *Studies in the History and Philosophy of Biological and Biomedical Sciences* 33, no. 4 (2002): 667–694, here 672.
138. Landecker, "New Times," 673.
139. Landecker examines this formulation with regard to the rise of microcinematography in "Creeping, Drinking, Dying: The Cinematic Portal and the Microscopic World of the Twentieth-Century Cell," *Science in Context* 24, no. 3 (September 2011): 381–416; and "The Life of Movement: From Microcinematography to Live-Cell Imaging," *Journal of Visual Culture* 11, no. 3 (December 2012): 378–399.
140. Harrison, "Further Experiments," 121.
141. Hermann Braus, "Experimentelle Beiträge zur Frage nach der Entwickelung peripherer Nerven," *Anatomischer Anzeiger* 26, nos. 17/18 (1 April 1905): 433–479.
142. Ross Granville Harrison, "Experiments in Transplanting Limbs and Their Bearing Upon the Problems of the Development of Nerves," *Journal of Experimental Zoology* 4, no. 2 (June 1907): 239–281, here 241.
143. Ross Granville Harrison, "Observations on the Living Developing Nerve Fiber," *Anatomical Record* 1, no. 5 (1 June 1907): 116–118, here 116. Harrison describes the method and its results more fully in "The Outgrowth of the Nerve Fiber as a Mode of Protoplasmic Movement," *Journal of Experimental Zoology* 9, no. 4 (December 1910): 787–846.
144. Leo Loeb is known to have accomplished in vivo tissue culture as early as 1897. See Lewis Philip Rubin, "Leo Loeb's Role in the Development of Tissue Culture," *Clio Medica* 12, no. 1 (1977): 33–56. See also H. M. Carleton, "Tissue Culture: A Critical Summary," *British Journal of Experimental Biology* 1, no. 1 (October 1923): 131–151.
145. On this experiment, see the announcements by Alexis Carrel and Montrose T. Burrows, including "Cultivation of Adult Tissues and Organs Outside of the

Body," *Journal of the American Medical Association* 55, no. 16 (15 October 1910): 1379–1381; "Cultivation of Sarcoma Outside of the Body: A Second Note," *Journal of the American Medical Association* 55, no. 18 (29 October 1910): 1554; "Human Sarcoma Cultivated Outside of the Body: A Third Note," *Journal of the American Medical Association* 55, no. 20 (12 November 1910): 1732; and "Cultivation of Tissues in Vitro and Its Technique," *Journal of Experimental Medicine* 13, no. 3 (1 March 1911): 387–396.

146. The definitive argument for this break in biological representation, to which I am indebted, is Hannah Landecker, *Culturing Life: How Cells Became Technologies* (Cambridge, Mass.: Harvard University Press, 2007); see also her "Technologies of Living Substance: Tissue Culture and Cellular Life in Twentieth Century Biomedicine" (PhD diss., Massachusetts Institute of Technology, 1999).

147. This shift in the discipline's attention was not just a result of representational issues but also to a certain extent due to philosophical discomfort as a result of Bergson's critiques. For a good example of a biologist wrestling with these issues, albeit much later, see P. Lecomte du Noüy, *Biological Time*, with a foreword by Alexis Carrel (New York: Macmillan, 1937).

148. Billings, "Concepts," 301–302.

149. Braus, "Mikro-Kino-Projektionen"; this was reprinted in slightly revised form in *Wiener medizinische Wochenschrift* 61, no. 44 (1911): 2809–2812.

150. Braus, "Mikro-Kino-Projektionen," 472–473.

151. Steven Shapin, "Pump and Circumstance: Robert Boyle's Literary Technology," *Social Studies of Science* 14, no. 4 (November 1984): 481–520. See also Steven Shapin and Simon Schaffer, *Leviathan and the Air-Pump: Hobbes, Boyle, and the Experimental Life* (Princeton, N.J.: Princeton University Press, 1985).

152. Braus, "Mikro-Kino-Projektionen," 473.

153. Braus, "Mikro-Kino-Projektionen," 474 (emphasis in original).

154. Braus, "Mikro-Kino-Projektionen," 474.

155. Walter Benjamin, "Little History of Photography," in *Walter Benjamin: Selected Writings*, vol. 2, *1927–1934*, ed. Michael W. Jennings, Howard Eiland, and Gary Smith, trans. Rodney Livingstone and others (Cambridge, Mass.: Belknap, 1999), 507–530, here 512.

156. For another example of the statistical use of film frames, see Scott Curtis, "'Tangible as Tissue': Arnold Gesell, Infant Behavior, and Film Analysis," *Science in Context* 24, no. 3 (September 2011): 417–442.

2. BETWEEN OBSERVATION AND SPECTATORSHIP

1. Paul Valéry, *Idée Fixe*, trans. David Paul (New York: Pantheon, 1965), 22.
2. For similar debates, see Michael Chanan, *The Dream That Kicks: The Prehistory and Early Years of Cinema in Britain* (London and Boston: Routledge & Kegan Paul, 1980); Richard Abel, *The Ciné Goes to Town: French Cinema, 1896–1914* (Berkeley: University of California Press, 1994); Yuri Tsivian, *Early Cinema in*

Russia and Its Cultural Reception, ed. Richard Taylor, trans. Alan Bodger (London and New York: Routledge, 1994); Lee Grieveson, *Policing Cinema: Movies and Censorship in Early-Twentieth-Century America* (Berkeley: University of California Press, 2004).

3. Sir James Paget, "An Address on the Utility of Scientific Work in Practice," *British Medical Journal* (15 October 1887): 811–814, here 811.
4. On physicians' adoption of scientific method, see W. F. Bynum, *Science and the Practice of Medicine in the Nineteenth Century* (Cambridge: Cambridge University Press, 1994); on the resistance of doctors to science, see John Harley Warner, "Ideals of Science and Their Discontents in Late Nineteenth-Century American Medicine," *Isis* 82, no. 3 (September 1991): 454–478.
5. Michel Foucault, *The Birth of the Clinic: An Archaeology of Medical Perception*, trans. A. M. Sheridan Smith (New York: Vintage, 1973); Michael Hau, "The Holistic Gaze in German Medicine, 1890–1930," *Bulletin of the History of Medicine* 74, no. 3 (Fall 2000): 495–524.
6. Scott Curtis, "Still/Moving: Digital Imaging and Medical Hermeneutics," in *Memory Bytes: History, Technology, and Digital Culture*, ed. Lauren Rabinovitz and Abraham Geil (Durham, N.C.: Duke University Press, 2004), 218–254.
7. Like chapter 1, this chapter will focus on research films, but for more on the early medical education film in Germany, see Waldemar Schweisheimer, *Die Bedeutung des Films für soziale Hygiene und Medizin* (Munich: Müller, 1920). On early American medical education films, see Martin Pernick, *The Black Stork: Eugenics and the Death of "Defective" Babies in American Medicine and Motion Pictures Since 1915* (New York: Oxford University Press, 1999); and esp. Kirsten Ostherr's valuable studies, *Cinematic Prophylaxis: Globalization and Contagion in the Discourse of World Health* (Durham, N.C.: Duke University Press, 2005); and *Medical Visions: Producing the Patient Through Film, Television, and Imaging Technologies* (New York: Oxford University Press, 2013).
8. The most comprehensive survey of early literature on medical research films remains Anthony R. Michaelis, *Research Films in Biology, Anthropology, Psychology, and Medicine* (New York: Academic, 1955). Specific statistics about Germany's output compared with other countries can be found on p. 326. See also Adolf Nichtenhauser, "History of Motion Pictures in Medicine" (unpublished manuscript, MS C 380, History of Medicine Division, National Library of Medicine, Bethesda, Md.). The definitive contemporary treatment of the subject is Lisa Cartwright, *Screening the Body: Tracing Medicine's Visual Culture* (Minneapolis: University of Minnesota Press, 1995).
9. A particularly vivid indication of the rise of German radiology is the change in the rosters of editorial board advisors of the major radiological journals. *The Archives of the Roentgen Ray* (London), e.g., had no one from Germany on its board in 1907, but it had seven members from Germany and Austria by 1913.
10. On Germany's research infrastructure, see, e.g., Margit Szöllösi-Janze, "Science and Social Space: Transformations in the Institutions of *Wissenschaft* from the Wilhelmine Empire to the Weimar Republic," *Minerva* 43 (2005): 339–360.

11. The best survey of the early literature in Germany is Martin Weiser, *Medizinische Kinematographie* (Dresden and Leipzig: Theodor Steinkopff, 1919). See also Karl Wilhelm Wolf-Czapek, *Die Kinematographie: Wesen, Entstehung und Ziele des lebenden Bildes* (Berlin: Union Deutsche Verlagsgesellschaft, 1908; 2d enl. ed., 1911); and relevant sections in Wolf-Czapek, ed., *Angewandte Photographie in Wissenschaft und Technik* (Berlin: Union Deutsche Verlagsgesellschaft, 1911); Hans Lehmann, *Die Kinematographie: Ihren Grundlagen und ihre Anwendungen* (Leipzig: Teubner, 1911); Oswald Polimanti, "Der Kinematograph in der biologischen und medizinischen Wissenschaft," *Naturwissenschaftliche Wochenschrift* 10, no. 49 (3 December 1911): 769–774; and Polimanti, "Die Anwendung der Kinematographie in den Naturwissenschaften, der Medizin und im Unterricht," in *Wissenschaftliche Kinematographie*, by Franz Paul Liesegang, with Karl Kieser and Oswald Polimanti (Düsseldorf: Liesegang, 1920), 257–310.

12. Weiser, *Medizinische Kinematographie*, 62.

13. Robert A. Nye, *Crime, Madness, and Politics in Modern France: The Medical Concept of National Decline* (Princeton, N.J.: Princeton University Press, 1984). See esp. his final chapter on the United Kingdom and Germany.

14. Robert Gaupp, "Der Kinematograph vom medizinischen und psychologischen Standpunkt," in *Der Kinematograph als Volkunterhaltungsmittel*, by Robert Gaupp and Konrad Lange (Munich: Dürer-Bund-Flugschrift zur Ausdruckskultur 100, 1912), 9 (emphasis in original).

15. The definitive statement on the role of experiment in medicine is Claude Bernard, *Introduction à l'étude de la médecine expérimentale* (Paris: Baillière, 1865), translated as *An Introduction to the Study of Experimental Medicine* by Henry Copley Greene (New York: Macmillan, 1927). See also Bynum, *Science and the Practice of Medicine in the Nineteenth Century*; and Warner, "Ideals of Science."

16. For overviews of the origins of medical photography, see Alison Gernsheim, "Medical Photography in the Nineteenth Century," *Medical and Biological Illustration* (London) 11, no. 2 (April 1961): 85–92; Renata Taurek, *Die Bedeutung der Photographie für die medizinische Abbildung im 19. Jahrhundert* (Cologne: Hansen, 1980); Daniel M. Fox and Christopher Lawrence, *Photographing Medicine: Images and Power in Britain and America Since 1840* (New York: Greenwood, 1988); Andreas-Holger Maehle, "The Search for Objective Communication: Medical Photography in the Nineteenth Century," in *Non-verbal Communication in Science Prior to 1900*, ed. Renato G. Mazzolini (Florence: Olschki, 1993), 563–586; Monique Sicard, Robert Pujade, and Daniel Wallach, *À corps et à raison: Photographies médicales, 1840–1920* (Paris: Marval, 1995); and Gunnar Schmidt, *Anamorphotische Körper: Medizinische Bilder vom Menschen im 19. Jahrhundert* (Cologne: Böhlau, 2001). On the impact of photography on medical practice, see Stanley Joel Reiser, *Medicine and the Reign of Technology* (Cambridge: Cambridge University Press, 1978).

17. Ludwig Braun, *Über Herzbewegung und Herzstoss* (Jena: Fischer, 1898). On Braun, see Nichtenhauser, "History of Motion Pictures in Medicine," 35–38; Michaelis, *Research Films*, 133; Cartwright, *Screening the Body*, 20–24; and Peter Geimer,

"Living and Non-living Pictures," in *Undead: Relations Between the Living and the Lifeless*, ed. Peter Geimer (Berlin: Max-Planck-Institut für Wissenschaftsgeschichte, 2003), 39–51. For other early attempts to film the action of the heart, see Étienne-Jules Marey, *Movement*, trans. Eric Pritchard (London: Heinemann, 1895), chap. 16; and Charles François-Franck, "La chronophotographie simultanée du coeur et des courbes cardiographiques chez les mammifères," *Comptes rendus hebdomadaires des séances et mémoires de la Société de Biologie* 54 (8 November 1902): 1193–1197. For other uses of film in experimental physiology, see, e.g., Martin Philippson, *L'autonomie et la centralisation dans le système nerveux des animaux: étude de physiologie expérimentale et compare* (Brussels: Falk, 1905); and esp. the work of Oswald Polimanti, "Über Ataxie cerebralen und cerebellaren Ursprungs," *Archiv für Physiologie* (1909): 123–134; and "Zur Physiologie der Stirnlappen," *Archiv für Physiologie* (1912): 337–342.

18. On the nature of scientific experiment, see Hans-Jörg Rheinberger, *Toward a History of Epistemic Things: Synthesizing Proteins in the Test Tube* (Stanford, Calif.: Stanford University Press, 1997); Ian Hacking, *Representing and Intervening: Introductory Topics in the Philosphy of Natural Science* (Cambridge: Cambridge University Press, 1983); and Hans Radder, esp. *The Material Realization of Science* (Assen and Maastricht: Van Gorcum, 1988), 59–69; *In and About the World: Philosophical Studies of Science and Technology* (Albany: State University of New York, 1996), 11–20; and his edited volume, *The Philosophy of Scientific Experimentation* (Pittsburgh, Pa.: University of Pittsburgh Press, 2003). See also Theodore Arabatzis, "Experiment," in *The Routledge Companion to Philosophy of Science*, ed. Stathis Psillos and Martin Curd (London and New York: Routledge, 2008), 159–170.

19. On the notion of "working objects," see Lorraine Daston and Peter Galison, "The Image of Objectivity," *Representations* 40 (Fall 1992): 81–128; and *Objectivity* (New York: Zone, 2007).

20. Steven Shapin, "Pump and Circumstance: Robert Boyle's Literary Technology," *Social Studies of Science* 14, no. 4 (November 1984): 481–520. See also Steven Shapin and Simon Schaffer, *Leviathan and the Air-Pump: Hobbes, Boyle, and the Experimental Life* (Princeton, N.J.: Princeton University Press, 1985).

21. See the bibliography in Otto Glasser, *Wilhelm Conrad Roentgen and the Early History of the Roentgen Rays* (Springfield, Ill.: Thomas, 1934).

22. Excellent reviews of the subject include Hans A. Jarre, "Roentgen Cinematography," in *The Science of Radiology*, ed. Otto Glasser (Springfield, Ill.: Thomas, 1933), 198–209; and L. J. Ramsey, "Early Cineradiography and Cinefluorography," *History of Photography* 7, no. 4 (October–December 1983): 311–322. See also Monika Dommann, *Durchsicht, Einsicht, Vorsicht: Eine Geschichte der Röntgenstrahlen, 1896–1963* (Zürich: Chronos, 2003). For early cinema and the X-ray, see Cartwright, *Screening the Body*; and Solveig Jülich, "Seeing in the Dark: Early X-Ray Imaging and Cinema," in *Moving Images: From Edison to the Webcam*, ed. John Fullerton and Astrid Söderberg Widding (Sydney: Libbey, 2000), 47–58.

23. John Macintyre, "X-Ray Records for the Cinematograph," *Archives of Skiagraphy* 1, no. 2 (April 1897): 37; and the report of his screening for "Ladies Night" of the Royal College of Surgeons of England in "The Royal Society Conversazione," *Lancet* 149 (19 June 1897): 1706.
24. P. H. Eykman, "Der Schlingact, dargestellt nach Bewegungsphotographien mittelst Röntgen-Strahlen," *Pflüger's Archiv für die gesammte Physiologie des Menschen und der Thiere* 99 (1903): 513–571.
25. Virgilio Tosi, *Cinema Before Cinema: The Origins of Scientific Cinematography*, trans. Sergio Angelini (London: British Universities Film & Video Council, 2005), 170.
26. C. Kästle, H. Rieder, and J. Rosenthal, "Ueber kinematographisch aufgenommene Röntgenogramme (Bio-Röntgenographie) der inneren Organe des Menschen," *Münchener medizinische Wochenschrift* 56, no. 6 (9 February 1909): 280–283; "The Bioroentgenography of the Internal Organs," *Archives of the Roentgen Ray* 15, no. 1 (June 1910): 3–12.
27. Lewis Gregory Cole, "The Gastric Motor Phenomena Demonstrated with the Projecting Kinetoscope," *American Quarterly of Roentgenology* 3, no. 4 (1912): 1–11.
28. Franz M. Groedel, "The Present State of Roentgen Cinematography and Its Results as to the Study of the Movements of the Inner Organs of the Human Body," *Interstate Medical Journal* 22 (March 1915): 281–290, here 290. Of his numerous texts on the topic, see esp. "Roentgen Cinematography and Its Importance in Medicine," *British Medical Journal* (24 April 1909): 1003; and his three-part series "Die Technik der Röntgenkinematographie," *Deutsche medizinische Wochenschrift* 35 (11 March 1909): 434–435; 39 (6 February 1913): 270–271; and 39 (24 April 1913): 798–799. See also W. Bruce Fye, "Franz M. Groedel," *Clinical Cardiology* 23, no. 2 (February 2000): 133–134.
29. On diagnosis, see Carlo Ginzburg, "Clues: Roots of an Evidential Paradigm," in *Clues, Myths, and the Historical Method*, trans. John and Ann C. Tedeschi (Baltimore: Johns Hopkins University Press, 1989), 96–125; or Caroline Whitbeck, "What Is Diagnosis? Some Critical Reflections," *Metamedicine* 2, no. 3 (October 1981): 319–329.
30. On doubts about the clinical application of X-rays, see Andrew Warwick, "X-Rays as Evidence in German Orthopedic Surgery, 1895–1900," *Isis* 96, no. 1 (March 2005): 1–24.
31. Friedrich Dessauer, *Die neuesten Fortschritte in der Röntgenphotographie* (Leipzig: Nemnich Verlag, 1912), 15. For a more optimistic view, see Carl Bruegel, "Bewegungsvorgänge am pathologischen Magen auf Grund röngenkinematographischer Untersuchungen," *Münchener medizinische Wochenschfrift* 60, no. 4 (28 January 1913): 179–181.
32. André Lomon and Jean Comandon, "Radiocinématographie par la photographie des écrans intensificateurs," *La Presse Médicale* 35 (3 May 1911): 359.
33. Jarre, "Roentgen Cinematography," 202.
34. For a more complete survey, see K. Podoll and J. Lüning, "Geschichte des wissenschaftlichen Films in der Nervenheilkunde in Deutschland 1895–1929,"

Fortschritte der Neurologie, Psychiatrie 66 (1998): 122–132; and Geneviève Aubert "From Photography to Cinematography: Recording Movement and Gait in a Neurological Context," *Journal of the History of the Neurosciences* 11, no. 3 (2002): 255–264. See also Juliet Clare Wagner, "Twisted Bodies, Broken Minds: Film and Neuropsychiatry in the First World War" (PhD diss., Harvard University, 2009).

35. Paul Schuster, "Vorführung pathologischer Bewegungscomplexe mittelst des Kinematographen und Erläuterung derselben," *Verhandlungen der Gesellschaft deutscher Naturforscher und Ärzte* 69, part 1 (1898): 196–199. For more on Schuster and his context, see Bernd Holdorff, "Die privaten Polikliniken für Nervenkranke vor und nach 1900" and "Zwischen Hirnforschung, Neuropsychiatrie und Emanzipation zur klinischen Neurologie bis 1933," in *Geschichte der Neurologie in Berlin*, ed. Bernd Holdorff and Rolf Winau (Berlin and New York: de Gruyter, 2001), 127–137, and 157–174.

36. Gheorghe Marinescu, "Les troubles de la marche dans l'hémiplégie organique étudiés à l'aide du cinématographe," *La semaine Médicale* (1899): 225–228; see also Alexandru C. Barboi, Christopher G. Goetz, and Radu Musetoiu, "The Origins of Scientific Cinematography and Early Medical Applications," *Neurology* 62 (June 2004): 2082–2086.

37. See Albert Londe's report of his work with Richer in Albert Londe, *Notice sur les titres et travaux scientifiques* (Paris: Masson, 1911).

38. Walter Greenough Chase, "The Use of the Biograph in Medicine," *Boston Medical and Surgical Journal* 153, no. 21 (23 November 1905): 571–572. See also Cartwright, *Screening the Body*, chap. 3.

39. Arthur Van Gehuchten, "Coup de couteau dans la moelle lombaire. Essai de physiologie pathologique," *Le Névraxe* 9 (1907): 208–232. See also Geneviève Aubert, "Arthur Van Gehuchten Takes Neurology to the Movies," *Neurology* 59 (November 2002): 1612–1618.

40. Emil Kraepelin, "Demonstration von Kinematogrammen," *Centralblatt für Nervenheilkunde und Psychiatrie* 32 (1909): 689.

41. Hans Hennes, "Die Kinematographie im Dienste der Neurologie und Psychiatrie, nebst Beschreibung einiger selteneren Bewegungsstörungen," *Medizinische Klinik* 6, no. 51 (18 December 1910): 2010–2014.

42. See Polimanti titles cited in note 11.

43. T. H. Weisenburg, "Moving Picture Illustrations in Medicine, with Special Reference to Nervous and Mental Diseases," *Journal of the American Medical Association* 59, no. 26 (28 December 1912): 2310–2312.

44. This section on live demonstration draws from my essay, "Photography and Medical Observation," in *The Educated Eye: Visual Pedagogy in the Life Sciences*, ed. Nancy Anderson and Michael R. Dietrich (Hanover, N.H.: Dartmouth College Press, 2012), 68–93.

45. Clare [Blake] to Dear Pater, Vienna, 9 November [1865], Clarence John Blake Papers, Francis A. Countway Library of Medicine, Harvard University, quoted in John Harley Warner, *Against the Spirit of System: The French Impulse in Nineteenth-Century American Medicine* (Princeton, N.J.: Princeton University Press, 1998), 304.

46. Clare to Sister Agnes, Vienna, 29 March 1869, Clarence John Blake Papers, Francis A. Countway Library of Medicine, Harvard University, quoted in Warner, *Against the Spirit of System*, 311.
47. Hennes, "Die Kinematographie im Dienste der Neurologie," 2012, quoted in Friedrich A. Kittler, *Gramophone, Film, Typewriter*, trans. and with an introduction by Geoffrey Winthrop-Young and Michael Wutz (Stanford, Calif.: Stanford University Press, 1999), 145 (emphasis in original).
48. See Cartwright, *Screening the Body*, chap. 3: "An Etiology of the Neurological Gaze."
49. See Reiser, *Medicine and the Reign of Technology*. See also Joel D. Howell, *Technology in the Hospital: Transforming Patient Care in the Early Twentieth Century* (Baltimore: Johns Hopkins University Press, 1995).
50. For discussion of the projection of images in medical education, see A. Wassermann, "Die medizinische Fakultät," in *Die Universitäten im Deutschen Reich*, ed. W. Lexis (Berlin: Asher, 1904), 146–147; or Erwin Christeller, "Die Bedeutung der Photographie für den pathologisch-anatomischen Unterricht und die pathologisch-anatomische Forschung," *Berliner klinische Wochenschrift* 55, no. 17 (29 April 1918): 399–401; for broader overviews, see Sigmund Theodor Stein, *Das Licht im Dienste wissenschaftlicher Forschung: Handbuch der Anwendung des Lichtes und der Photographie in der Natur- und Heilkunde* (Leipzig: Spamer, 1877); Sigmund Theodor Stein, *Die optische Projektionskunst im dienste der exakten Wissenschaften: ein Lehr- und Hilfsbuch zur unterstützung des naturwissenschaftlichen Unterrichts* (Halle an der Saale, Germany: Knapp, 1887); see also Henning Schmidgen, "Pictures, Preparations, and Living Processes: The Production of Immediate Visual Perception (*Anschauung*) in Late-19th-Century Physiology," *Journal of the History of Biology* 37, no. 3 (October 2004): 477–513.
51. Schuster, "Vorführung pathologischer Bewegungscomplexe," 196–197.
52. See, e.g., Albert E. Stein, "Ueber medizinisch-photographische und -kinematographische Aufnahmen," *Deutsche medizinische Wochenschrift* 38 (20 June 1912): 1184–1186. For early reviews of the use of photography and cinematography for the study of pathological movement, with implications for therapy, see Ernst Jendrassik, "Klinische Beiträge zum Studium der normalen und pathologischen Gangarten," *Deutsche Archiv für klinische Medizin* 70 (1901): 81–132; and James Fränkel, "Kinematographische Untersuchung des normalen Ganges und einiger Gangstörungen," *Zeitschrift für orthopädische Chirurgie* 20 (1908): 617–646.
53. An example of a German report on Comandon's work that stresses the importance of his films as both exploratory and documentary is "Die Kinematographie des Unsichtbaren," *Prometheus* 21, no. 1054 (5 January 1910): 218–220.
54. J. Frey, "Report of the Photographic Department of Bellevue Hospital for the Year 1869," in *Tenth Annual Report of the Commissioners of Public Charities and Correction of the City of New York for the Year 1869* (Albany: van Benthuysen, 1870), 85. www.artandmedicine.com/ogm/1869.html.
55. Hennes, "Die Kinematographie im Dienste der Neurologie und Psychiatrie," 2014. For other calls for a central agency to handle medical or scientific films,

see Robert Kutner, "Die Bedeutung der Kinematographie für medizinische Forschung und Unterricht sowie für die volkshygienische Belehrung," *Zeitschrift für Ärztliche Fortbildung* 8, no. 8 (15 April 1911): 249–251; or Weiser, *Medizinische Kinematographie*, 4. This call was eventually answered after World War I with the formation of the Kulturfilm section of UFA. See Universum-Film A. G., *Das medizinische Filmarchiv bei der Kulturabteilung der Universum-Film A.G.* (Berlin: Gahl, 1919).

56. Franz Goerke, "Proposal for Establishing an Archive for Moving Pictures (1912)," trans. Cecilie L. French and Daniel J. Leab, *Historical Journal of Film, Radio and Television* 16, no. 1 (March 1996): 9–12, here 10. Originally published as "Vorschlag zur Einrichtung eines Archives für Kino-films" in *Der Deutsche Kaiser im Film. Zum 25 jährigen Regierungs-Jubiläum Seiner Majestät des Deutschen Kaisers Königs von Preußen Wilhelm II*, ed. Paul Klebinder (Berlin: Klebinder, 1912), 63–68.

57. Goerke, "Proposal for Establishing an Archive," 9–10.

58. Eugène Doyen, "Le cinématograph et l'enseignement de la chirurgie," *Revue critique de médecine et de chirurgie* 1, no. 1 (15 August 1899): 1–6; partially translated as "The Cinematograph and the Teaching of Surgery," *British Gynæcological Journal* 15 (1899): 579–586, here 581. On Doyen, see Robert Didier, *Le Docteur Doyen: Chirurgien de la Belle Époque* (Paris: Librairie Maloine, 1962); and esp. the work of Thierry Lefebvre, including "Le cas étrange du Dr Doyen, 1859–1916," *Archives* 29 (February 1990): 1–12; "Le Dr Doyen, un précurseur," in *Le cinéma et la science*, ed. Alexis Martinet (Paris: CNRS Éditions, 1994), 70–77; "Die Trennung der Siamesischen Zwillinge Doodica und Radica durch Dr. Doyen," *KINtop* 6 (1997): 97–101; and *La chair et le celluloid: Le cinéma chirurgical du Docteur Doyen* (Brionne: Jean Doyen éditeur, 2004).

59. Although the films were not mentioned in the *British Medical Journal*'s proceedings of the July 1898 meeting, there is a letter to the editor that remarks on the strong impression they made: G. P. Coldstream, "The Cinematoscope as an Aid in Teaching," *British Medical Journal* (3 September 1898): 658. On Doyen's films, see Thierry Lefebvre, "La collection des films du Dr Doyen," *1895* 17 (December 1994): 100–114; and Tiago Baptista, " 'Il faut voir le maître': A Recent Restoration of Surgical Films by E.-L.Doyen (1859–1916)," *Journal of Film Preservation* 70 (November 2005): 42–50. After 1907, French film manufacturer Eclipse distributed Doyen's films throughout Europe (see Lefebvre, *La chair et le celluloid*, 67–76).

60. Doyen, "Le cinématograph et l'enseignement de la chirurgie," 3.

61. For a catalog of Doyen's films to that point, see Eugène Louis Doyen, *L'enseignement de la technique opératoire par les projections animées* (Paris: Société générale des cinématographes Eclipse, ca. 1911).

62. For Doyen's detailed discussion of the medical case, the surgical technique, and its ethical aftermath, see Eugène Louis Doyen, *La cas des xiphopages hindoues Radica et Doodica* (Paris: Bourse de Commerce, 1904). This pamphlet also reprints a heated exchange of letters between Doyen and a Dr. Legrain, who accused Doyen

of besmirching the profession with the film. See also Lefebvre, "Die Trennung der Siamesischen Zwillinge." On the Parnaland incident, see Lefebvre, *La chair et le celluloid*, 39–59.

63. The concern about spectacle in medicine was more pronounced in France—probably due to the Doyen controversy—but there were grumblings from England and Germany as well, especially with regard to Doyen's films. See "A Surgical Showman," *British Medical Journal* (19 January 1907): 163, which reports from Germany and elsewhere about screenings of Doyen films that left doctors disgusted; at one of his demonstrations, "it is said that he was hissed at Brussels."

64. See the report in "Verwandte Gebiete," *Zentralblatt für Röntgenstrahlen, Radium und verwandte Gebiete* 1, no. 2 (1910): 78–80.

65. Kutner was a proponent of medical photography in general, having taken some of the earliest images from a cytoscope with famed urologist Max Nitze. See Max Nitze, *Kystophotographischer Atlas* (Wiesbaden: Bergmann, 1894). See also Harry W. Herr, "Max Nitze, the Cystoscope and Urology," *Journal of Urology* 176, no. 4 (October 2006): 1313–1316. Kutner was also a strong advocate of continuing education for physicians, having founded the journal *Zeitschrift für Ärztliche Fortbildung*.

66. See also James Fränkel, "Kinematographische Demonstration," *Verhandlungen der freien Vereinigung der Chirurgen Berlins* 20, part 1 (1907): 12–13.

67. Karl Reicher, "Kinematographie in der Neurologie," *Verhandlungen der Gesellschaft deutscher Naturforscher und Ärzte* 79, part 2 (1907): 235–236.

68. See "Special Correspondence: Berlin," *British Medical Journal* (5 March 1910): 598, for another report of the evening. In England, Dr. William Stirling was Kutner's counterpart as a champion of the educational value of medical films. See reports of Stirling's presentations of a variety of films in medicine and biology in "Medical News," *Lancet* 177 (27 May 1911): 1470; and *Lancet* 182 (11 October 1913): 1083–1084.

69. Kutner, "Die Bedeutung der Kinematographie," 250.

70. Alongside this discourse of "seeing as" (film as an extension of direct perception and the moving image as a substitute for the thing itself) there was another discourse of "seeing differently" (film as a technology for representing things in ways the naked eye could not perceive). Educators regarded both features of cinema to be pedagogically useful, while film theorists such as Jean Epstein and Béla Balàzs prioritized the latter. This difference also corresponds to the "documentary" and "exploratory" functions outlined so far.

71. Doyen, "The Cinematograph and the Teaching of Surgery," 580–581 (translation modified).

72. A good overview is Jennifer Karns Alexander, *The Mantra of Efficiency: From Waterwheel to Social Control* (Baltimore: Johns Hopkins University Press, 2008). See also Evelyn Cobley, *Modernism and the Culture of Efficiency: Ideology and Fiction* (Toronto and Buffalo, N.Y.: University of Toronto Press, 2009).

73. Paul Starr, *The Social Transformation of American Medicine* (New York: Basic, 1982), 146.

74. George Rosen, "The Efficiency Criterion in Medical Care, 1900–1920," *Bulletin of the History of Medicine* 50, no. 1 (Spring 1976): 28–44; Margarete Arndt and Barbara Bigelow, "Toward the Creation of an Institutional Logic for the Management of Hospitals: Efficiency in the Early Nineteen Hundreds," *Medical Care Research and Review* 63, no. 3 (June 2006): 369–394.
75. See the issue devoted to the "Conference on Hospital Standardization," *Bulletin of the American College of Surgeons* 3, no. 1 (1917); or Frank B. Gilbreth, "Scientific Management in the Hospital," *Modern Hospital* 3 (1914): 321–324.
76. My essay on "Photography and Medical Observation" (see note 44) likewise examines correspondences between formal features of still photography and practices/ideals of medical observation.
77. Ludwig Braun, "Ueber den Werth des Kinematographen für die Erkenntniss der Herzmechanik," *Verhandlungen der Gesellschaft deutscher Naturforscher und Ärzte* 69, part 1 (1898): 185–186. This list is included and elaborated in his *Über Herzbewegung und Herzstoss* (Jena: Fischer, 1898).
78. Kelly Wilder, *Photography and Science* (London: Reaktion, 2009), 23.
79. Gernsheim, "Medical Photography in the Nineteenth Century," 87.
80. Cartwright, *Screening the Body*, 38, 48.
81. Ludwik Fleck, "Some Specific Features of the Medical Way of Thinking [1927]," in *Cognition and Fact: Materials on Ludwik Fleck*, ed. Robert S. Cohen and Thomas Schnelle (Dordrecht, Netherlands, and Boston: Reidel, 1986), 39–46, here 39–40. Foucault also discusses the importance of the series in medical thinking in *The Birth of the Clinic*, 97. On the importance of comparing pathological states to find some sense of consistency, see Georges Canguilhem, *The Normal and the Pathological*, trans. Carolyn R. Fawcett in collaboration with Robert S. Cohen (New York: Zone, 1991), 51. On medical logic, see Friedrich Oesterlen, *Medical Logic*, trans. G. Whitley (London: Sydenham Society, 1855); Frederick P. Gay, "Medical Logic," *Bulletin of the History of Medicine* 7 (1939): 6–27; Lester S. King, "Medical Logic," *Journal of the History of Medicine and Allied Sciences* 33, no. 3 (July 1978): 377–385; and King, *Medical Thinking: A Historical Preface* (Princeton, N.J.: Princeton University Press, 1982).
82. Bernike Pasveer, "Representing or Mediating: A History and Philosophy of X-Ray Images in Medicine," in *Visual Cultures of Science: Rethinking Representational Practices in Knowledge Building and Science Communication*, ed. Luc Pauwels (Lebanon, N.H.: Dartmouth College Press/University Press of New England, 2006), 41–62.
83. For an example of the use of series photography to track the development of smallpox in a patient over several days, see Samuel A. Powers, *Variola: A Series of Twenty-One Heliotype Plates Illustrating the Progressive Stages of the Eruption* (Boston: Samuel A. Powers, 1882). My thanks to Mark Rowland for pointing me to this text. For examples of different angles of a patient during the same session, see Albert Londe's photographs of female patients documented in the journal *Nouvelle iconographie de la Salpêtrière* (Paris: Masson, 1888–1918).

84. For an interesting discussion of X-ray cinematography's diagnostic value, see the exchange between Drs. Fränkel, von Bergmann, Levy-Dorn, Albu, and Kutner in Albert Fränkel, "Röntgendiagnosen und Röntgenfehldiagnosen beim Magenkarzinom; diagnostischer Fortschritt durch Röntgenkinographie." *Zentralblatt für Röntgenstrahlen, Radium und verwandte Gebiete* 3, no. 4 (1912): 149–150.
85. Warwick, "X-Rays as Evidence in German Orthopedic Surgery, 1895–1900."
86. Photographs were often used as illustrations exchanged between attendees of a lecture. An example, picked more or less at random, is Adolf Magnus-Levy, "Ueber Organ-Therapie beim endemischen Kretinismus," *Verhandlungen der Berliner medicinischen Gesellschaft* 34, part 2 (1903): 350–357. See esp. the discussion of this presentation on 22 July 1903 in part 1, pp. 246–249. No photos are published with the paper, but they discuss the photographs that were passed around among the audience. Many such uses of photographs can be found in the *Verhandlungen* and similar proceedings.
87. Cartwright, *Screening the Body*, 36.
88. Georges Didi-Huberman, *Invention of Hysteria: Charcot and the Photographic Iconography of the Salpêtrière*, trans. Alisa Hartz (Cambridge, Mass.: MIT Press, 2003), 24–25.
89. For more on this topic, see Scott Curtis, "Photography and Medical Observation."
90. Foucault, *Birth of the Clinic*, 109.
91. Foucault, *Birth of the Clinic*, esp. 107–123.
92. Neurologists such as Schuster needed movement to recognize the disorder—which would otherwise be lost in the individual frames—so they consistently slowed (but did not stop) the image in projection. See Schuster, "Vorführung pathologischer Bewegungskomplexe."
93. Many researchers emphasized this feature of motion picture technology, but for an extended discussion of the applications of film's temporal malleability, with interesting responses from the expert audience, see the report of Herr v. Grützner's presentation at the Tübingen Society of Medical and Natural Scientists: "Medizinisch-Naturwissenschaftlicher Verein Tübingen," *Münchener medizinische Wochenschrift* 56, no. 3 (19 January 1909): 154–155.
94. Thomas Laycock, *Lectures on the Principles and Methods of Medical Observation and Research* (Philadelphia: Blanchard and Lea, 1857), 64.
95. Rudolph Virchow, *Post-mortem Examinations*, trans. T. P. Smith, 3d ed. (Philadelphia: Blakiston, 1895), 12.
96. Foucault, *Birth of the Clinic*, xiv.
97. Foucault, *Birth of the Clinic*, xiii (emphasis added).
98. Foucault, *Birth of the Clinic*, esp. 107–122.
99. Theodor Billroth, *The Medical Sciences in the German Universities: A Study in the History of Civilization*, trans. William H. Welch (New York: Macmillan, 1924), 52–53. Originally published as *Über das Lehren und Lernen de medicinischen Wissenschaften an den Universitäten der deutschen Nation nebst allgemeinen Bemerkungen über Universitäten; eine culturhistorische Studie* (Vienna: Gerold, 1876).

100. Foucault, *Birth of the Clinic*, xiii.
101. Hau, "The Holistic Gaze in German Medicine, 1890–1930," 503. For more on the resistance of clinicians to laboratory methods in medicine, see Russell C. Maulitz, "Physician Versus Bacteriologist: The Ideology of Science in Clinical Medicine," in *The Therapeutic Revolution: Essays in the Social History of American Medicine*, ed. Morris J. Vogel and Charles E. Rosenberg (Philadelphia: University of Pennsylvania Press, 1979), 91–107.
102. Foucault, *Birth of the Clinic*, 121.
103. Lorraine Daston, "On Scientific Observation," *Isis* 99, no. 1 (2008): 97–110. See also Lorraine Daston and Elizabeth Lunbeck, eds., *Histories of Scientific Observation* (Chicago: University of Chicago Press, 2011).
104. There are perhaps some connections here to be made between the glance of the connoisseur and theories of cinephilia. See Christian Keathley, *Cinephilia and History; or, The Wind in the Trees* (Bloomington: Indiana University Press, 2006).
105. Again, see Curtis, "Still/Moving: Digital Imaging and Medical Hermeneutics," for more on the relationship between film and medical hermeneutics.
106. Which should remind us of similar patterns of showmanship in early entertainment film described by Tom Gunning in "An Aesthetic of Astonishment: Early Film and the (In)Credulous Spectator," *Art and Text* 34 (Spring 1989): 31–45.
107. My thanks to Christian Quendler for stating this so succinctly for me.
108. Erwin Risak, *Der klinische Blick*, 7th and 8th eds. (Vienna: Springer, 1943), 4.
109. Carl F. Flemming, *Pathologie und Therapie der Psychosen* (Berlin: Hirschwald, 1859), 281–282.
110. For a good overview of the relationship between clinical observation and scientific methods, see Kenneth D. Keele, *The Evolution of Clinical Methods in Medicine* (London: Pitman, 1963). On the debates about clinical observation and the sphygmomanometer specifically, see Jeremy Booth, "A Short History of Blood Pressure Measurement," *Proceedings of the Royal Society of Medicine* 70, no. 11 (November 1977): 793–799; Reiser, *Medicine and the Reign of Technology*, 101–106; and Hughes Evans, "Losing Touch: The Controversy Over the Introduction of Blood Pressure Instruments Into Medicine," *Technology and Culture* 34, no. 4 (October 1993): 784–807.
111. Paget, "An Address on the Utility of Scientific Work in Practice," 811.
112. Karl Wilhelm Wolf-Czapek, "Die Kinematographie im medizinische Unterricht," *Jahrbuch für Photographie und Reproduktionstechnik* 22 (1908): 58–59, here 58.
113. A close review of the primary literature on scientific film shows that analysis and synthesis were always considered two sides of the same coin. See Étienne-Jules Marey, *Movement*, trans. Eric Pritchard (London: Heinemann, 1895), esp. chap. 18: "Synthetic Reconstruction of the Elements of an Analyzed Movement." See also Hannah Landecker, "Microcinematography and the History of Science and Film," *Isis* 97, no. 1 (2006): 121–132; and Oliver Gaycken, "'The Swarming of Life': Moving Images, Education, and Views Through the Microscope," *Science in Context* 24, no. 3 (September 2011): 361–380.

114. On the theoretical implications of experimental apparatuses, see Davis Baird, *Thing Knowledge: A Philosophy of Scientific Instruments* (Berkeley: University of California Press, 2004).
115. More could be said about the relationship between control of and submission to or pleasure in the scientific image. A good place to start would be Anne Secord, "Botany on a Plate: Pleasure and the Power of Pictures in Promoting Early Nineteenth-Century Scientific Knowledge," *Isis* 93, no. 1 (March 2002): 28–57.
116. Claudia Huerkamp, *Der Aufstieg der Arzte im 19. Jahrhundert: Vom gelehrten Stand zum professionellen Experten: Das Beispiel Preußens* (Göttingen: Vandenhoeck & Ruprecht, 1985). See also Ute Frevert, *Krankheit als politisches Problem 1770–1880. Soziale Unterschichten in Preußen zwischen medizinischer Polizei und staatlicher Sozialversicherung* (Göttingen: Vandenhoeck & Ruprecht, 1984).
117. Huerkamp stresses this medicalization of culture in her essay, "The Making of the Modern Medical Profession, 1800–1914: Prussian Doctors in the Nineteenth Century," in *German Professions, 1800–1950*, ed. Geoffrey Cocks and Konrad H. Jarausch (New York and Oxford: Oxford University Press, 1990), 66–84.
118. On the social prestige and authority of physicians, see Alfons Labisch, *Homo Hygienicus: Gesundheit und Medizin in der Neuzeit* (Frankfurt and New York: Campus, 1992); Michael H. Kater, "Professionalization and Socialization of Physicians in Wilhelmine and Weimar Germany," *Journal of Contemporary History* 20 (1985): 677–701; Paul Weindling, "Bourgeois Values, Doctors and the State: The Professionalization of Medicine in Germany 1848–1933," in *The German Bourgeoisie*, ed. David Blackbourn and Richard J. Evans (London and New York: Routledge, 1991), 198–223; Charles E. McClelland, "Modern German Doctors: A Failure of Professionalization?" in *Medicine and Modernity: Public Health and Medical Care in Nineteenth- and Twentieth-Century Germany*, ed. Manfred Berg and Geoffrey Cocks (Cambridge and New York: Cambridge University Press, 1997), 81–97. For a discussion of this phenomenon from the viewpoint of medical ethics, see Robert M. Veatch, "Generalization of Expertise," *Hastings Center Studies* 1, no. 2 (1973): 29–40.
119. On scientists, especially, as *Kulturträger*, see Fritz K. Ringer, *The Decline of the German Mandarins: The German Academic Community, 1890–1933* (Cambridge, Mass.: Harvard University Press, 1969), 6; and Russell McCormmach, "On Academic Scientists in Wilhelmian Germany," in *Science and Its Public: The Changing Relationship*, ed. Gerald Horton and William A. Blanpied (Dordrecht, Netherlands, and Boston: Reidel, 1976), 157–171.
120. Paul Weindling, "Public Health in Germany," in *The History of Public Health and the Modern State*, ed. Dorothy Porter (Amsterdam and Atlanta, Ga.: Rodopi, 1994), 119–131. See also Weindling, *Health, Race, and German Politics Between National Unification and Nazism, 1870–1945* (Cambridge and New York: Cambridge University Press, 1989).
121. Eric J. Engstrom, "Emil Kraepelin: Psychiatry and Public Affairs in Wilhelmine Germany," *History of Psychiatry* 2, no. 6 (June 1991): 111–132. See also Engstrom's *Clinical Psychiatry in Imperial Germany: A History of Psychiatric Practice*

(Ithaca, N.Y.: Cornell University Press, 2003); and Emil Kraepelin, *Memoirs*, ed. H. Hippius, G. Peters, D. Ploog in collaboration with P. Hoff and A. Kreuter, trans. Cheryl Wooding-Deane (Berlin and New York: Springer-Verlag, 1987). Kraepelin was also swept up by the degeneration craze described below: see Emil Kraepelin, "Zur Entartungsfrage," *Zentralblatt für Nervenheilkunde und Psychiatrie* 31 (1908): 745–751, translated as "On the Question of Degeneration," *History of Psychiatry* 18, no. 3 (2007): 399–404.

122. According to Andreas Killen, as early as 1912 the Reich Health Office started to collect materials documenting the educational benefits of scientific films and the health risks of commercial cinema. Andreas Killen, "Psychiatry, Cinema, and Urban Youth in Early-Twentieth-Century Germany," *Harvard Review of Psychiatry* 14, no. 1 (2006): 38–43.

123. Robert Gaupp, *Psychologie des Kindes* (Leipzig: Teubner, 1910), and "Das Pathologische in Kunst und Literatur," *Deutsche Revue* 36, no. 2 (April 1911): 11–23. For an especially explicit statement of the physician's duty to society, see Gaupp, "Der Arzt als Erzieher seines Volkes," *Medicinisches Correspondenz-Blatt* 89, no. 32 (9 August 1919): 295–296. On Gaupp, see William Mayer, "Robert Gaupp," *American Journal of Psychiatry* 108, no. 10 (April 1952): 724–725.

124. Max Nordau, *Degeneration* (London: Appleton, 1895), 43. Originally published as *Entartung*, 2 vols. (Berlin: Duncker, 1892–1893). Nordau, an Austro-Hungarian physician living in Paris, was a prolific writer of fiction and cultural criticism as well as a foreign correspondent for German-language newspapers in Berlin, Vienna, and Budapest.

125. For contemporary discussions of nervousness, see, e.g., Wilhelm Heinrich Erb, *Ueber die wachsende Nervosität unserer Zeit* (Heidelberg: Universitäts Buchdruckerei von J. Hörning, 1893); Auguste Forel, *Hygiene der Nerven und des Geistes im gesunden und kranken Zustande* (Stuttgart: Moritz, 1903); and Robert Gaupp, "Die Nervosität unserer Zeit im Lichte der Wissenschaft," *Medicinisches Correspondenz-Blatt* 77, no. 31 (3 August 1907): 633–639. On nervousness and modernity in Germany, see Joachim Radkau, *Das Zeitalter der Nervosität: Deutschland zwischen Bismarck und Hitler* (Munich: Hanser, 1998); Andreas Killen, *Berlin Electropolis: Shock, Nerves, and German Modernity* (Berkeley: University of California Press, 2006); Michael Cowan, *Cult of the Will: Nervousness and German Modernity* (University Park: Pennsylvania State University Press, 2008).

126. On Lombroso and Nordau, see Charles Bernheimer, "Decadent Diagnostics," in *Decadent Subjects: The Idea of Decadence in Art, Literature, Philosophy, and Culture of the Fin de Siècle in Europe*, ed. T. Jefferson Kline and Naomi Schor (Baltimore: Johns Hopkins University Press, 2002), 139–162. For more on Nordau, see George L. Mosse, "Max Nordau and His Degeneration," in Max Nordau, *Degeneration* (New York: Fertig, 1968), xiii–xxxvi; Thomas Anz, "Gesundheit, Krankheit und literarische Norm: Max Nordaus 'Entartung' als Paradigma pathologisierender Kunstkritik," in *Gesund oder Krank?: Medizin, Moral und Ästhetik in der deutschen Gegenwartsliteratur* (Stuttgart: Metzler, 1989), 33–52; Hans-Peter Söder, "Disease and Health as Contexts of Modernity: Max Nordau as

a Critic of Fin-de-Siècle Modernism," *German Studies Review* 14, no. 3 (October 1991): 473–487; Christoph Schulte, *Psychopathologie des Fin de Siècle: Der Kulturkritiker, Arzt und Zionist Max Nordau* (Frankfurt: Fischer Taschenbuch, 1997); Céline Kaiser, *Rhetorik der Entartung: Max Nordau und die Sprache der Verletzung* (Bielefeld, Germany: Transcript, 2007).

127. To be fair, even those friendly to modern art and culture often saw it in similar, especially primitivist terms. See Doris Kaufmann, "'Pushing the Limits of Understanding': The Discourse on Primitivism in German *Kulturwissenschaften*, 1880–1930," *Studies in History and Philosophy of Science* 39 (2008): 434–443.

128. Söder, "Disease and Health as Contexts of Modernity," 474.

129. Surveys of cultural pessimism in Germany include Fritz Stern, *The Politics of Cultural Despair: A Study in the Rise of the Germanic Ideology* (Berkeley: University of California Press, 1961); and Ringer, *The Decline of the German Mandarins*.

130. Ike Spier, "Die sexuelle Gefahr des Kinos," *Die neue Generation* 8 (1912): 192–198, here 192 (emphasis in original). I will discuss the battle against *Schundfilms* ("trash films") in chap. 3, but for representative works, see Albert Hellwig, *Schundfilms: Ihr Wesen, ihre Gefahren und ihre Bekämpfung* (Halle an der Saale, Germany: Waisenhaus, 1911); Max Grempe, "Gegen die Frauenverblödung im Kino," *Gleichheit* 23, no. 5 (1912): 70–72; Malwine Rennert, "Die Zaungäste des Lebens im Kino," *Bild und Film* 4, no. 11 (1914/1915): 217–218; and, for a reasonable rebuttal, Joseph Landau, "Mechanisierte Unsterblichkeit," in *Der Deutsche Kaiser im Film. Zum 25jährigen Regierungs-Jubiläum Seiner Majestät des Deutschen Kaisers Königs von Preußen Wilhelm II*, ed. Paul Klebinder (Berlin: Klebinder, 1912), 18–22.

131. Paul Schenk, "Der Kinematograph und die Schule," *Aerztliche Sachverständigen-Zeitung* 14, no. 15 (1 August 1908): 312–313. For an entertaining "experiment" in which a writer submits three men to hours of continuous, flickering projection with predictable results, see Naldo Felke, "Die Gesundheitsschädlichkeit des Kinos," *Die Umschau* 17, no. 1 (1 January 1913): 254–255.

132. A survey of the (mostly French) discussion of "flicker" in early cinema can be found in Thierry Lefebvre, "Flimmerndes Licht: Zur Geschichte der Filmwahrnehmung im frühen Kino," *KINtop* 5 (1996): 71–80.

133. For excellent surveys of this trend, see Killen, "Psychiatry, Cinema, and Urban Youth in Early-Twentieth-Century Germany"; and Killen, "The Scene of the Crime: Psychiatric Discourses on the Film Audience in Early Twentieth Century Germany," in *Film 1900: Technology, Perception, Culture*, ed. Annemone Ligensa and Klaus Kreimeier (New Barnet, U.K.: Libbey, 2009) 99–111.

134. Albert Hellwig, "Über die schädliche Suggestivkraft kinematographischer Vorführung," *Aerztliche Sachverständigen-Zeitung* 20, no. 6 (15 March 1914): 122; I will discuss Hellwig's work in more depth in chap. 3, but for a taste of his reliance on medical terminology and audiences, see Hellwig, "Zur Psychologie kinematographischer Vorführungen," *Zeitschrift für Psychotherapie und medizinische Psychologie* 6 (1916): 88–120; and "Hypnotismus und

Kinematograph," *Zeitschrift für Psychotherapie und medizinische Psychologie* 6 (1916): 310–315.

135. O. Götze, "Jugendpsyche und Kinematograph," *Zeitschrift für Kinderforschung* 16 (1911): 418 (emphasis in original).
136. Thierry Lefebvre quotes a similar, French objection from 1913 to film's temporal malleability: "The cinema, with its rapid unfolding, its somewhat brutal speed of images which follow one another, distort the slow and progressive work of nature. Here is a film showing a seed which suddenly sprouts, becomes stem, flower, fruit all in just a couple of seconds. Nature does not do this; nature 'does not jump,' as told to us by the old philosophy." "The Scientia Production (1911–1914): Scientific Popularization Through Pictures," *Griffithiana* no. 47 (May 1993): 137–153, here 145.
137. A good survey is Hartmut Rosa and William E. Scheuerman, eds., *High-Speed Society: Social Acceleration, Power, and Modernity* (University Park: Pennsylvania State University Press, 2009).
138. Nordau, *Degeneration*, 40. For more on Nordau's critique of the speed of modernity, see Günther A. Höfler, "La naissance de la 'nervosité' issue de l'esprit de la modernité technologique. Dégénérescence et nomadisation chez Max Nordau et Adolph Wahrmund," in *Max Nordau (1849–1923): Critique de la Dégénérescence, Médiateur Franco-Allemand, Père Fondateur du Sionisme*, ed. Delphine Bechtel, Dominique Bourel, and Jacques Le Rider (Paris: Cerf, 1996), 149–160.
139. Nordau, *Degeneration*, 42.
140. Nordau, *Degeneration*, 55.
141. Jonathan Crary, *Suspensions of Perception: Attention, Spectacle, and Modern Culture* (Cambridge, Mass.: MIT Press, 1999).
142. Crary, *Suspensions of Perception*, 25. See also Friedrich Nietzsche's discussion of attention and the will to mastery in *Beyond Good and Evil*, sect. 19, in *Basic Writings of Nietzsche*, trans. Walter Kaufmann (New York: Modern Library, 1966), 215–217.
143. Crary, *Suspensions of Perception*, 17. If "attention" were a difficult concept to define, at least it remained ideologically productive through the twentieth century; the same cannot be said for "will" or "volition," which lost currency after around 1900. See G. E. Berrios and M. Gili, "Will and Its Disorders: A Conceptual History," *History of Psychiatry* 6 (1995): 87–104.
144. Gaupp, "Der Kinematograph vom medizinischen und psychologischen Standpunkt," 9 (emphasis in original).
145. Adolf Sellmann, "Das Geheimnis des Kinos," *Bild und Film* 1, nos. 3–4 (1912): 65–67, here 66 (emphasis in original).
146. Wilhelm Stapel, "Der homo cinematicus," *Deutsches Volkstum* 21 (October 1919): 319–320, here 319.
147. See the articles by Andreas Killen (notes 122 and 133 above), as well as Stefan Andriopoulos, *Possessed: Hypnotic Crimes, Corporate Fiction, and the Invention of Cinema* (Chicago: University of Chicago Press, 2008); and Rae Beth Gordon, *Why the French Love Jerry Lewis: From Cabaret to Early Cinema* (Stanford, Calif.: Stanford University Press, 2001).

148. Stefan Andriopoulos, "Spellbound in Darkness: Hypnosis as an Allegory of Early Cinema," *Germanic Review* 77, no. 2 (Spring 2002): 102–116, here 103.
149. Leopold Laquer, "Über die Schädlichkeit kinematographischer Veranstaltungen für die Psyche des Kindesalters," *Aerztliche Sachverständigen-Zeitung* 27, no. 11 (1 June 1911): 221.
150. Laquer, "Über die Schädlichkeit kinematographischer Veranstaltungen," 222.
151. The most famous of these was the case of the Borbacher Knabenmord, in which a young man accused of killing a little boy recounted the films he saw leading up to the crime. This case was a touchstone for medical and reformist literature on cinema through the 1920s. See Killen, "The Scene of the Crime," 104; and Hellwig, "Über die schädliche Suggestivkraft kinematographischer Vorführung."
152. Gaupp, "Der Kinematograph vom medizinischen und psychologischen Standpunkt," 9 (emphasis in original).
153. Gordon, *Why the French Love Jerry Lewis*, 128.
154. Gustave Le Bon, *The Crowd: A Study of the Popular Mind* (Atlanta, Ga.: Cherokee, 1982), 16. Originally published as *Psychologie des foules* (Paris: Alcan, 1895). Translated and published in German as *Psychologie der Massen* (Leipzig: Klinkhardt, 1908).
155. Le Bon, *The Crowd*, 21. For a version of this argument applied to film audiences, see Hermann Duenschmann, "Kinematograph und Psychologie der Volksmenge. Eine sozialpolitische Studie," *Konservative Monatsschrift* 69, no. 9 (June 1912): 920–930.
156. Le Bon, *The Crowd*, 29.
157. Didi-Huberman, *Invention of Hysteria*.
158. Léon Chertok and Isabelle Stengers, *A Critique of Psychoanalytic Reason: Hypnosis as a Scientific Problem from Lavoisier to Lacan*, trans. Martha Noel Evans (Stanford, Calif.: Stanford University Press, 1992). For a detailed study of the historical relationship between hypnosis and psychoanalysis, see Andreas Mayer, *Sites of the Unconscious: Hypnosis and the Emergence of the Psychoanalytical Setting* (Chicago: University of Chicago Press, 2013).
159. Albert Moll, *Hypnotism* (London: Scott, 1890), 333 (emphasis added). Originally published as *Der Hypnotismus* (Berlin: Fischer's Medicinische, 1889). To be fair, it should be noted that this particular application of hypnosis is relatively uncommon during the nineteenth and early twentieth centuries, when the therapeutic technique is used overwhelmingly for somatic ailments. For an example, see W. P. Carr, "Suggestion as Used and Misused in Curing Disease," in *Hypnotism and Hypnotic Suggestion*, ed. E. Virgil Neal and Charles S. Clark (Rochester: New York State Publishing, 1900), 5–17. For more on the experimental and therapeutic uses of hypnosis, see Mayer, *Sites of the Unconscious*.
160. [Henri-Étienne] Beaunis, "L'expérimentation en psychologie par le somnambulisme provoqué," *Revue philosophique* 10, no. 7 (1885): 2 (emphasis in original).
161. Immanuel Kant, "What Is Enlightenment?," in *German Aesthetic and Literary Criticism*, ed. David Simpson (Cambridge: Cambridge University Press, 1984), 29–34, here 30.

3. THE TASTE OF A NATION

1. "Die Bremer Lehrerinnen und die Kinogefahr," *Die Lehrerin* 30 (1913): 156, quoted in Albert Hellwig, *Kind und Kino* (Langensalza: Beyer, 1914), 71.
2. Stephen Kern discusses the bourgeois attitude toward sexuality in *Anatomy and Destiny: A Cultural History of the Human Body* (Indianapolis: Bobbs-Merrill, 1975). Cf. Peter Gay, *The Education of the Senses*, vol. 1 (New York: Oxford University Press, 1984); and Michel Foucault, *The History of Sexuality*, trans. Robert Hurley (New York: Vintage, 1990).
3. Konrad Lange, *Die künstlerische Erziehung der deutschen Jugend* (Darmstadt: Bergstraeßer, 1893), 12.
4. Pierre Bourdieu, *Distinction* (Cambridge, Mass.: Harvard University Press, 1984), 190 (emphasis in original).
5. A more complete treatment of this theme can be found in Patrice Petro, *Joyless Streets: Women and Melodramatic Representation in Weimar Germany* (Princeton, N.J.: Princeton University Press, 1989). Ann Douglas discusses a similar concern in the United States in *The Feminization of American Culture* (New York: Knopf, 1977).
6. See Josef Chytry, *The Aesthetic State: A Quest in Modern German Thought* (Berkeley and Los Angeles: University of California Press, 1989).
7. Norbert Elias, *The Civilizing Process*, vol. 1, *The History of Manners*, trans. Edmund Jephcott (New York: Pantheon, 1978), 19. For other discussions of *Kultur* and *Zivilisation* in the German context, see Fritz Ringer, *The Decline of the German Mandarins: The German Academic Community, 1890–1933* (Cambridge, Mass.: Harvard University Press, 1969); and Jeffrey Herf, *Reactionary Modernism: Technology, Culture, and Politics in Weimar and the Third Reich* (Cambridge and New York: Cambridge University Press, 1984). See also Jörg Fisch, "Zivilisation, Kultur," in *Geschichtliche Grundbegriffe. Historisches Lexikon zur politisch-sozialen Sprache in Deutschland*, ed. Otto Brunner, Werner Conze, and Reinhardt Koselleck (Stuttgart: Klett, 1992), 7: 679–774.
8. Raymond Geuss, "Kultur, Bildung, Geist," *History and Theory* 35, no. 2 (May 1996): 151–164, here 153.
9. Another dichotomy, Ferdinand Tönnies's *Gemeinschaft* (community) and *Gesellschaft* (society), struck a similarly antimodernist tone. Contrasting the unity of the small, rural community, which he felt was disappearing in the modern industrial transformation, with the alienation and fragmentation of the metropolis, Tönnies was perhaps more elegiac than staunchly antimodernist. Still, the tendency to describe or criticize modern bourgeois society through reference to a precapitalist past was described by Georg Lukács as "romantic anti-capitalism," capturing the deeply ambivalent, contradictory character of nineteenth-century reactions to industrialization. See Tönnies, *Gemeinschaft und Gesellschaft* (Leipzig: Fues, 1887), translated by Charles Loomis as *Community and Society* (East Lansing: Michigan State University Press, 1957); Georg Lukács,

"Über den Dostojewski-Nachlass," *Moskauer Rundschau* 17 (22 March 1931): 4; Robert Sayre and Michael Löwy, "Figures of Romantic Anti-capitalism," *New German Critique* 32 (Spring/Summer 1984): 42–92. For a good, general overview of Germany's ambivalent reaction to modernity, see Kenneth D. Barkin, "The Crisis of Modernity, 1887–1902," in *Imagining Modern German Culture, 1889–1910*, ed. Françoise Forster-Hahn (Washington, D.C.: National Gallery of Art, 1996), 19–35.

10. Dennis Sweeney, "Reconsidering the Modernity Paradigm: Reform Movements, the Social and the State in Wilhelmine Germany," *Social History* 31, no. 4 (November 2006): 405–434, here 406. See also Kevin Repp, *Reformers, Critics, and the Paths of German Modernity: Anti-Politics and the Search for Alternatives, 1890–1914* (Cambridge, Mass.: Harvard University Press, 2000); and Andrew Lees, *Cities, Sin, and Social Reform in Imperial Germany* (Ann Arbor: University of Michigan Press, 2002).

11. Repp, *Reformers, Critics, and the Paths of German Modernity*, 278.

12. This chapter began life as "The Taste of a Nation: Training the Senses and Sensibility of Cinema Audiences in Imperial Germany," *Film History* 6, no. 4 (Winter 1994): 445–469.

13. On reform movements in general, see Norman Rich, *The Age of Nationalism and Reform, 1850–1890* (New York: Norton, 1970), 103–122; Maureen A. Flanagan, *America Reformed: Progressives and Progressivisms, 1890s–1920s* (New York: Oxford University Press, 2007); or Judith F. Stone, *The Search for Social Peace: Reform Legislation in France, 1890–1914* (Albany: State University of New York Press, 1985).

14. See, e.g., such general surveys as Hans-Ulrich Wehler, *The German Empire, 1871–1918* (Leamington Spa, U.K.: Berg, 1985); or David Blackbourn, *History of Germany, 1780–1918: The Long Nineteenth Century*, 2d ed. (Malden, Mass.: Blackwell, 2003). See also studies of German urbanization, such as Jürgen Reulecke, *Geschichte der Urbanisierung in Deutschland* (Frankfurt: Suhrkamp, 1985); or surveys of European modernity, such as Andrew Lees and Lynn Lees, eds., *The Urbanization of European Society in the Nineteenth Century* (Lexington, Mass.: Heath, 1976); and Hans Jürgen Teuteberg, ed., *Urbanisierung im 19. und 20. Jahrhundert: historische und geographische Aspekte* (Cologne: Böhlau, 1983).

15. On reform in Germany, in addition to Repp (*Reformers, Critics, and the Paths of German Modernity*) and Lees and Lees (*The Urbanization of European Society*), see Rüdiger vom Bruch, *Wissenschaft, Politik und öffentliche Meinung: Gelehrtenpolitik im Wilhelminischen Deutschland, 1890–1914* (Husum, Germany: Matthiesen, 1980); Jürgen Reulecke, *Sozialer Frieden durch soziale Reform: Der Centralverein für das Wohl der Arbeitenden Klassen in der Frühindustrialisierung* (Wuppertal: Hammer, 1983); and vom Bruch, ed., *Weder Kommunismus noch Kapitalismus: Bürgerliche Sozialreform in Deutschland vom Vormärz bis zur Ära Adenauer* (Munich: Beck, 1985). On educational reform in particular, see Christa Berg, ed., *Handbuch der deutschen Bildungsgeschichte* (Munich: Beck, 1991); and Wolfgang Scheibe, *Die Reformpädagogische Bewegung, 1900–1932: Eine einführende Darstellung*, 9th ed. (Weinheim and Basel: Beltz, 1984). Some

have rightly argued that, despite the implications of the concept of "reform," we should be careful not to view the educational or social theories and practices that came out of this period as complete breaks with tradition. See Jürgen Oelkers, *Reformpädagogik. Eine kritische Dogmengeschichte* (Weinheim and Munich: Juventa, 1989).

16. A good survey of late nineteenth-century *Kulturkritik* is David L. Gross, "*Kultur* and Its Discontents: The Origins of a 'Critique of Everyday Life' in Germany, 1880–1925," in *Essays on Culture and Society in Modern Germany*, ed. Gary D. Stark and Bede Karl Lackner (College Station: Texas A&M University Press, 1982), 70–97.

17. *Verhandlung über Fragen des höheren Unterrichts, Berlin 4. bis 17. Dezember 1890* (Berlin, 1891), 770, quoted in James C. Albisetti, *Secondary School Reform in Imperial Germany* (Princeton, N.J.: Princeton University Press, 1983), 4. For an especially compelling discussion of the debates about the value of Greek ideals in imperial Germany, see Suzanne L. Marchand, *Down from Olympus: Archaeology and Philhellenism in Germany, 1750–1970* (Princeton, N.J.: Princeton University Press, 1996).

18. See, e.g., Wilhelm Frei, *Landerziehungsheime: Darstellung und Kritik einer modernen Reformschule* (Leipzig: Klinkhardt, 1902); or Herbert Bauer, *Zur Theorie und Praxis der ersten deutschen Landerziehungsheime: Erfahrungen zur Internats- und Ganztagserziehung aus den Hermann-Lietz-Schulen* (Berlin: Volk und Wissen, 1961).

19. See Georg Kerschensteiner, "Begriff der Arbeitsschule," in *Die deutsche Reformpädagogik*, ed. Wilhelm Flitner and Gerhard Kudritzki (Düsseldorf and Munich: Küpper, 1961), 222–238.

20. Essays expressing the themes of "the modern," "the healthy," and "the national" might be, respectively: Hermann Kienzl, "Theater und Kinematograph," *Der Strom* 1, no. 7 (October 1911): 219–221; Robert Gaupp, "Die Gefahren des Kino," *Süddeutsche Monatshefte* 9, no. 9 (1911/1912): 363–366; and Albert Hellwig, "Kinematograph und Zeitgeschichte," *Die Grenzboten* 72, no. 39 (1913): 612–620. These and other representative essays can be found in Jörg Schweinitz, ed., *Prolog vor dem Film: Nachdenken über ein neues Medium, 1909–1914* (Leipzig: Reclam, 1992). Essays from this period are also collected in Anton Kaes, ed., *Kino-Debatte: Texte zum Verhältnis von Literatur und Film 1909–1929* (Tübingen: Niemeyer, 1978); and Fritz Güttinger, ed., *Kein Tag ohne Kino: Schriftsteller über den Stummfilm* (Frankfurt: Deutsches Filmmuseum, 1984). Among the numerous commentaries, see especially Kaes's introduction, revised and translated as "Literary Intellectuals and the Cinema: Charting a Controversy (1909–1929)," *New German Critique* 40 (Winter 1987): 7–34; Heide Schlüpmann, *Unheimlichkeit des Blicks: Das Drama des frühen deutschen Kinos* (Frankfurt: Stroemfeld/Roter Stern, 1990), 189–243; and Sabine Hake, *The Cinema's Third Machine: Writing on Film in Germany, 1907–1933* (Lincoln: University of Nebraska Press, 1993), 27–42.

21. Repp (*Reformers, Critics, and the Paths of German Modernity*) and Sweeney ("Reconsidering the Modernity Paradigm") emphasize imperial Germany's diversity

of approaches to the problems of modernity, as well as the deep ambivalence toward the modern that most elites felt. On the other hand, for histories of ideas that stress the retrograde elements of German society and the reactionary responses that, for some, foreshadow National Socialism, see Fritz Stern, *The Politics of Cultural Despair: A Study in the Rise of the Germanic Ideology* (Berkeley and Los Angeles: University of California Press, 1961); or George L. Mosse, *The Crisis of German Ideology: Intellectual Origins of the Third Reich* (New York: Grosset & Dunlap, 1964).

22. Paul Schultze-Naumburg, *Die Kultur des weiblichen Körpers als Grundlage der Frauenkleidung*, quoted in Kern, *Anatomy*, 15. Schultze-Naumburg shifted easily from advocating "natural clothing" to supporting art fashioned after natural bodies; during the Third Reich he was an architect of the campaign against "degenerate" art. See Kern, *Anatomy*, 223–226.

23. Carl Heinrich Stratz, *Die Frauenkleidung und ihre natürliche Entwicklung* (Stuttgart: Enke, 1900).

24. See Schlüpmann, *Unheimlichkeit*, 8–25; Beth Irwin Lewis, "*Lustmord*: Inside the Windows of the Metropolis," in *Berlin: Culture and Metropolis*, ed. Charles W. Haxthausen and Heidrun Suhr (Minneapolis: University of Minnesota Press, 1990), 111–140; and the essays included in J. Edward Chamberlin and Sander L. Gilman, eds., *Degeneration: The Dark Side of Progress* (New York: Columbia University Press, 1985).

25. Along with Kaes ("Literary Intellectuals and the Cinema"), Schlüpmann (*Unheimlichkeit*), and Hake (*Third Machine*), see Miriam Hansen, "Early Silent Cinema: Whose Public Sphere?," *New German Critique* 29 (Spring/Summer 1983): 147–184. On women and reform in Germany, see Christoph Sachße, *Mütterlichkeit als Beruf: Sozialarbeit, Sozialreform und Frauenbewegung, 1871–1929* (Frankfurt: Suhrkamp, 1986); and Ann Taylor Allen, *Feminism and Motherhood in Germany, 1800–1914* (New Brunswick, N.J.: Rutgers University Press, 1991).

26. Susanne Asche, "Fürsorge, Partizipation und Gleichberechtigung—die Leistungen der Karlsruherinnen für die Entwicklung zur Großstadt (1859–1914)," in *Karlsruher Frauen, 1715–1945: Eine Stadtgeschichte* (Karlsruhe: Badenia, 1992), 171–256.

27. Eventually exemplified by Oswald Spengler's *The Decline of the West*, trans. Charles Francis Atkinson (New York: Knopf, 1926–28). There are numerous commentaries, but see, e.g., Ringer, *The Decline of the German Mandarins*; Klaus Vondung, "Zur Lage der Gebildeten in der wilhelminischen Zeit," in *Das wilhelminische Bildungsbürgertum Zur Sozialgeschichte seiner Ideen*, ed. Klaus Vondung (Göttingen: Vandenhoeck & Ruprecht, 1976), 20–33; or Charles E. McClelland, "The Wise Man's Burden: The Role of Academicians in Imperial German Culture," in Stark and Lackner, *Essays on Culture and Society in Modern Germany*, 45–69. An excellent expression of the perceived loss of cultural authority of experts in an age of mass culture and mediocre scientists is Ernst Meumann, "Wilhelm Wundt. Zu seinem achtzigsten Geburtstag," *Deutsche Rundschau* 38, no. 11 (August 1912): 193–224.

28. On the vital role of feminist activists in shaping the direction of reform movements and the public sphere in general, see Allen, *Feminism and Motherhood in Germany*,

1800–1914; and Kathleen Canning, *Languages of Labor and Gender: Female Factory Work in Germany, 1850–1914* (Ithaca, N.Y.: Cornell University Press, 1996).

29. Mirjam Storim, "'Einer, der besser ist, als sein Ruf': Kolportageroman und Kolportagebuchhandel um 1900 und die Haltung der Buchbranche," in *Schund und Schönheit: Populäre Kultur um 1900*, ed. Kaspar Maase and Wolfgang Kaschuba (Cologne: Böhlau, 2001), 252–282, here 255–256. See also Rudolf Schenda, *Die Lesestoffe der kleinen Leute: Studien zur populären Literatur im 19. und 20. Jahrhundert* (Munich: Beck, 1976).

30. Luke Springman, "Poisoned Hearts, Diseased Minds, and American Pimps: The Language of Censorship in the *Schund und Schmutz* Debates," *German Quarterly* 68, no. 4 (Autumn 1995): 408–429, here 413.

31. Corey Ross, *Media and the Making of Modern Germany: Mass Communications, Society, and Politics from the Empire to the Third Reich* (New York: Oxford University Press, 2008), 66.

32. Kaspar Maase, "Krisenbewußtsein und Reformorientierung: Zum Deutungshorizont der Gegener der modernen Populärkünste 1880–1918," in Maase and Kaschuba, *Schund und Schönheit: Populäre Kultur um 1900*, 290–342. See also his "Struggling About 'Filth and Trash': Educationalists and Children's Culture in Germany Before the First World War," *Paedagogica Historica* 34, no. 1 (1998): 8–28.

33. Professor Dr. Friß Johannesson, "Das Lesen der Jugend außerhalb der Schule," *Die Hochwacht* no. 2 (November 1911), quoted in Kara L. Ritzheimer, "Protecting Youth from 'Trash': Anti-*Schund* Campaigns in Baden, 1900–1933," PhD diss. (State University of New York–Binghamton, 2007), 25.

34. Class warfare, however, was not unknown in these campaigns, especially given the historical coincidence of the rise of mass entertainment and the rise of the Social Democratic Party of Germany (SPD), which many (such as Karl Brunner) saw as uncoincidental and fought both with equal vigor. For more, see Ross, *Media and the Making of Modern Germany*.

35. A fine articulation of Brunner's position with regard to film (and the SPD) is *Der Kinematograph von heute—eine Volksgefahr* (Berlin: Vaterländischen Schriftenverbandes, 1913).

36. Other trade periodicals included *Der deutsche Lichtspiel-Theater-Besitzer* (Berlin, 1909–1914), *Erste Internationale Film-Zeitung* (Berlin, 1908–1920), *Film und Lichtbild* (Stuttgart, 1912–1914), and *Die Lichtbild-Bühne* (Berlin, 1908–1940). Helmut H. Diederichs provides a more complete survey of the trade press in his *Anfänge deutscher Filmkritik* (Stuttgart: Fischer, 1986). On *Der Kinematograph* in particular, see Thomas Schorr, "Die Film- und Kinoreformbewegung und die Deutsche Filmwirtschaft. Eine Analyse des Fachblatts *Der Kinematograph* (1907–1935) unter pädagogischen und publizistischen Aspekten," PhD diss. (Universität der Bundeswehr, Munich, 1990). Hake also discusses the trade press in *Third Machine*, 3–26.

37. C. H. Dannmeyer, *Bericht der Kommission für "Lebende Photographien"* (Hamburg: Kampen, 1907), 27–28.

38. Hellwig, *Kind und Kino*, 22. Hellwig wrote much on *Schundfilms* and censorship, including "Die Beziehungen zwischen Schundliteratur, Schundfilms und Verbrechen," *Archiv für Kriminal-Anthropologie und Kriminalistik* 51, no. 1 (24 January 1913): 1–32; "Die maßgebenden Grundsätze für Verbote von Schundfilms nach geltendem und künstigem Rechte," *Verwaltungsarchiv* 21 (1913): 405–455; and *Die Filmzensur: Eine rechtsdogmatische und rechtpolitische Erörterung* (Berlin: Frankenstein, 1914).
39. See Eileen Bowser, *The Transformation of Cinema, 1907–1915* (New York: Scribner, 1990), 37–52. See also J. A. Lindstrom, "'Getting a Hold Deeper in the Life of the City': Chicago Nickelodeons, 1905–1914." PhD diss. (Northwestern University, 1998); Lee Grieveson, *Policing Cinema: Movies and Censorship in Early-Twentieth-Century America* (Berkeley: University of California Press, 2004); and Jennifer Lynn Peterson, *Education in the School of Dreams: Travelogues and Early Nonfiction Film* (Durham, N.C.: Duke University Press, 2013), esp. chap. 3.
40. Dannmeyer, *Bericht der Kommission*, 39.
41. "Die Eröffnung des Reform-Kinematographentheater," *Der Kinematograph* no. 32 (7 August 1907). *Der Kinematograph* was not paginated. For more on the sometimes stuffy discussion of ventilation, see a translation of the American Society of Heating and Ventilation Engineers' "Report of Committee on Standards for Ventilation Legislation for Motion Picture Show Places," in *Gesundheits-Ingenieur* 36, no. 22 (31 May 1913): 409–410; and a German response, Konrad Meier, "Vorschriften über Lüftung von Kinotheatern," *Gesundheits-Ingenieur* 36, no. 26 (28 June 1913): 483–484.
42. "Ein kurzer Rückblick auf die erste Woche des Reform-Kinematographen-Theaters," *Der Kinematograph* no. 33 (14 August 1907).
43. "Kinematographische Reformvereinigung," *Der Kinematograph* no. 43 (23 October 1907).
44. For more on *Kino-Kommissions*, see Sabine Lenk and Frank Kessler, "The Institutionalization of Educational Cinema: The Case of the *Kinoreformbewegung* in Germany," in *The Institutionalization of Educational Cinema: Educational Cinemas in North America and Europe in the 1910s and 1920s*, ed. Marina Dahlquist and Joel Frykholm (Bloomington: Indiana University Press, forthcoming). See also Rudolf W. Kipp, *Bilddokumente zur Geschichte des Unterrichtsfilms* (Grünwald, Germany: Institut für Film und Bild in Wissenschaft und Unterricht, 1975), 13–15.
45. "Kinematographische Reformvereinigung." Oskar Kalbus, one of the driving forces behind UFA's *Kulturfilm* division, later intimated that these donations were not uncontroversial: "Although this association was soon sharply criticized because of the close relationship between Lemke and the French film industry, it can nevertheless take credit for having given the first important impetus for the introduction of film in schools." Kalbus, "Abriß einer Geschichte der deutschen Lehrfilmbewegung," in *Das Kulturfilmbuch*, ed. Edgar Beyfuß and Alexander Kossowsky (Berlin: Chryselius'scher, 1924), 1–13, here 3.
46. Hermann Lemke, "Die Verwertung und Nutzbarmachung neuer Film-Ideen—Künstlerische Films," *Der Kinematograph* no. 57 (29 January 1908).

47. Ludwig Brauner, "Die Kino-Ausstellung in Berlin," *Der Kinematograph* no. 104 (25 December 1908).
48. Indeed, by this time the relations between the exhibitors and the reformers and trade journals were downright hostile. See "Die Kino-Austellung und 'Wir,'" *Erste Internationale Film-Zeitung* 6, no. 50 (14 December 1912): 52.
49. Hermann Häfker, "Eine Reise an die Quellen der Kinematographie," *Der Kinematograph* no. 163 (9 February 1910); and 172 (13 April 1910).
50. Hermann Lemke, "Volkstümliche Reisebeschreibungen," *Der Kinematograph* no. 34 (21 August 1907).
51. *Der Kinematograph* no. 258 (6 December 1911).
52. Paul Samuleit and Emil Borm, *Der Kinematograph als Volks- und Jugendbildungsmittel* (Berlin: Gesellschaft für Verbreitung von Volksbildung, 1912), 23–24, quoted in Hake, *Third Machine*, 36 (translation modified).
53. Hake, *Third Machine*, 36–38.
54. Unwilling to rely just on production companies, by 1909 Lemke hoped to create a cost-sharing distribution cooperative among interested schools. See Hermann Lemke, *Praktische Forderungen für die Verwertung der Kinematographie im Unterricht* (Friedenau: Schule und Technik, 1909). Georg Victor Mendel agreed and followed up with a plan to open a "purely scientific [*wissenschaftlichen*] theater": Georg Victor Mendel, *Kinematographie und Schule: Plan zur Gründung eines rein wissenschaftlichen Theaters für Kinematographie und Projektion* (Berlin: privately printed, 1909).
55. The legal discourse on cinema in Germany is far too vast to even attempt a survey here. Albert Hellwig's reviews are the best place to start, however: *Rechtsquellen des öffentlichen Kinematographenrechts* (M. Gladbach [Mönchengladbach]: Volksvereins, 1913); and *Öffentliches Lichtspielrecht* (M. Gladbach: Volksvereins, 1921). Other contemporary surveys include Bruno May, *Das Recht des Kinematographen* (Berlin: Falk, 1912); and Hans Müller-Sanders, "Die Kinematographenzensur in Preußen," PhD diss. (Badischen Ruprecht-Karls-Universität, Heidelberg, 1912). See also Gary D. Stark, "Cinema, Society, and the State," in Stark and Lackner, *Essays on Culture and Society in Modern Germany*, 122–166; and Kaspar Maase, "Massenkunst und Volkserziehung: Die Regulierung von Film und Kino im deutschen Kaiserreich," *Archiv für Sozialgeschichte* 41 (2001): 39–77.
56. Hellwig, *Öffentliches Lichtspielrecht*, 32–33. Not all regulations applied to the same theaters at the same time, of course. For an excellent case study of the variety of local tactics, see Amelie Duckwitz, Martin Loiperdinger, and Susanne Theisen, "'Kampf dem Schundfilm!': Kinoreform and Jugendschutz in Trier," *KINtop* 9 (2000): 53–63.
57. My presentation of the GVV is indebted to Schorr, "Die Film- und Kinoreformbewegung," 81–94; and Horst Dräger, *Die Gesellschaft für Verbreitung von Volksbildung: Eine historisch-problemgeschichtliche Darstellung von 1871–1914* (Stuttgart: Klett, 1975), 226–237.
58. Dräger, *Gesellschaft für Verbreitung*, 236.

59. Willi Warstat and Franz Bergmann, *Kino und Gemeinde* (M. Gladbach: Volksvereins, 1913), 114–116.
60. Heiner Schmitt, *Kirche und Film: Kirchliche Filmarbeit in Deutschland von ihren Anfängen bis 1945* (Boppard: Boldt, 1979), 41. For more on the role of Catholicism in this sphere, see Margaret Stieg Dalton, *Catholicism, Popular Culture, and the Arts in Germany, 1880–1933* (Notre Dame, Ind.: University of Notre Dame Press, 2005).
61. The "monopoly" system, established in Germany between 1910 and 1911, allowed distributors to acquire sole rights to a film and pass this exclusivity to cinema managers in the form of local exhibition rights. The theater owner's local monopoly enabled him to charge more and make, for the first time in Germany, a considerable profit. See Corinna Müller, *Frühe deutsche Kinematographie: formale, wirtschaftliche und kulturelle Entwicklungen 1907–1912* (Stuttgart and Weimar: Metzler, 1994), 126–158; as well as her essay, "The Emergence of the Feature Film in Germany Between 1910 and 1911," in *Before Caligari: German Cinema, 1895–1920*, ed. Paolo Cherchi Usai and Lorenzo Codelli (Madison: University of Wisconsin Press, 1990), 94–113.
62. The best survey of the role of the *Lichtbilderei* in the reform movement is Diederichs, *Anfänge deutscher Filmkritik*, 84–88.
63. Dräger, *Gesellschaft für Verbreitung*, 234–235.
64. Volker Schulze, "Frühe kommunale Kinos und die Kinoreformbewegung in Deutschland bis zum Ende des ersten Weltkriegs," *Publizistik* 22, no. 1 (January–March 1977): 61–71.
65. Minutes from the meeting of the community representatives of Eickel, 14 May 1912 (archive of the City of Wanne-Eickel), quoted in Schulze, "Frühe kommunale Kinos," 64.
66. Rudolf Pechel in *Literarischen Echo* 16 (1913/1914): 582, quoted in Ludwig Greve, Margot Pehle, and Heidi Westhoff, eds., *Hätte ich das Kino! Die Schriftsteller und der Stummfilm* (Munich: Kösel, 1976), 68. Pechel reviewed Willy Rath's *Kino und Bühne* (M. Gladbach: Volksvereins, 1913).
67. Arthur Mellini, "Die ganze Richtung passt uns nicht!" *Lichtbild-Bühne* 5 (4 February 1911): 3–4, quoted in Karen J. Kenkel, "The Nationalisation of the Mass Spectator in Early German Film," in *Celebrating 1895: The Centenary of Cinema*, ed. John Fullerton (Sydney: Libbey, 1998), 155–162, here 158.
68. Max Kullmann, "Die Entwicklung des deutschen Lichtspieltheater," PhD diss. (University of Nuremberg, 1935), quoted in Hake, *Third Machine*, 27. Kullmann quoted a film theater owner.
69. Siegfried Kracauer, *From Caligari to Hitler* (Princeton, N.J.: Princeton University Press, 1947), 19.
70. Hake, *Third Machine*, 28.
71. Helmut Kommer, *Früher Film und späte Folgen: Zur Geschichte der Film- und Fernseherziehung* (Berlin: Basis, 1979).
72. Hermann Lemke, *Die Kinematographie der Gegenwart, Vergangenheit und Zukunft* (Leipzig: Demme, 1911), 24.

73. Many scholars have stressed the connection between "the masses" and "the feminine" as an indication of the anxieties and spirit of the age. This line of reasoning is indeed extremely significant, but the connection between "the masses" and "children" (as a similarly charged rhetorical construction) deserves a closer look. On the masses as feminine, see esp. Susanna Barrows, *Distorting Mirrors: Visions of the Crowd in Late Nineteenth-Century France* (New Haven: Yale University Press, 1981). For another, parallel examination of German reformers on cinema, children, and the masses, see Karen J. Kenkel, "The Adult Children of Early Cinema," *Women in German Yearbook* (2000): 137–160.
74. Lorenz Pieper, "Kino und Drama," *Bild und Film* 1, no. 1 (1912): 5.
75. Georg Lukács, "Thoughts Toward an Aesthetic of the Cinema," trans. Janelle Blankenship, *Polygraph* 13 (2001): 13–18, here 16 (emphasis in original). Originally published as "Gedanken zu einer Ästhetik des 'Kino,'" *Frankfurter Zeitung* 251 (10 September 1913): 1–2.
76. This phrase and *vom Kinde aus* are attributed to Hamburg pedagogue Johannes Gläser, one of many who popularized and realized Key's suggestions. See Scheibe, *Die Reformpädagogische Bewegung, 1900–1932*, 65.
77. Ellen Key, "Erziehung," in *Das Jahrhundert des Kindes* (Berlin, 1905), in Flitner and Kudritzki, *Die deutsche Reformpädagogik*, 52–54, here 52.
78. Stephen Kern, "Freud and the Emergence of Child Psychology, 1880–1910," PhD diss. (Columbia University, 1970), 264.
79. Charles Darwin, *The Descent of Man* (London, 1871), quoted in Kern, "Freud," 212.
80. The importance of "primitivism"—of which Darwin was a prime but not uncommon example—for this connection between children (or the feminine) and the masses cannot be underestimated. For its role in shaping turn of the century cultural agendas in Germany, see Doris Kaufmann, "'Pushing the Limits of Understanding': The Discourse on Primitivism in German Kulturwissenschaften, 1880–1930," *Studies in History and Philosophy of Science* 39 (2008): 434–443. For examinations in relation to cinema, see Assenka Oksiloff, *Picturing the Primitive: Visual Culture, Ethnography, and Early German Cinema* (New York: Palgrave, 2001); and Beth Corzo-Duchardt, "Primal Screen: Primitivism and American Silent Film Spectatorship," PhD diss. (Northwestern University, 2013).
81. Gustave Le Bon, *The Crowd: A Study of the Popular Mind* (Atlanta, Ga.: Cherokee, 1982), 16. Hereafter cited parenthetically. Originally published as *Psychologie des foules* (Paris: Alcan, 1895). Translated and published in German as *Psychologie der Massen* (Leipzig: Klinkhardt, 1908). See also Robert A. Nye, *The Origins of Crowd Psychology: Gustave Le Bon and the Crisis of Mass Democracy in the Third Republic* (London and Beverly Hills, Calif.: Sage, 1975)..
82. Erika Apfelbaum and Gregory R. McGuire, "Models of Suggestive Influence and the Disqualification of the Social Crowd," in *Changing Conceptions of Crowd Mind and Behavior*, ed. C. F. Graumann and S. Moscovici (New York and Berlin: Springer, 1986), 27–50.

83. See Walter Serner, "Kino und Schaulust," in Schweinitz, *Prolog vor dem Film*, 208–214. Originally published in *Die Schaubühne* 9, nos. 34/35 (1913): 807–811. Chapter 4 will discuss *Schaulust* and this essay in more detail.
84. Dannmeyer, *Bericht der Kommission*, 27–28.
85. Emilie Altenloh, *Zur Soziologie des Kino: Die Kino-Unternehmung und die sozialen Schichten ihrer Besucher* (Jena: Diederichs, 1914), 91.
86. Altenloh, *Zur Soziologie des Kino*, 65.
87. Albert Hellwig, "Über die schädliche Suggestivkraft kinematographischer Vorführung," *Aerztliche Sachverständigen-Zeitung* 20, no. 6 (15 March 1914): 122. Hellwig was reviewing and citing from an article by Italian psychiatrist Giuseppe d'Abundo, "Sopra alcuni particolari effetti delle projezioni cinematografiche nei nevrotici," *Rivista Italiana di Neuropatologia, Psichiatria ed Elettroterapia* 4, no. 10 (October 1911): 433–442; for another German review, see "Kinematograph als Krankheitsstifter," in *Fortschritte der Medizin* 30 (1912): 302. For more on Italian uses of cinematography in the human sciences, see Silvio Alovisio, *L'occhio sensibile. Cinema e scienze della mente nell'Italia del primo Novecento. Con una antologia di testi d'epoca* (Turin: Edizioni Kaplan, 2013).
88. Anson Rabinbach, *The Human Motor: Energy, Fatigue, and the Origins of Modernity* (Berkeley and Los Angeles: University of California Press, 1990), 157.
89. Hermann Lemke, "Die kinematographische Reformpartei, ihre Aufgaben und Ziele," *Der Kinematograph* no. 42 (16 October 1907).
90. Albert Hellwig, *Schundfilms: Ihr Wesen, ihre Gefahren und ihre Bekämpfung* (Halle an der Saale, Germany: Waisenhaus, 1911), 33, quoted in Hake, *Third Machine*, 39.
91. Friedrich Schiller, *On the Aesthetic Education of Man*, trans. Elizabeth M. Wilkinson and L. A. Willoughby (Oxford: Oxford University Press, 1967), 161.
92. "Die Kultur-Arbeit des Kinematographen-Theaters," *Die Lichtbild-Bühne* 2, no. 41 (4 February 1909).
93. Erwin Ackerknecht, *Das Lichtspiel im Dienste der Bildungspflege: Handbuch für Lichtspielreformer* (Berlin: Weidmannsche, 1918), 66.
94. August Julius Langbehn, "Rembrandt als Erzieher," in *Die Kunsterziehungsbewegung*, ed. Hermann Lorenzen (Bad Heilbrunn, Germany: Klinkhardt, 1966), 7–17. Hereafter cited parenthetically. Stern's *The Politics of Cultural Despair* provides the standard account of Langbehn's place in history.
95. For more on the reverberations of Langbehn's essay through Wilhelmine Germany, see Corona Hepp, *Avantgarde: Moderne Kunst, Kulturkritik und Reformbewegungen nach der Jahrhundertwende* (Munich: Deutscher Taschenbuch, 1987).
96. Alfred Lichtwark, "Der Deutsche der Zukunft," in Flitner and Kudritzki, *Die deutsche Reformpädagogik*, 99–110, here 104 (emphasis in original). Hereafter cited parenthetically. Lichtwark and Langbehn were acquaintances; Lichtwark introduced Langbehn to the work of Rembrandt in 1887. See Gisela Wilkending, *Volksbildung und Pädagogik "vom Kinde aus": Eine Untersuchung zur Geschichte der Literaturpädagogik in den Anfängen der Kunsterziehungsbewegung* (Weinheim, Germany: Beltz, 1980), 79–85. For more on Lichtwark, see Julius Gebhard, *Alfred Lichtwark und die Kunsterziehungsbewegung in Hamburg*

(Hamburg: Hoffmann und Campe, 1947); Hans Präffcke, *Der Kunstbegriff Alfred Lichtwarks* (Hildesheim, Zürich, and New York: Olms, 1986); Carolyn Kay, *Art and the German Bourgeoisie: Alfred Lichtwark and Modern Painting in Hamburg, 1886–1914* (Toronto and Buffalo, N.Y.: University of Toronto Press, 2002); but esp. Jennifer Jenkins, *Provincial Modernity: Local Culture and Liberal Politics in Fin-de-Siècle Hamburg* (Ithaca, N.Y.: Cornell University Press, 2003).

97. Jenkins, *Provincial Modernity*, 76.
98. Konrad Lange, "Das Wesen der künstlerischen Erziehung," in Lorenzen, *Kunsterziehungsbewegung*, 21–26, here 22. For more on the art education movement, see Peter Joerissen, *Kunsterziehung und Kunstwissenschaft im wilhelminischen Deutschland, 1871–1918* (Cologne and Vienna: Böhlau, 1979).
99. Lange, *Die künstlerische Erziehung der deutschen Jugend*, 10.
100. Lange, "Das Wesen der künstlerischen Erziehung," 26.
101. Alfred Lichtwark, "Die Aufgaben der Kunsthalle: Antrittsrede den 9. December 1886," In *Drei Programme*, 2d ed. (Berlin: Cassirer, 1902), 11–31, here 29.
102. Eckard Schaar, "Zustände," in Alfred Lichtwark, *Erziehung des Auges: Ausgewählte Schriften*, ed. Eckard Schaar (Frankfurt: Fischer, 1991), 8, quoted in Jenkins, *Provincial Modernity*, 64.
103. Alfred Lichtwark, *Die Bedeutung der Amateur-Photographie* (Halle an der Saale, Germany: Knapp, 1894), 1.
104. Alfred Lichtwark, "Museen als Bildungsstätten," *Der Deutsche der Zukunft* (Berlin: Cassirer, 1905), 89–107.
105. Alfred Lichtwark, *Übungen in der Betrachtung von Kunstwerken* (Dresden: Kühtmann, 1900), 17.
106. See, for instance, John Dewey, *Art as Experience* (New York: Putnam, 1984).
107. See, e.g., Walter Geisel, *Wie ich mit meinen Jungens Kunstwerke betrachte* (Glückstadt: Geisel, 1904); Paul Quensel, *Meisterbilder und Schule: Anregungen zu praktischen Versuchen* (Munich: Kunstwart-Verl, 1905); Leipziger Lehrerverein, ed., *Bildbetrachtungen: Arbeiten aus der Abteilung für Kunstpflege des Leipziger Lehrervereins* (Leipzig: Teubner, 1906); and Ulrich Diem, *Bildbetrachtung: Eine Wegleitung für Kunstfreunde* (St. Gallen: Fehr'sche, 1919). *Bildbetrachtung* was even more popular as a teaching method after World War II.
108. Heinrich Wolgast, "Die Bedeutung der Kunst für die Erziehung," in Lorenzen, *Die Kunsterziehungsbewegung*, 17–20, here 19.
109. For a European history of the movement, along with a clear explication of the principles of *Bild- und Kunstbetrachtung*, see Ludwig Praehauser, *Erfassen und Gestalten: Die Kunsterziehung als Pflege formender Kräfte* (Salzburg: Müller, 1950).
110. Pestalozzi, *How Gertrude Teaches Her Children* [1801], ed. Ebenezer Cooke, trans. Lucy E. Holland and Frances C. Turner, 2d ed. (Syracuse, N.Y.: Bardeen, 1898), tenth letter, 220 (emphasis in original).
111. Clive Ashwin, "Pestalozzi and the Origins of Pedagogical Drawing," *British Journal of Educational Studies* 29, no. 2 (June 1981): 138–151, here 146.
112. Keiichi Takaya, "The Method of *Anschauung*: From Johann H. Pestalozzi to Herbert Spencer," *Journal of Educational Thought* 37, no. 1 (2003): 77–99, here 84.

113. Robert Ulich, "Pestalozzi, Johann Heinrich," in *The Encyclopedia of Philosophy*, vol. 6, ed. Paul Edwards (New York: Macmillan, 1967), 121–122.
114. Melanie Judith Keene, "Object Lessons: Sensory Science Education, 1830–1870," PhD diss. (University of Cambridge, 2008), 51–54.
115. Ulich, "Pestalozzi, Johann Heinrich," 122.
116. Ashwin, "Pestalozzi and the Origins of Pedagogical Drawing," 146.
117. Christopher Owen Ritter, "Re-presenting Science: Visual and Didactic Practice in Nineteenth-Century Chemistry," PhD diss. (University of California, Berkeley), 2001, 128.
118. W. T. Harris, editor's preface to *Herbart's ABC of Sense Perception and Minor Pedagogical Works*, by Johann Friedrich Herbart, ed. and trans. William J. Eckoff (New York: Appleton, 1896), vii.
119. Herbert Spencer, *Education: Intellectual, Moral, and Physical* [1861] (New York: Appleton, 1896), quoted in Takaya, "The Method of *Anschauung*," 81.
120. The best survey of its dissemination in Germany is Gottlieb Gustav Deussing, "Der Anschauungsunterricht in der deutschen Schule von Comenius bis zur Gegenwart," PhD diss. (Universität Jena, 1884).
121. On *Anschauungsunterricht* in the natural sciences, see Massimiano Bucchi, "Images of Science in the Classroom: Wallcharts and Science Education, 1850–1920," *British Journal for the History of Science* 31, no. 2 (1998): 161–184; Lynn K. Nyhart, "Science, Art, and Authenticity in Natural History Displays," in *Models: The Third Dimension of Science*, ed. Soraya de Chadarevian and Nick Hopwood (Stanford, Calif.: Stanford University Press, 2004), 307–335; Nyhart, *Modern Nature: The Rise of the Biological Perspective in Germany* (Chicago: University of Chicago Press, 2009), esp. chap. 5, "The 'Living Community' in the Classroom"; Henning Schmidgen, "Pictures, Preparations, and Living Processes: The Production of Immediate Visual Perception (*Anschauung*) in Late-19th-Century Physiology," *Journal of the History of Biology* 37, no. 3 (October 2004): 477–513; and Schmidgen, "1900—The Spectatorium: On Biology's Audiovisual Archive," *Grey Room* 43 (2011): 42–65.
122. I have examined the trope of efficiency in visual education in "The Efficiency of Images: Educational Effectiveness and the Modernity of Motion Pictures," in *The Visual Culture of Modernism*, SPELL: Swiss Papers in English Language and Literature 26, ed. Deborah L. Madsen and Mario Klarer (Tübingen: Narr, 2011), 41–59; and "Dissecting the Medical Training Film," in *Beyond the Screen: Institutions, Networks and Publics of Early Cinema*, ed. Marta Braun et al. New Barnet, U.K.: Libbey, 2012), 161–167.
123. Carl Jacobj, "Anschauungsunterricht und Projektion," *Zeitschrift für wissenschaftliche Mikroskopie und mikroskopische Technik* 36, no. 4 (1919): 273–314, here 275, quoted in Schmidgen, "1900—The Spectatorium," 51.
124. Ashwin, "Pestalozzi and the Origins of Pedagogical Drawing," 146 (emphasis in original).
125. For a contemporary assessment of the gap between reform ideals and actual practice, see I. L. Kandel, "Germany," in *Comparative Education: Studies of the*

Educational Systems of Six Modern Nations, ed. Peter Sandiford (London and Toronto: Dent, 1918), 121–130; or, to a lesser extent, the apologetic William S. Learned, *An American Teacher's Year in a Prussian Gymnasium* (New York: Educational Review, 1911).

126. "Besuch kinematographischer Vorführung durch Schüler höherer Lehranstalten (Breslau 1910)" and "Besuch der Kinematographentheater durch Schüler und Schülerinnen sowie durch die Zöglinge der Seminare und Präparandenanstalten (Berlin 1912)" in *Dokumente zur Geschichte der Schulfilmbewegung in Deutschland*, ed. Fritz Terveen (Emsdetten: Lechte, 1959), 16–17.

127. Otfrid von Hanstein, *Kinematographie und Schule. Ein Vorschlag zur Reform des Anschauungs-Unterrichts* (Berlin: Lichtspiele Mozartsaal, 1911), 3. Other major statements about the educational use of film before World War I include Samuleit and Borm, *Der Kinematograph als Volks- und Jugendbildungsmittel*; Adolf Sellmann, *Der Kinematograph als Volkserzieher?* (Langensalza, Germany: Beyer, 1912); Friedrich Murawski, *Die Kinematographie und ihre Beziehungen zu Schule und Unterricht* (Dresden: Bieyl and Kaemmerer, 1914); Sellmann, *Kino und Volksbildung* (M. Gladbach: Volksvereins Verlag, 1914); and esp. Sellmann, *Kino und Schule* (M. Gladbach: Volksvereins Verlag, 1914).

128. For a friendly assessment, see Wilhelm Richter, "Der Kinematograph als naturwissenschaftliches Anschauungsmittel," *Naturwissenschaftliche Wochenschrift* 12, no. 52 (28 December 1913): 817–820.

129. K. Rüswald, "Der Film im Erdkundlichen und Naturwissenschaftlichen Unterricht," in Terveen, *Dokumente zur Geschichte der Schulfilmbewegung in Deutschland*, 43–44. A fine summary of the arguments forwarded against educational uses of film can be found in H. Graupner, "Unterrichtshygiene," in *Handbuch der deutschen Schulhygiene*, ed. Hugo Selter (Dresden and Leipzig: Steinkopff, 1914), 174–321, esp. his section on "Kinematograph und Unterrichtshygiene," 302–307.

130. Sellmann, *Kino und Schule*, 15 (emphasis in original).

131. Richard Kretz, "Die Anwendung der Photographie in der Medicin," *Wiener klinische Wochenschrift* 7, no. 44 (1 November 1894): 832.

132. Paul Knospe, *Der Kinematograph im Dienste der Schule. Unter besonderer Berücksichtigung des erdkundlichen Unterrichts* (Halle an der Saale, Germany: Waisenhaus, 1913), 9. There are many more examples, but see also Bastian Schmid, "Kinematographie und Schule," *Die Naturwissenschaften* 1, no. 6 (7 February 1913): 145–146.

133. Oskar Kalbus summarizes and quotes from such complaints in *Der Deutsche Lehrfilm in der Wissenschaft und im Unterricht* (Berlin: Heymanns, 1922), 6.

134. Mendel, *Kinematographie und Schule*.

135. "Zur Eröffnung des Ernemann-Kino in Dresden," *Der Kinematograph* no. 134 (21 July 1909). It was not unusual for film equipment manufacturers in Germany to have exhibition storefronts that were open to the public in a quasi-museum-like setting. See Deac Rossell, "Beyond Messter: Aspects of Early Cinema in Berlin," *Film History* 10 (1998): 52–69.

136. "Die Dresdner 'Kosmographia,'" *Bild und Film* 1, no. 1 (1912): 19. See also Uli Jung, "Film für Lehre und Bildung," in *Geschichte des dokumentarischen Films in Deutschland*, vol. 1, ed. Uli Jung and Martin Loiperdinger (Stuttgart: Reclam, 2005), 333–340.
137. Those venues were the Ernemann-Kino (1909) and the Kosmographia (1910) in Dresden; the Reform-Kino in Braunschweig (1910); the Reformtheater in Bremen (1911); the Gemeindekino in Eickel (1912); the Germania Saal in Hagen (1912); the Musterlichtbildbühne in Altona (1912); and the Urania in Stettin (1914).
138. Hermann Bredtmann, "Kinematographie und Schule," *Pädagogisches Archiv* 56, no. 3 (1914): 154–163, here 161 and 163.
139. On Lemke, see Schorr, "Die Film- und Kinoreformbewegung," 56ff; Müller, *Frühe deutsche Kinematographie*, 71–76; and Müller, "Der frühe Film, das frühe Kino und seine Gegner und Befürworter," in *Schund und Schönheit: Populäre Kultur um 1900*, ed. Kaspar Maase und Wolfgang Kaschuba (Cologne, Weimar, and Vienna: Böhlau, 2001), 62–91.
140. See the technophilia of Hermann Lemke's *Durch die Technik zur Schulreform. Zwei modern-technische Lehrmethoden und Veranschaulichungsmittel in der Schule der Zukunft* (Leipzig: Demme, 1911), in which he predicts that the combination of films and phonographs will make teachers obsolete. See also Sellmann's vision of a "complete revolution" in teaching once projectors are universally installed; *Kino und Schule*, 39.
141. Hermann Lemke, *Die kinematographische Unterrichtsstunde* (Leipzig: Demme, 1911), 5.
142. Müller, "Der frühe Film, das frühe Kino und seine Gegner und Befürworter," 75. Sellmann held similar workshops in Eickel, but they also did not last in the long run. See "Bericht über eine Besprechung der Kinokommssion des Westfälischen Landgemeindetages anläßlich der Eröffnungsfeier des Gemeindelichtspielhauses in Eickel," *Bild und Film* 2, no. 3 (1912): 70–71. See also Lenk and Kessler, "The Institutionalization of Educational Cinema."
143. Actually, Herbart was not this clear or consistent, so these steps are the result of refinements by later Herbartians such as Wilhelm Rein, esp. *Pädagogik im Grundriß* (1890), 4th ed. (Leipzig: Göschen, 1907), 109. For a helpful discussion of Herbart and Herbartianism, see Harold B. Dunkel, *Herbart and Education* (New York: Random House, 1969).
144. Miriam Hansen, *Babel and Babylon: Spectatorship in American Silent Film* (Cambridge, Mass.: Harvard University Press, 1991), 93.
145. Helmut H. Diederichs, "Naturfilm als Gesamtkunstwerk: Hermann Häfker und sein 'Kinetographie'-Konzept," *Augenblick* 8 (1990): 37–60. If Häfker is known at all to English-speaking readers, it is through Kracauer's characterization of him in *From Caligari to Hitler* as the man who "praised war as the salvation from the evils of peace" (28). Häfker saw World War I mainly as an opportunity for the state to take control of cinema and put his plans into action. While there is no doubt that Häfker was conservative, nationalistic, and blind to the horrors of war, it would be unfair to depict him as a warmonger with the prefascist tendencies implied by

Kracauer. Häfker earned a "heart attack" in a concentration camp for his resistance to the Nazi government. All biographical information comes from Diederichs's article and his entry on Häfker in *Cinegraph*, ed. Hans-Michael Bock (Munich: edition text + kritik, 1984). My presentation of Häfker is indebted to these essays and my conversations with Diederichs.

146. Hermann Häfker, "Zur Dramaturgie der Bilderspiele," *Der Kinematograph* no. 32 (7 August 1907).
147. Hermann Häfker, "Meisterspiele," *Der Kinematograph* no. 56 (22 January 1908).
148. Häfker, "Zur Dramaturgie der Bilderspiele."
149. Hermann Häfker, *Kino und Kunst* (M. Gladbach: Volksvereins, 1913), 5.
150. See, for instance, Robert Gaupp and Konrad Lange, *Der Kinematograph als Volksunterhaltungsmittel* (Munich: Callwey, 1912); R. Stigler, "Über das Flimmern der Kinematographen," *Archiv für die gesamte Physiologie des Menschen und der Tiere* (Bonn) 123 (1908): 224–232; or Naldo Felke, "Die Gesundheitsschädlichkeit des Kinos," *Die Umschau* 17, no. 1 (1 January 1913): 254–255.
151. Rabinbach, *Human Motor*, 21.
152. Häfker, "Meisterspiele."
153. Häfker, *Kino und Kunst*, 9. Hereafter cited parenthetically.
154. Ernst Schultze, *Der Kinematograph als Bildungsmittel* (Halle an der Saale, Germany: Waisenhaus, 1911), 118.
155. Max Brethfeld, "Neue Versuche, die Kinematographie für die Volksbildung und Jugenderziehung zu verwerten," *Neue Bahnen* (1910): 422, quoted in Diederichs, "Naturfilm," 41.
156. Häfker, *Kino und Kunst*, 61. Hereafter cited parenthetically.
157. On the Urania lecture hall influence, see Gerhard Ebel and Otto Lührs, "Urania—eine Idee, eine Bewegung, eine Institution wird 100 Jahre alt," in *100 Jahre Urania: Wissenschaft heute für morgen* (Berlin: Urania Berlin, 1988), 15–74. There are also structural similarities between Häfker's presentations and the presentation of early cinema's passion plays. See Charles Musser, *The Emergence of Cinema: The American Screen to 1907* (New York: Scribner, 1990), 208–218.
158. Jonathan Crary offers a concise overview of the attention psychologists paid to attention in his "Unbinding Vision," *October* 68 (Spring 1994): 21–44, and offers a more lengthy treatment in *Suspensions of Perception: Attention, Spectacle, and Modern Culture* (Cambridge, Mass.: MIT Press, 1999). With regard to the perceived increase in sensual diversions, see, e.g., George M. Beard, *American Nervousness: Its Causes and Consequences* (New York: Putnam, 1881); and Tom Lutz, *American Nervousness, 1903: An Anecdotal History* (Ithaca, N.Y.: Cornell University Press, 1991).
159. See Martin Jay, *Downcast Eyes: The Denigration of Vision in Twentieth-Century French Thought* (Berkeley and Los Angeles: University of California Press, 1993).
160. Schiller, *On the Aesthetic Education of Man*, 183 (emphasis in original)
161. Walter Benjamin, "The Work of Art in the Age of Its Technological Reproducibility (Third Version)," in *Walter Benjamin, Selected Writings*, vol. 4, *1938–1940*, ed. Howard Eiland and Michael W. Jennings, trans. Edmund Jephcott and others (Cambridge, Mass.: Belknap, 2003), 251–283.

162. Schiller, *On the Aesthetic Education of Man*, 183.
163. Immanuel Kant, *Critique of Judgement*, trans. James Creed Meredith (Oxford: Oxford University Press, 1952), 150–154, §40.
164. Schiller, *On the Aesthetic Education of Man*, 215 (emphasis in original). For my discussion of aesthetics and ideology, I am indebted to Terry Eagleton, *The Ideology of the Aesthetic* (Cambridge, Mass.: Blackwell, 1990).

4. THE PROBLEM WITH PASSIVITY

1. Walter Benjamin, "The Work of Art in the Age of Its Technological Reproducibility (Second Version)," in *Walter Benjamin: Selected Writings*, vol. 3, *1935–1938*, ed. Howard Eiland and Michael W. Jennings, trans. Edmund Jephcott, Howard Eiland, and others (Cambridge, Mass.: Belknap, 2002), 101–133, here 109.
2. For representative essays, see the following anthologies: Anton Kaes, ed., *Kino-Debatte: Texte zum Verhältnis von Literatur und Film, 1909–1929* (Tübingen: Niemeyer, 1978); Ludwig Greve, Margot Pehle, and Heidi Westhoff, eds., *Hätte ich das Kino! Die Schriftsteller und der Stummfilm* (Munich: Kösel, 1976); Fritz Güttinger, ed., *Kein Tag ohne Kino: Schriftsteller über den Stummfilm* (Frankfurt: Deutsches Filmmuseum, 1984); Jörg Schweinitz, ed., *Prolog vor dem Film: Nachdenken über ein neues Medium, 1909–1914* (Leipzig: Reclam, 1992); Helmut H. Diederichs, ed., *Geschichte der Filmtheorie: Kunsttheoretische Texte von Méliès bis Arnheim* (Frankfurt: Suhrkamp, 2004); and Richard W. McCormick and Alison Guenther-Pal, eds., *German Essays on Film* (New York: Continuum, 2004).
3. The best overview of the early German film industry remains Corinna Müller, *Frühe deutsche Kinematographie: formale wirtschaftliche und kulturelle Entwicklungen, 1907–1912* (Stuttgart and Weimar: Metzler, 1994). On early German cinema in general, see the following anthologies: Paolo Cherchi Usai and Lorenzo Codelli, eds., *Before Caligari: German Cinema, 1895–1920* (Madison: University of Wisconsin Press, 1990); Thomas Elsaesser with Michael Wedel, eds., *A Second Life: German Cinema's First Decades*, (Amsterdam: Amsterdam University Press, 1996); and Thomas Elsaesser and Michael Wedel, eds., *Kino der Kaiserzeit: Zwischen Tradition und Moderne* (Munich: edition text + kritik, 2002). On the *Autorenfilm*, see esp. Helmut H. Diederichs, *Anfänge deutscher Filmkritik* (Stuttgart: Fischer, 1986); Diederichs, "The Origins of the *Autorenfilm*," in Cherchi Usai and Codelli, *Before Caligari*, 380–401; and Leonardo Quaresima, "Dichter, Heraus! The *Autorenfilm* and German Cinema of the 1910s," *Griffithiana* 38/39 (October 1990): 101–120.
4. Anton Kaes, "Literary Intellectuals and the Cinema: Charting a Controversy (1909–1929)," *New German Critique* 40 (Winter 1987): 7–34, here 7–8.
5. Sabine Hake, *The Cinema's Third Machine: Writing on Film in Germany, 1907–1933* (Lincoln: University of Nebraska Press, 1993), 63.
6. Peter Jelavich, "'Am I Allowed to Amuse Myself Here?': The German Bourgeoisie Confronts Early Film," in *Germany at the Fin de Siècle: Culture, Politics, and*

Ideas, ed. Suzanne Marchand and David Lindenfeld (Baton Rouge: Louisiana State University Press, 2004), 227–249, here 247.
7. Helmut H. Diederichs, "Kino und die Wortkünste: Zur Diskussionen der deutschen literarischen Intelligenz 1910 bis 1915," *KINtop* 13 (2004): 9–23. See also Diederichs, "Frühgeschicht deutscher Filmtheorie. Ihre Entstehung und Entwicklung bis zum Ersten Weltkrieg," unpublished Habilitationsschrift (J. W. Goethe-Universität Frankfurt am Main, 1996).
8. Stefanie Harris, *Mediating Modernity: German Literature and the "New" Media, 1895–1930* (University Park: Pennsylvania State University Press, 2009).
9. Heinz-B. Heller, *Literarische Intelligenz und Film: Zu Veränderungen der ästhetischen Theorie und Praxis unter dem Eindruck des Films 1910–1930 in Deutschland* (Tübingen: Niemeyer, 1985). See also Thomas Koebner, "Der Film als neue Kunst: Reaktionen der literarischen Intelligenz: Zur Theorie des Stummfilms (1911–1924)," in *Literaturwissenschaft-Medienwissenschaft*, ed. Volker Canaris and Helmut Kreuzer (Heidelberg: Quelle und Meyer, 1977), 1–31. While they do not focus as much on the relationship between literature and film, Miriam Hansen and Heide Schlüpmann set the agenda for the study of this era in other ways, which will be discussed later in the chapter: Miriam Hansen, "Early Silent Cinema: Whose Public Sphere?," *New German Critique* 29 (Spring/Summer 1983): 147–184; Heide Schlüpmann, *Unheimlichkeit des Blicks: Das Drama des frühen deutschen Kinos* (Frankfurt: Stroemfeld/Roter Stern, 1990).
10. This applies, of course, to descriptions of the *beautiful*, rather than to those of the *sublime*, which often emphasized an immediate overwhelming of the imagination. My thanks to Dan Morgan for reminding me of this distinction.
11. Friedrich Schiller, *On the Aesthetic Education of Man*, trans. Elizabeth M. Wilkinson and L. A. Willoughby (Oxford: Oxford University Press, 1967), 123.
12. See, for example, Robert Vischer's description of the difference between *Sehen* and *Schauen* ("seeing" and "scanning") in *On the Optical Sense of Form* (1873), in *Empathy, Form, and Space: Problems in German Aesthetics, 1873–1893*, ed. and trans. Harry Francis Mallgrave and Eleftherios Ikonomou (Santa Monica, Calif.: Getty Center for the History of Art and the Humanities, 1994), 89–124, esp. 93–95. See also Adolf Hildebrand's elaboration of this idea in *The Problem of Form in the Fine Arts* (1893), in Mallgrave and Ikonomou, *Empathy, Form, and Space*, 227–279, esp. 229–232.
13. For a complete survey, see Steve Odin, *Artistic Detachment in Japan and the West: Psychic Distance in Comparative Aesthetics* (Honolulu: University of Hawaii Press, 2001).
14. Emilie Altenloh, *Zur Soziologie des Kino: Die Kino-Unternehmung und die sozialen Schichten ihrer Besucher* (Jena: Diederichs, 1914), 91.
15. William Egginton, "Intimacy and Anonymity, or How the Audience Became a Crowd," in *Crowds*, ed. Jeffrey T. Schnapp and Matthew Tiews (Stanford, Calif.: Stanford University Press, 2006), 97–110.
16. Jacques Rancière, *The Emancipated Spectator*, trans. Gregory Elliott (London and New York: Verso, 2009), 2.

17. Carolin Duttlinger, "Between Contemplation and Distraction: Configurations of Attention in Walter Benjamin," *German Studies Review* 30, no. 1 (February 2007): 33–54.
18. Dudley Andrew, "Film and Society: Public Rituals and Private Space," in *Exhibition: The Film Reader*, ed. Ina Rae Hark (New York: Routledge, 2001), 161–172, here 165. Originally published in *East-West Film Journal* 1, no. 1 (1986): 7–22.
19. A good introduction to the broader social problem, minus cinema, is Hartmut Rosa and William E. Scheuerman, eds., *High-Speed Society: Social Acceleration, Power, and Modernity* (University Park: Pennsylvania State University Press, 2009).
20. Georg Kleibömer, "Kinematograph und Schuljugend," *Der Kinematograph*, no. 124 (12 May 1909) (emphasis in original in this and subsequent quotations).
21. Karl Hans Strobl, "Der Kinematograph," in *Kein Tag ohne Kino: Schriftsteller über den Stummfilm*, ed. Fritz Güttinger (Frankfurt: Deutsches Filmmuseum, 1984), 50–54, here 52. Originally published in *Die Hilfe* 17, no. 9 (2 March 1911).
22. Ph. Sommer, "Zur Psychologie des Kinematographen," *Der Kinematograph* no. 227 (3 May 1911).
23. Wilhelm Stapel, "Der homo cinematicus," *Deutsches Volkstum* 21 (October 1919): 319–320, here 319.
24. Schiller, *On the Aesthetic Education of Man*, 79. Hereafter cited parenthetically.
25. Arthur Schopenhauer, *The World as Will and Representation*, trans. E. F. J. Payne (New York: Dover, 1966), vol. 1, §38, p. 197.
26. Josef Chytry, *The Aesthetic State: A Quest in Modern German Thought* (Berkeley and Los Angeles: University of California Press, 1989). See also David Aram Kaiser, *Romanticism, Aesthetics, and Nationalism* (Cambridge: Cambridge University Press, 1999), esp. his chapter on "Schiller's Aesthetic State."
27. Egon Friedell, "Prolog vor dem Film," in Kaes, *Kino-Debatte*, 42–47, here 43. Originally published in *Blätter des Deutschen Theaters* 2 (1912): 509–511.
28. Hermann Kienzl, "Theater und Kinematograph," in Schweinitz, *Prolog vor dem Film*, 230–234, here 231. Originally published in *Der Strom* 1, no. 7 (October 1911): 219–221.
29. Walter Hasenclever, "Der Kintopp als Erzieher: Eine Apologie," in Kaes, *Kino-Debatte*, 47–49, here 48. Originally published in *Revolution* 1, no. 4 (1 December 1913): n.p.
30. Lou Andreas-Salomé, *In der Schule bei Freud: Tagebuch eines Jahres 1912/1913*, ed. Ernst Pfeiffer (Zürich: Niehans, 1958), 102–103.
31. Hermann Duenschmann, "Kinematograph und Psychologie der Volksmenge. Eine sozialpolitische Studie," *Konservative Monatsschrift* 69, no. 9 (June 1912): 920–930, here 924.
32. René Descartes, "The Search for Truth," *The Philosophical Writings of Descartes*, trans. John Cottingham, Robert Stoothoff, and Dugald Murdoch, vol. 2 (Cambridge and New York: Cambridge University Press, 1984–1991), 400–420, here 406.
33. John Locke, *An Essay Concerning Human Understanding*, ed. and with an introduction by Peter H. Nidditch (Oxford: Oxford University Press, 1975), 152, book 2, chap. 10, sect. 5.

34. Locke, *Essay*, 607, book 4, chap. 7, sect. 16.
35. Lex Newman, "Ideas, Pictures, and the Directness of Perception in Descartes and Locke," *Philosophy Compass* 4, no. 1 (2009): 134–154.
36. Ulrich Rauscher, "Die Kino-Ballade," in Güttinger, *Kein Tag ohne Kino*, 143–149, here 148. Originally published in *Der Kunstwart* 26, no. 13 (1 April 1913): 1–6.
37. Georges Duhamel, *Scènes de la vie future* (Paris: Mercure de France, 1930), 52, quoted in Walter Benjamin, "The Work of Art in the Age of Its Technological Reproducibility (Third Version)," in *Walter Benjamin: Selected Writings*, vol. 4, *1938–1940*, trans. Edmund Jephcott and others, ed. Howard Eiland and Michael W. Jennings (Cambridge, Mass.: Belknap, 2003), 267.
38. Georg Simmel, "The Metropolis and Mental Life," in *The Sociology of Georg Simmel*, ed. and trans. Kurt H. Wolff (New York: Free Press, 1950), 409–424; Sigmund Freud, *Beyond the Pleasure Principle*, trans. James Strachey (London: Hogarth, 1950).
39. Joseph Landau, "Mechanisierte Unsterblichkeit," in *Der Deutsche Kaiser im Film*, ed. Paul Klebinder (Berlin: Klebinder, 1912), 18–22, here 20.
40. Max Brod, "Kinematographentheater," in Kaes, *Kino-Debatte*, 39–41, here 41. Originally published in *Die neue Rundschau* 20, no. 2 (February 1909): 319–320.
41. "Neuland für Kinematographentheater," in Kaes, *Kino-Debatte*, 41. Originally published in *Lichtbild-Bühne* 3 (September 1910): 3.
42. Strobl, "Der Kinematograph," in *Kein Tag ohne Kino*, 51.
43. Juliet Koss describes the gentleman art historian in terms of a unified, stable subjectivity in "On the Limits of Empathy," *Art Bulletin* 88, no. 1 (March 2006): 139–157; see also Koss, *Modernism After Wagner* (Minneapolis: University of Minnesota Press, 2010).
44. Harry Francis Mallgrave and Eleftherios Ikonomou, "Introduction," *Empathy, Form, and Space*, 10–11. Herbart's aesthetics were widely influential in the nineteenth century (but not as influential as his pedagogical ideas, as we saw in the previous chapter). See his 1813 *Lehrbuch zur Einleitung in die Philosophie*, third edition (Königsberg: Unzer, 1834) or his 1831 *Kurze Encyklopädie der Philosophie* (Hamburg: Voss, 1884). Robert Zimmermann later expanded Herbartian aesthetics into an equally influential comprehensive system devoted to the "the science of form." See his *Allgemeine Aesthetik als Formwissenschaft* (Vienna: Braumüller, 1865).
45. Many of Schopenhauer's arguments in "Supplements to the Third Book" take into account questions of perception and physiology, esp. chap. 30, "On the Pure Subject of Knowing," and chap. 39, "On the Metaphysics of Music." Schopenhauer, *The World as Will and Representation*, vol. 2, pp. 367–375 and 447–457, respectively.
46. Looking for "laws" of beauty as a scientist might look for laws of nature, Fechner measured hundreds of paintings to find a statistical, scientific basis for the perfect format. See Gustav Fechner, *Vorschule der Aesthetik* (Leipzig: Breitkopf & Härtel, 1897 [1876]). Likewise, Wundt, the father of experimental psychology, tried to link pleasure in the perception of forms to the physiological structure of the eye by noting the ease with which the eye traced the contours of various forms.

See Wilhelm Wundt, *Grundzüge der physiologischen Psychologie*, 5th ed. (Leipzig: Engelmann, 1902–1903 [1874]), 1: 486ff. For a discussion of the implications of the work of these and other researchers for modern subjectivity, see Jonathan Crary, *Techniques of the Observer: On Vision and Modernity in the Nineteenth Century* (Cambridge, Mass.: MIT Press, 1990). For more on the relationship between the sciences, especially physiology, and aesthetics, see Robert Michael Brain, "The Pulse of Modernism: Experimental Physiology and Aesthetic Avant-Gardes Circa 1900," *Studies in History and Philosophy of Science* 39, no. 3 (2008): 393–417.

47. Mallgrave and Ikonomou, "Introduction," 2.
48. Vischer, *On the Optical Sense of Form*, 89–124. Hereafter cited parenthetically.
49. Gustav Jahoda, "Theodor Lipps and the Shift from 'Sympathy' to 'Empathy,'" *Journal of the History of the Behavioral Sciences* 4, no. 2 (2005): 151–163; and Susan Lanzoni, "Empathy in Translation: Movement and Image in the Psychological Laboratory," *Science in Context* 25, no. 3 (September 2012): 301–327.
50. The early "formal-analytic" approach to art, so important for the formation of the discipline of art history, is exemplified by Conrad Fiedler, *Über die Berurtheilung von Werken der bildenden Kunst* (Leipzig: Hirzel, 1876). Translated by Henry Schaefer-Simmern and Fulmer Mood as *On Judging Works of Visual Art* (Berkeley: University of California Press, 1949).
51. The *Zeitschrift für Ästhetik und allgemeine Kunstwissenschaft* (Journal for Aesthetics and Art History), which began in 1906 under the editorship of Max Dessoir, was the leading forum in Germany for discussions of *Einfühlung* and its implications for aesthetics and reception.
52. These questions were not limited to Germany, as many ideas and approaches spread to the United Kingdom, France, and the United States during the nineteenth century. Representative English-language examples of psychological aesthetics include Herbert Spencer, *The Principles of Psychology*, 2 vols. (London: Williams and Norgate, 1855); Grant Allen, *Physiological Aesthetics* (London: King, 1877); James Sully, "Pleasure of Visual Form," *Mind: A Quarterly Review of Psychology and Philosophy* 5, no. 18 (April 1880): 181–201; and Vernon Lee and C. Anstruther-Thomson, *Beauty and Ugliness and Other Studies in Psychological Aesthetics* (London and New York: Lane, 1912). On Lee, see esp. Hilary Fraser, "Women and the Ends of Art History: Vision and Corporeality in Nineteenth-Century Critical Discourse," *Victorian Studies* 42, no. 1 (Autumn 1998): 77–100. Applications of these ideas to film include Hugo Münsterberg, *The Photoplay: A Psychological Study* (London and New York: Appleton, 1916); Victor Oscar Freeburg, *Pictorial Beauty on the Screen* (New York: Macmillan, 1923); and Frances Taylor Patterson, *Scenario and Screen* (New York: Harcourt, Brace, 1928). Contemporary analyses of this trend in English-language (film) aesthetics include Laura Marcus, *The Tenth Muse: Writing About Cinema in the Modernist Period* (Oxford and New York: Oxford University Press, 2007); Lynda Nead, *The Haunted Gallery: Painting, Photography, Film c. 1900* (New Haven and London: Yale University Press, 2007); and Kaveh Askari, *Making Movies into Art: Picture Craft from the Magic Lantern to Early Hollywood* (London: British Film Institute, 2014), esp. chap. 3.

53. Joseph Imorde, "Einfühlung in der Kunstgeschichte," in *Einfühlung. Zur Geschichte und Gegenwart eines ästhetischen Konzepts*, ed. Robin Curtis and Gertrud Koch (Paderborn, Germany: Fink, 2009), 127–142.
54. In addition to my own essay on "Einfühlung und frühe deutsche Filmtheorie," in Curtis and Koch, *Einfühlung. Zur Geschichte und Gegenwart eines ästhetischen Konzepts*, 61–84, see Robin Curtis, "Einfühlung and Abstraction in the Moving Image: Historical and Contemporary Reflections," *Science in Context* 25, no. 3 (September 2012): 425–446; and Robert Michael Brain, "Self-Projection: Hugo Münsterberg on Empathy and Oscillation in Cinema Spectatorship," in the same issue, pp. 329–353.
55. Even when one of the preeminent names in aesthetic and *Einfühlung* theory, Max Dessoir, was given the opportunity to discuss the phenomenon of film from a theoretical standpoint, he chose, rather uninterestingly, to extol the virtues of words for literature and denounce the lack of words in silent film. See "Kino und Buchhandel," in Schweinitz, *Prolog vor dem Film*, 284–285.
56. Walter von Molo, "Im Kino," in Schweinitz, *Prolog vor dem Film*, 28–39, here 31. Originally published in *Velhagen & Klasings Monatshefte* 26, no. 8 (April 1912): 618–627.
57. Alfred Polgar, "Das Drama im Kinematographen," in *Kein Tag ohne Kino*, 56–61, here 60. Originally published in *Der Strom* 1, no. 2 (May 1911).
58. Vivian Sobchack, *The Address of the Eye: A Phenomenology of Film Experience* (Princeton, N.J.: Princeton University Press, 1992).
59. Hildebrand, *The Problem of Form in the Fine Arts*, 261.
60. Polgar, "Das Drama im Kinematographen," 59.
61. Mallgrave and Ikonomou, "Introduction," 23.
62. Ernst Bloch, "Die Melodie im Kino oder immanente und transzendentale Musik," in Schweinitz, *Prolog vor dem Film*, 326–334, here 328–329. Originally published in *Die Argonauten* 1, no. 2 (1914): 84–85. This translation, while largely my own, borrows some phrases from Ernst Bloch, "On Music in the Cinema," in *Literary Essays*, trans. Andrew Joron and others (Stanford, Calif.: Stanford University Press, 1998), 157–158.
63. See also Hildebrand, *The Problem of Form in the Fine Arts*, 229. Hildebrand further equates *Schauen* with *Abtasten* ("probing" or "touching").
64. See Hildebrand, *The Problem of Form in the Fine Arts*, 261, for a concurring opinion: "What we simply call the life of nature is actually the animation of nature through the imagination."
65. August Schmarsow, "The Essence of Architectural Creation" (1893), in Mallgrave and Ikonomou, *Empathy, Form, and Space*, 281–297, here 287 (emphasis in original). Hereafter cited parenthetically.
66. For a survey of his output, see Niels W. Bokhove and Karl Schuhmann, "Bibliographie der Schriften von Theodor Lipps," *Zeitschrift für philosophische Forschung* 45, no. 1 (January–March 1991): 112–130.
67. Theodor Lipps, "Einfühlung und ästhetischer Genuß," *Die Zukunft* 54, no. 14 (20 January 1906): 100–114, here 101.

68. Lipps, "Einfühlung und ästhetischer Genuß," 103.
69. Theodor Lipps, "Einfühlung, innere Nachahmung, und Organempfindungen," *Archiv für die gesamte Psychologie* 1, nos. 2/3 (1903): 185–204, here 201, 202, 203. Translated as "Empathy, Inner Imitation, and Sense-Feeling," in *A Modern Book of Aesthetics*, ed. Melvin M. Rader (New York: Holt, 1935), 291–304, here 302–303 (translation modified).
70. Theodor Lipps, *Ästhetik. Psychologie des Schönen und der Kunst. Erster Teil: Grundlegung der Ästhetik* (Hamburg and Leipzig: Voss, 1903), 148.
71. Other writers on *Einfühlung* explored the relation between empathy and bodily response, however tentatively. See Karl Groos, "Das ästhetische Miterleben und die Empfindungen aus dem Körperinnern," *Zeitschrift für Ästhetik und allgemeine Kunstwissenschaft* 4 (1909): 161–182; Vernon Lee, "Weiteres über Einfühlung und ästhetisches Miterleben," *Zeitschrift für Ästhetik und allgemeine Kunstwissenschaft* 5 (1910): 145–190; and later, Johannes Volkelt, *System der Ästhetik, Erster Band: Grundlegung der Ästhetik*, 2d ed. (Munich: Beck, 1927), esp. 186–201, in which he discusses motor sensations as way of facilitating *Einfühlung*. See also Christian G. Allesch, *Geschichte der psychologischen Ästhetik* (Göttingen: Verlag für Psychologie, 1987) for a complete discussion of this topic.
72. *Schaulust* has been translated as "scopophilia" by Freud's translators, but this rendering gives it a clinical connotation that neither Freud nor Serner intended. It literally means "the desire to look" or "the sexual pleasure in looking," but "voyeurism" tends to narrow its meaning as well. Not having an adequate English word at hand, I will simply refer to the concept in German. (On the inadequacies of the standard translation of Freud, see Bruno Bettelheim, *Freud and Man's Soul* [New York: Knopf, 1983]).
73. Walter Serner, "Kino und Schaulust," in Schweinitz, *Prolog vor dem Film*, 208–214, here 208. Originally published in *Die Schaubühne* 9, nos. 34/35 (1913): 807–811.
74. Serner, "Kino und Schaulust," 210.
75. Serner, "Kino und Schaulust," 211.
76. Tom Gunning, "The Cinema of Attractions: Early Film, Its Spectator and the Avant-Garde," in *Early Cinema: Space, Frame, Narrative*, ed. Thomas Elsaesser with Adam Barker (London: British Film Institute, 1990), 56–62.
77. Lipps, "Einfühlung, innere Nachahmung, und Organempfindungen," 195 (emphasis in original). Translated as "Empathy, Inner Imitation, and Sense-Feeling," in Rader, *A Modern Book of Aesthetics*, 300–301 (translation modified).
78. Lipps, "Einfühlung und ästhetischer Genuß," 113.
79. Charles Musser, "A Cinema of Contemplation, A Cinema of Discernment: Spectatorship, Intertextuality and Attractions in the 1890s," in *The Cinema of Attractions Reloaded*, ed. Wanda Strauven (Amsterdam: Amsterdam University Press, 2006), 159–179.
80. Miriam Hansen, "Early Silent Cinema: Whose Public Sphere?," *New German Critique* 29 (Spring/Summer 1983): 147–184; Heide Schlüpmann, *Unheimlichkeit des Blicks: Das Drama des frühen deutschen Kinos* (Frankfurt: Stroemfeld/Roter Stern, 1990).

81. Diderot, *Salons*, III, ed. Jean Seznec and Jean Adhémar (Oxford, 1963), 134–135, quoted in Michael Fried, *Absorption and Theatricality: Painting and Beholder in the Age of Diderot* (Chicago: University of Chicago Press, 1980), 125.
82. Richard Huelsenbeck, *Wozu Dada. Texte 1916–1936* (Giessen: Anabas, 1994), 35, quoted in David C. Durst, *Weimar Modernism: Philosophy, Politics, and Culture in Germany 1918–1933* (Lanham, Md.: Lexington, 2004), 48. Durst's book was instrumental in crafting my argument about the politics of contemplation.
83. Siegfried Kracauer, "Cult of Distraction: On Berlin's Picture Palaces," in *The Mass Ornament: Weimar Essays*, trans. and ed. Thomas Y. Levin (Cambridge, Mass.: Harvard University Press, 1995), 323–328. Hereafter cited parenthetically.
84. Walter Benjamin, "Surrealism: The Last Snapshot of the European Intelligensia," in *Walter Benjamin: Selected Writings*, vol. 2, *1927–1934*, ed. Michael W. Jennings, Howard Eiland, and Gary Smith, trans. Rodney Livingstone and others (Cambridge, Mass.: Belknap, 1999), 207–221, here 213.
85. Benjamin, "The Work of Art in the Age of Its Technological Reproducibility (Third Version)"; and Benjamin, "The Work of Art in the Age of Its Technological Reproducibility (Second Version)," 101–136.
86. Benjamin, "The Work of Art in the Age of Its Technological Reproducibility (Second Version)," 119.
87. Benjamin, "Theory of Distraction," in *The Work of Art in the Age of Its Technological Reproducibility, and Other Writings on Media*, ed. Michael W. Jennings, Brigid Doherty, and Thomas Y. Levin, trans. Edmund Jephcott, Rodney Livingstone, Howard Eiland, and others (Cambridge, Mass.: Belknap, 2008), 56–57 (emphasis added).
88. Of the many commentaries on Benjamin and Kracauer on distraction, see esp. Miriam Bratu Hansen, *Cinema and Experience: Siegfried Kracauer, Walter Benjamin, and Theodor W. Adorno* (Berkeley: University of California Press, 2012); and Paul North, *The Problem of Distraction* (Stanford, Calif.: Stanford University Press, 2012).
89. The standard postwar discussion of this topic is Hannah Arendt, *The Human Condition* (Chicago: University of Chicago Press, 1958).
90. Georg Lukács, *History and Class Consciousness*, trans. Rodney Livingstone (Cambridge, Mass.: MIT Press, 1971), 319.
91. Lukács, "Preface to the New Edition [1967]," *History and Class Consciousness*, xviii.
92. Georg Lukács, *The Theory of the Novel*, trans. Anna Bostock (Cambridge, Mass.: MIT Press, 1971), 135.
93. Lukács, *The Theory of the Novel*, 118.
94. Georg Lukács, "Gedanken zu einer Aesthetik des 'Kino,'" *Frankfurter Zeitung* 251 (10 September 1913): 1–2. There is also an earlier version, "Gedanken zu einer Aesthetik des 'Kino,'" which appeared in the German-Hungarian journal *Pester Lloyd* (16 April 1911): 44–46. I will be using Janelle Blankenship's excellent translation, "Thoughts Toward an Aesthetic of the Cinema," *Polygraph* 13 (2001): 13–18. Hereafter cited parenthetically.

95. See especially Tom Levin, "From Dialectical to Normative Specificity: Reading Lukács on Film," *New German Critique* 40 (Winter 1987): 35–61; and Janelle Blankenship, "Futurist Fantasies: Lukács's Early Essay 'Thoughts Toward an Aesthetic of the Cinema,'" *Polygraph* 13 (2001): 21–36.
96. Janelle Blankenship notes that this formulation recalls Bergson's concept of *durée*, which might have decisively shaped Lukács's later work. Blankenship "Futurist Fantasies," 22.
97. Levin, "From Dialectical to Normative Specificity," 35–61.
98. We must here note that this designation of cinema—or any form for that matter—as utopian is very provisional in Lukács's work. Lukács's early work sometimes endorsed the possibility of a glimpse in art of unalienated, authentic life and, at other times, foreclosed that possibility, depending on his mode of analysis, metaphysical or historical. See György Márkus, "Life and the Soul: The Young Lukács and the Problem of Culture," in *Lukács Revalued*, ed. Agnes Heller (London: Blackwell, 1983), 1–26.
99. Franz Pfemfert, "Kino als Erzieher," in Schweinitz, *Prolog*, 165–169. Originally published in *Die Aktion* 1, no. 18 (19 June 1911): 560–563.
100. Kurt Pinthus, *Das Kinobuch* (Frankfurt: Fischer Taschenbuch, 1983), 27.
101. Benjamin, "Garlanded Entrance: On the 'Sound Nerves' Exhibition at the Gesundheitshaus Kreuzberg," in Jennings, Doherty, and Levin, eds., *The Work of Art in the Age of Its Technological Reproducibility, and Other Writings on Media*, 60–66, here 62.
102. Duttlinger, "Between Contemplation and Distraction," 51.
103. Jocelyn Szczepaniak-Gillece, "Machines for Seeing: Cinema, Architecture, and Mid-century American Spectatorship," PhD diss. (Northwestern University, 2013).

CONCLUSION

1. For a history of the *Kulturfilm*, see Oskar Kalbus, *Pionere des Kulturfilms: Ein Beitrag zur Geschichte des Kulturfilmschaffens in Deutschland* (Karlsruhe: Neue Verlags-Gesellschaft, 1956).
2. For a clear explanation of *dispositif* as a historiographical concept, see Frank Kessler, "La cinématographie comme dispositif (du) spectaculaire," *CiNéMAS* 14, no. 1 (2003): 21–34; and "The Cinema of Attractions as *Dispositif*," in *The Cinema of Attractions Reloaded*, ed. Wanda Strauven (Amsterdam: Amsterdam University Press, 2006), 57–69.
3. Tom Gunning, "Systematizing the Electric Message: Narrative Form, Gender, and Modernity in *The Lonedale Operator*," in *American Cinema's Transitional Era: Audiences, Institutions, Practices*, ed. Charlie Keil and Shelley Stamp (Berkeley: University of California Press, 2004), 15–50. See also Tom Gunning, "Film History and Film Analysis: The Individual Film in the Course of Time," *Wide Angle* 12, no. 3 (July 1990): 4–19.

4. Ludwik Fleck, *Genesis and Development of a Scientific Fact*, ed. Thaddeus J. Trenn and Robert K. Merton, trans. Fred Bradley and Thaddeus J. Trenn (Chicago: University of Chicago Press, 1979); and *Cognition and Fact: Materials on Ludwik Fleck*, ed. Robert S. Cohen and Thomas Schnelle (Dordrecht, Netherlands, and Boston: Reidel, 1986).
5. Nicolas Rasmussen follows the development of a representational technology and its disciplinary conventions in *Picture Control: The Electron Microscope and the Transformation of Biology in America, 1940–1960* (Stanford, Calif.: Stanford University Press, 1997).
6. See my "Vergrösserung und das mikroskopische Erhabene," *Zeitschrift für Medienwissenschaft* 5 (2011): 96–110. See also Hannah Landecker, "Cellular Features: Microcinematography and Film Theory," *Critical Inquiry* 31, no. 4 (2005): 903–937.
7. Hannah Landecker, "Creeping, Drinking, Dying: The Cinematic Portal and the Microscopic World of the Twentieth-Century Cell," *Science in Context* 24, no. 3 (September 2011): 381–416.
8. Hans-Jörg Rheinberger, *Toward a History of Epistemic Things: Synthesizing Proteins in the Test Tube* (Stanford, Calif.: Stanford University Press, 1997).
9. These ideas about linked solutions and mutant genes I owe to George Kubler, *The Shape of Time: Remarks on the History of Things* (New Haven: Yale University Press, 1962).
10. A particularly compelling statement of this dynamic is in Andreas Gailus, "Of Beautiful and Dismembered Bodies: Art as Social Discipline in Schiller's *On the Aesthetic Education of Man*," in *Impure Reason: Dialectic of Enlightenment in Germany*, ed. W. Daniel Wilson and Robert C. Holub (Detroit, Mich.: Wayne State University Press, 1993), 146–165, here 158–159.

BIBLIOGRAPHY

PRIMARY SOURCES

Ackerknecht, Erwin. *Das Lichtspiel im Dienste der Bildungspflege: Handbuch für Lichtspielreformer*. Berlin: Weidmannsche, 1918.
Allen, Grant. *Physiological Aesthetics*. London: King, 1877.
Altenloh, Emilie. *Zur Soziologie des Kino: Die Kino-Unternehmung und die sozialen Schichten ihrer Besucher*. Jena: Diederichs, 1914.
American Society of Heating and Ventilation Engineers. "Report of Committee on Standards for Ventilation Legislation for Motion Picture Show Places." *Gesundheits-Ingenieur* 36, no. 22 (31 May 1913): 409–410.
Andreas-Salomé, Lou. *In der Schule bei Freud: Tagebuch eines Jahres 1912/1913*. Edited by Ernst Pfeiffer. Zürich: Niehans, 1958.
"An unsere Abonnenten!" *Film und Lichtbild* 2, no. 5 (1913): 84.
"A Surgical Showman." *British Medical Journal* (19 January 1907): 163.
Beard, George M. *American Nervousness: Its Causes and Consequences*. New York: Putnam, 1881.
Beaunis, [Henri-Étienne]. "L'expérimentation en psychologie par le somnambulisme provoqué." *Revue philosophique* 10, no. 7 (1885): 2.

Bénard, Henri. "Formation de centres de giration à l'arrière d'un obstacle en movement." *Comptes rendus de l'Académie des Sciences* 147 (1908): 839–842.

———. "Les tourbillons cellulaires dans une nappe liquide. II. Procédés mécaniques et optiques d'examens, lois numériques des phénomènes." *Revue générale des sciences pures et appliquées* 11 (1900): 1309–1328.

Benjamin, Walter. "Garlanded Entrance: On the 'Sound Nerves' Exhibition at the Gesundheitshaus Kreuzberg." In *The Work of Art in the Age of Its Technological Reproducibility, and Other Writings on Media*, ed. Michael W. Jennings, Brigid Doherty, and Thomas Y. Levin, trans. Edmund Jephcott, Rodney Livingstone, Howard Eiland, and others, 60–66. Cambridge, Mass.: Belknap, 2008.

———. "Little History of Photography." In *Walter Benjamin: Selected Writings*, vol. 2, *1927–1934*, ed. Michael W. Jennings, Howard Eiland, and Gary Smith, trans. Rodney Livingstone and others, 507–530. Cambridge, Mass.: Belknap, 1999.

———. "On Some Motifs in Baudelaire." In *Walter Benjamin: Selected Writings*, vol. 4, *1938–1940*, ed. Howard Eiland and Michael W. Jennings, trans. Edmund Jephcott and others, 313–355. Cambridge, Mass.: Belknap, 2003.

———. "Surrealism: The Last Snapshot of the European Intelligensia." In *Walter Benjamin: Selected Writings*, vol. 2, *1927–1934*, ed. Michael W. Jennings, Howard Eiland, and Gary Smith, trans. Rodney Livingstone and others, 207–221. Cambridge, Mass.: Belknap, 1999.

———. "Theory of Distraction." In *The Work of Art in the Age of Its Technological Reproducibility, and Other Writings on Media*, ed. Michael W. Jennings, Brigid Doherty, and Thomas Y. Levin, trans. Edmund Jephcott, Rodney Livingstone, Howard Eiland, and others, 56–57. Cambridge, Mass.: Belknap, 2008.

———. "The Work of Art in the Age of Its Technological Reproducibility (Second Version)." In *Walter Benjamin: Selected Writings*, vol. 3, *1935–1938*, ed. Howard Eiland and Michael W. Jennings, trans. Edmund Jephcott, Howard Eiland, and others, 101–133. Cambridge, Mass.: Belknap, 2002.

———. "The Work of Art in the Age of Its Technological Reproducibility (Third Version)." In *Walter Benjamin: Selected Writings*, vol. 4, *1938–1940*, ed. Howard Eiland and Michael W. Jennings, trans. Edmund Jephcott and Others, 251–283. Cambridge, Mass.: Belknap, 2003.

Bergmann, Franz. "Der Kinematograph im Hochschulunterricht." *Bild und Film* 2, no. 2 (1912/1913): 48.

Bergson, Henri. *Creative Evolution*. Translated by Arthur Mitchell. New York: Random House, 1944.

———. *An Introduction to Metaphysics*. Translated by T. E. Hulme. New York: Macmillan, 1955.

———. *Matter and Memory*. Translated by Nancy Margaret Paul and W. Scott Palmer. Cambridge, Mass.: Zone, 1988.

"Bericht über eine Besprechung der Kinokommssion des Westfälischen Landgemeindetages anläßlich der Eröffnungsfeier des Gemeindelichtspielhauses in Eickel." *Bild und Film* 2, no. 3 (1912): 70–71.

"Besuch kinematographischer Vorführung durch Schüler höherer Lehranstalten (Breslau 1910)" and "Besuch der Kinematographentheater durch Schüler und Schülerinnen sowie durch die Zöglinge der Seminare und Präparandenanstalten (Berlin 1912)." In *Dokumente zur Geschichte der Schulfilmbewegung in Deutschland*, ed. Fritz Terveen, 16–17. Emsdetten: Lechte, 1959.

Billroth, Theodor. *The Medical Sciences in the German Universities: A Study in the History of Civilization*. Translated by William H. Welch. New York: Macmillan, 1924. Originally published as *Über das Lehren und Lernen de medicinischen Wissenschaften an den Universitäten der deutschen Nation nebst allgemeinen Bemerkungen über Universitäten; eine culturhistorische Studie* (Vienna: Gerold, 1876).

Bloch, Ernst. "Die Melodie im Kino oder immanente und transzendentale Musik." In *Prolog vor dem Film: Nachdenken über ein neues Medium, 1909–1914*, ed. Jörg Schweinitz, 326–334. Leipzig: Reclam, 1992. Originally published in *Die Argonauten* 1, no. 2 (1914): 84–85.

———. "On Music in the Cinema." In *Literary Essays*, trans. Andrew Joron and others, 157–158. Stanford, Calif.: Stanford University Press, 1998.

Braun, Ludwig. *Über Herzbewegung und Herzstoss*. Jena: Fischer, 1898.

———. "Ueber den Werth des Kinematographen für die Erkenntniss der Herzmechanik." *Verhandlungen der Gesellschaft deutscher Naturforscher und Ärzte* 69, part 1 (1898): 185–186.

Braune, Wilhelm, and Otto Fischer. "Betrachtungen über die weiteren Ziele der Untersuchung und Überblick über die Bewegungen der unteren Extremitäten." *Abhandlungen der Mathematisch-Physischen Klasse der Königlich Sächsischen Gesellschaft der Wissenschaften* 26, no. 3 (1900): 85–170.

———. *Das Gesetz der Bewegungen an der Basis der mittleren Finger und im Handgelenk des Menschen*. Leipzig, 1887.

———. *Determination of the Moments of Inertia of the Human Body and Its Limbs*. Translated by Paul Macquet and Ronald Furlong. Berlin and New York: Springer, 1988. Originally published as *Bestimmung der Trägheitsmomente des menschlichen Koerpers und seiner Glieder*. Leipzig, 1892.

———. "Die Bewegung des Gesamtschwerpunktes und die äußeren Kräfte." *Abhandlungen der Mathematisch-Physischen Klasse der Königlich Sächsischen Gesellschaft der Wissenschaften* 25, no. 1 (1899): 1–130.

———. "Die Kinematik des Beinschwingens." *Abhandlungen der Mathematisch-Physischen Klasse der Königlich Sächsischen Gesellschaft der Wissenschaften* 28, no. 5 (1903): 319–418.

———. *The Human Gait*, trans. Paul Maquet and Ronald Furlong. Berlin and New York: Springer, 1987.

———. *On the Centre of Gravity of the Human Body as Related to the Equipment of the German Infantry Soldier*. Translated by Paul Macquet and Ronald Furlong. Berlin and New York: Springer, 1985. Originally published as *Über den Schwerpunkt des menschlichen Körpers mit Rücksicht auf die Ausrüstung des deutschen Infanteristen*. Leipzig, 1889.

———. "Über den Einfluß der Schwere und der Muskeln auf die Schwingungsbewegung des Beins." *Abhandlungen der Mathematisch-Physischen Klasse der Königlich Sächsischen Gesellschaft der Wissenschaften* 28, no. 7 (1904): 531–617.

———. "Über die Bewegung des Fußes und die auf denselben einwirkenden Kräfte." *Abhandlungen der Mathematisch-Physischen Klasse der Königlich Sächsischen Gesellschaft der Wissenschaften* 26, no. 7 (1901): 467–556.

———. "Versuche am unbelasteten und belasteten Menschen." *Abhandlungen der Mathematisch-Physischen Klasse der Königlich Sächsischen Gesellschaft der Wissenschaften* 21, no. 4 (1895): 151–322.

Brauner, Ludwig. "Die Kino-Ausstellung in Berlin." *Der Kinematograph* no. 104 (25 December 1908).

Braus, Hermann. "Experimentelle Beiträge zur Frage nach der Entwickelung peripherer Nerven." *Anatomischer Anzeiger* 26, nos. 17/18 (1 April 1905): 433–479.

———. "Mikro-Kino-Projektionen von in vitro gezüchteten Organanlagen." *Verhandlungen der Gesellschaft deutscher Naturforscher und Ärzte* 83, part 2 (1911): 472–475.

———. "Mikro-Kino-Projektionen von in vitro gezüchteten Organanlagen." *Wiener medizinische Wochenschrift* 61, no. 44 (1911): 2809–2812.

Bredtmann, Hermann. "Kinematographie und Schule." *Pädagogisches Archiv* 56, no. 3 (1914): 154–163.

Brod, Max. "Kinematographentheater." In *Kino-Debatte: Texte zum Verhältnis von Literatur und Film 1909–1929*, ed. Anton Kaes, 39–41. Tübingen: Niemeyer, 1978. Originally published in *Die neue Rundschau* 20, no. 2 (February 1909): 319–320.

Bruegel, Carl. "Bewegungsvorgänge am pathologischen Magen auf Grund röngenkinematographischer Untersuchungen." *Münchener medizinische Wochenschfrift* 60, no. 4 (28 January 1913): 179–181.

Brunner, Karl. *Der Kinematograph von heute—eine Volksgefahr*. Berlin: Vaterländischen Schriftenverbandes, 1913.

Carr, W. P. "Suggestion as Used and Misused in Curing Disease." In *Hypnotism and Hypnotic Suggestion*, ed. E. Virgil Neal and Charles S. Clark, 5–17. Rochester: New York State Publishing, 1900.

Carrel, Alexis, and Montrose T. Burrows. "Cultivation of Adult Tissues and Organs Outside of the Body." *Journal of the American Medical Association* 55, no. 16 (15 October 1910): 1379–1381.

———. "Cultivation of Sarcoma Outside of the Body: A Second Note." *Journal of the American Medical Association* 55, no. 18 (29 October 1910): 1554.

———. "Cultivation of Tissues in Vitro and Its Technique." *Journal of Experimental Medicine* 13, no. 3 (1 March 1911) 387–396.

———. "Human Sarcoma Cultivated Outside of the Body: A Third Note." *Journal of the American Medical Association* 55, no. 20 (12 November 1910): 1732.

Chase, Walter Greenough. "The Use of the Biograph in Medicine." *Boston Medical and Surgical Journal* 153, no. 21 (23 November 1905): 571–572.

Chevroton, L., and F. Vlès. "La cinématique de la segmentation de l'oeuf et la chronophotographie du développement de l'Oursin." *Comptes rendus hebdomadaires des séances de l'Academie des Sciences* 149 (8 November 1909): 806–809.

Christeller, Erwin. "Die Bedeutung der Photographie für den pathologisch-anatomischen Unterricht und die pathologisch-anatomische Forschung." *Berliner klinische Wochenschrift* 55, no. 17 (29 April 1918): 399–401.

Coldstream, G. P. "The Cinematoscope as an Aid in Teaching." *British Medical Journal* (3 September 1898): 658.

Cole, Lewis Gregory. "The Gastric Motor Phenomena Demonstrated with the Projecting Kinetoscope." *American Quarterly of Roentgenology* 3, no. 4 (1912): 1–11.

Comandon, Jean with Albert Dastre. "Cinématographie, à l'ultra-microscope, de microbes vivants et des particules mobiles." *Comptes rendus hebdomadaires des séances de l'Académie des Sciences* 149 (22 November 1909): 938–941.

"Conference on Hospital Standardization." *Bulletin of the American College of Surgeons* 3, no. 1 (1917).

Cotton, Aimé. "Recherches récentes sur les mouvements browniens." *La Revue du Mois* 5 (10 June 1908): 737–741.

Cranz, Carl. "Über einen ballistischen Kinematographen." *Deutscher Mechaniker-Zeitung* 18 (15 September 1909): 173–177.

Cros, Charles. "Inscription." In *Oeuvres completes*, ed. Louis Forestier and Pascal Pia, 135–136. Paris: Pauvert, 1964.

d'Abundo, Giuseppe. "Sopra alcuni particolari effetti delle projezioni cinematografiche nei nevrotici." *Rivista Italiana di Neuropatologia, Psichiatria ed Elettroterapia* 4, no. 10 (October 1911): 433–442.

Dannmeyer, C. H. *Bericht der Kommission für "Lebende Photographien."* Hamburg: Kampen, 1907.

Descartes, René. "The Search for Truth." *The Philosophical Writings of Descartes*, vol. 2. Translated by John Cottingham, Robert Stoothoff, and Dugald Murdoch, 400–420. Cambridge and New York: Cambridge University Press, 1984.

Dessauer, Friedrich. *Die neuesten Fortschritte in der Röntgenphotographie*. Leipzig: Nemnich, 1912.

Dewey, John. *Art as Experience*. New York: Putnam, 1984.

"Die Dresdner 'Kosmographia.'" *Bild und Film* 1, no. 1 (1912): 19.

"Die Eröffnung des Reform-Kinematographentheater." *Der Kinematograph* no. 32 (7 August 1907).

"Die Kinematographie des Unsichtbaren." *Prometheus* 21, no. 1054 (5 January 1910): 218–220.

"Die Kino-Austellung und 'Wir.'" *Erste Internationale Film-Zeitung* 6, no. 50 (14 December 1912): 52.

"Die Kultur-Arbeit des Kinematographen-Theaters." *Die Lichtbild-Bühne* 2, no. 41 (4 February 1909).

Diem, Ulrich. *Bildbetrachtung: Eine Wegleitung für Kunstfreunde*. St. Gallen: Fehr'sche, 1919.

Doyen, Eugène. "Le cinématograph et l'enseignement de la chirurgie." *Revue critique de médecine et de chirurgie* 1, no. 1 (15 August 1899): 1–6. Partially translated as "The Cinematograph and the Teaching of Surgery," *British Gynæcological Journal* 15 (1899): 579–586.

Doyen, Eugène Louis. *La cas des xiphopages hindoues Radica et Doodica*. Paris: Bourse de Commerce, 1904.

———. *L'enseignement de la technique opératoire par les projections animées*. Paris: Société générale des cinématographes Eclipse, ca. 1911.

Du Bois-Reymond, Emil. *Untersuchungen über thierische Elektricität*. Berlin: Reimer, 1848–1884.
Duenschmann, Hermann. "Kinematograph und Psychologie der Volksmenge. Eine sozialpolitische Studie." *Konservative Monatsschrift* 69, no. 9 (June 1912): 920–930.
"Ein kurzer Rückblick auf die erste Woche des Reform-Kinematographen-Theaters." *Der Kinematograph* no. 33 (14 August 1907).
"Ein neues wissenschaftliches Kino." *Film und Lichtbild* 1, no. 5 (1912): 62.
Einstein, Albert. "On the Movement of Small Particles Suspended in a Stationary Liquid Demanded by the Molecular-Kinetic Theory of Heat." In *Investigations on the Theory of the Brownian Movement*, ed. R. Fürth, trans. A. D. Cowper, 1–18. New York: Dover, 1956.
Erb, Wilhelm Heinrich. *Ueber die wachsende Nervosität unserer Zeit*. Heidelberg: Universitäts Buchdruckerei von J. Hörning, 1893.
Exner, Felix M. "Notiz zu Brown's Molecularbewegung." *Annalen der Physik* 2, no. 8 (1900): 843–847.
Eykman, P. H. "Der Schlingact, dargestellt nach Bewegungsphotographien mittelst Röntgen-Strahlen." *Pflüger's Archiv für die gesammte Physiologie des Menschen und der Thiere* 99 (1903): 513–571.
Fechner, Gustav. *Vorschule der Aesthetik*. Leipzig: Breitkopf & Härtel, 1897.
Felke, Naldo. "Die Gesundheitsschädlichkeit des Kinos." *Die Umschau* 17, no. 1 (1 January 1913): 254–255.
Fiedler, Conrad. *Über die Berurtheilung von Werken der bildenden Kunst*. Leipzig: Hirzel, 1876. Translated by Henry Schaefer-Simmern and Fulmer Mood as *On Judging Works of Visual Art* (Berkeley: University of California Press, 1949).
Fischer, Otto. *Theoretische Grundlagen für eine Mechanik der lebenden Körper*. Leipzig, 1906.
Flemming, Carl F. *Pathologie und Therapie der Psychosen*. Berlin: Hirschwald, 1859.
Forel, Auguste. *Hygiene der Nerven und des Geistes im gesunden und kranken Zustande*. Stuttgart: Moritz, 1903.
François-Franck, Charles. "Démonstrations de microphotographie instantanée et de chronomicrophotographie." *Comptes rendus hebdomadaires des séances et mémoires de la Société de Biologie* 62 (25 May 1907): 964–967.
———. "Études graphiques et photographiques de mécanique respiratoire comparée." *Comptes rendus hebdomadaires des séances et mémoires de la Société de Biologie* 61 (28 July 1906): 174–176.
———. "La chronophotographie simultanée du coeur et des courbes cardiographiques chez les mammifères." *Comptes rendus hebdomadaires des séances et mémoires de la Société de Biologie* 54 (8 November 1902): 1193–1197.
———. "Note sur quelques points de technique relatifs à la photographie et à la chronophotographie avec le magnésium à deflagration lente." *Comptes rendus hebdomadaires des séances et mémoires de la Société de Biologie* 55 (5 December 1903): 1538–1540.
Fränkel, Albert. "Röntgendiagnosen und Röntgenfehldiagnosen beim Magenkarzinom; diagnostischer Fortschritt durch Röntgenkinographie." *Zentralblatt für Röntgenstrahlen, Radium und verwandte Gebiete* 3, no. 4 (1912): 149–50.

Fränkel, James. "Kinematographische Demonstration." *Verhandlungen der freien Vereinigung der Chirurgen Berlins* 20, part 1 (1907): 12–13.

——. "Kinematographische Untersuchung des normalen Ganges und einiger Gangstörungen." *Zeitschrift für orthopädische Chirurgie* 20 (1908): 617–646.

Freeburg, Victor Oscar. *Pictorial Beauty on the Screen*. New York: Macmillan, 1923.

Frei, Wilhelm. *Landerziehungsheime: Darstellung und Kritik einer modernen Reformschule*. Leipzig: Klinkhardt, 1902.

Freud, Sigmund. *Beyond the Pleasure Principle*. Translated by James Strachey. London: Hogarth, 1950.

Frey, J. "Report of the Photographic Department of Bellevue Hospital for the Year 1869." In *Tenth Annual Report of the Commissioners of Public Charities and Correction of the City of New York for the Year 1869*, 85. Albany: van Benthuysen, 1870. www.artandmedicine.com/ogm/1869.html.

Friedell, Egon. "Prolog vor dem Film." In *Kino-Debatte: Texte zum Verhältnis von Literatur und Film 1909–1929*, ed. Anton Kaes, 42–47. Tübingen: Niemeyer, 1978. Originally published in *Blätter des Deutschen Theaters* 2 (1912): 509–511.

Gaupp, Robert. "Das Pathologische in Kunst und Literatur." *Deutsche Revue* 36, no. 2 (April 1911): 11–23.

——. "Der Arzt als Erzieher seines Volkes." *Medizinisches Correspondenz-Blatt* 89, no. 32 (9 August 1919): 295–296.

——. "Der Kinematograph vom medizinischen und psychologischen Standpunkt." In *Der Kinematograph als Volkunterhaltungsmittel*, by Robert Gaupp and Konrad Lange, 1–12. Munich: Dürer-Bund-Flugschrift zur Ausdruckskultur 100, 1912.

——. "Die Gefahren des Kino." In *Prolog vor dem Film: Nachdenken über ein neues Medium, 1909–1914*, ed. Jörg Schweinitz, 64–69. Leipzig: Reclam, 1992. Originally published in *Süddeutsche Monatshefte* 9, no. 9 (1911/1912): 363–366.

——. "Die Nervosität unserer Zeit im Lichte der Wissenschaft." *Medizinisches Correspondenz-Blatt* 77, no. 31 (3 August 1907): 633–639.

——. *Psychologie des Kindes*. Leipzig: Teubner, 1910.

Gaupp, Robert, and Konrad Lange, *Der Kinematograph als Volksunterhaltungsmitte*. Munich: Callwey, 1912.

Geisel, Walter. *Wie ich mit meinen Jungens Kunstwerke betrachte*. Glückstadt: Geisel, 1904.

Georges-Michel, Michel. "Henri Bergson nous parle au cinéma." *Le Journal* (20 February 1914): 7. Translated by Louis-Georges Schwartz as "Henri Bergson Talks to Us About Cinema." *Cinema Journal* 50, no. 3 (Spring 2011): 79–82.

Gilbreth, Frank B. *Motion Study: A Method for Increasing the Efficiency of the Workman*. New York: Van Nostrand, 1911.

——. "Scientific Management in the Hospital." *Modern Hospital* 3 (1914): 321–324.

Goerke, Franz. "Proposal for Establishing an Archive for Moving Pictures (1912)." Translated by Cecilie L. French and Daniel J. Leab. *Historical Journal of Film, Radio and Television* 16, no. 1 (March 1996): 9–12. Originally published as "Vorschlag zur Einrichtung eines Archives für Kino-films," in *Der Deutsche Kaiser im Film: zum 25jährigen Regierungs-Jubiläum Seiner Majestät des Deutschen Kaisers Königs von Preußen Wilhelm II*, ed. Paul Klebinder, 63–68 (Berlin: Klebinder, 1912).

Götze, O. "Jugendpsyche und Kinematograph." *Zeitschrift für Kinderforschung* 16 (1911): 418.

Gouy, Louis-Georges. "Le mouvement brownien et les mouvements moléculaires." *Revue générale des sciences pures et appliquées* 6, no. 1 (15 January 1895): 1–7.

Graupner, H. "Unterrichtshygiene." In *Handbuch der deutschen Schulhygiene*, ed. Hugo Selter, 174–321. Dresden and Leipzig: Theodor Steinkopff, 1914.

Grempe, Max. "Gegen die Frauenverblödung im Kino." *Gleichheit* 23, no. 5 (1912): 70–72.

Groedel, Franz M. "Die Technik der Röntgenkinematographie." *Deutsche medizinische Wochenschrift* 35 (11 March 1909): 434–435.

———. "Die Technik der Röntgenkinematographie." *Deutsche medizinische Wochenschrift* 39 (6 February 1913): 270–271.

———. "Die Technik der Röntgenkinematographie." *Deutsche medizinische Wochenschrift* 39 (24 April 1913): 798–799.

———. "The Present State of Roentgen Cinematography and Its Results as to the Study of the Movements of the Inner Organs of the Human Body." *Interstate Medical Journal* 22 (March 1915): 281–290.

———. "Roentgen Cinematography and Its Importance in Medicine." *British Medical Journal* (24 April 1909): 1003.

Groos, Karl. "Das ästhetische Miterleben und die Empfindungen aus dem Körperinnern." *Zeitschrift für Ästhetik und allgemeine Kunstwissenschaft* 4 (1909): 161–182.

Günther, Hanns. "Mikrokinematographische Aufnahmeapparate." *Film and Lichtbild* 1, no. 1 (1912): 4–6; 1, no. 2 (1912): 13–14.

Häfker, Hermann. "Eine Reise an die Quellen der Kinematographie." *Der Kinematograph* no. 163 (9 February 1910); 172 (13 April 1910).

———. *Kino und Kunst*. M. Gladbach [Mönchengladbach]: Volksvereins, 1913.

———. "Meisterspiele." *Der Kinematograph* no. 56 (22 January 1908).

———. "Zur Dramaturgie der Bilderspiele." *Der Kinematograph* no. 32 (7 August 1907).

Harris, W. T. Editor's preface to *Herbart's ABC of Sense Perception and Minor Pedagogical Works*, by Johann Friedrich Herbart. Translated and edited by William J. Eckoff, vii–xi. New York: Appleton, 1896.

Harrison, Ross Granville. "Experiments in Transplanting Limbs and Their Bearing Upon the Problems of the Development of Nerves." *Journal of Experimental Zoology* 4, no. 2 (June 1907): 239–281.

———. "Further Experiments on the Development of Peripheral Nerves." *American Journal of Anatomy* 5, no. 2 (31 May 1906): 121–131.

———. "Observations on the Living Developing Nerve Fiber." *Anatomical Record* 1, no. 5 (1 June 1907): 116–118.

———. "The Outgrowth of the Nerve Fiber as a Mode of Protoplasmic Movement." *Journal of Experimental Zoology* 9, no. 4 (December 1910): 787–846.

Hasenclever, Walter. "Der Kintopp als Erzieher: Eine Apologie." In *Kino-Debatte: Texte zum Verhältnis von Literatur und Film 1909–1929*, ed. Anton Kaes, 47–49. Tübingen: Niemeyer, 1978. Originally published in *Revolution* 1, no. 4 (1 December 1913): n.p.

Hellwig, Albert. "Die Beziehungen zwischen Schundliteratur, Schundfilms und Verbrechen." *Archiv für Kriminal-Anthropologie und Kriminalistik* 51, no. 1 (24 January 1913): 1–32.

———. "Die maßgebenden Grundsätze für Verbote von Schundfilms nach geltendem und künstigem Rechte." *Verwaltungsarchiv* 21 (1913): 405–455.

———. *Die Filmzensur: Eine rechtsdogmatische und rechtpolitische Erörterung.* Berlin: Frankenstein, 1914.

———. "Hypnotismus und Kinematograph." *Zeitschrift für Psychotherapie und medizinische Psychologie* 6 (1916): 310–315.

———. *Kind und Kino.* Langensalza: Beyer, 1914.

———. "Kinematograph und Zeitgeschichte." In *Prolog vor dem Film: Nachdenken über ein neues Medium, 1909–1914,* ed. Jörg Schweinitz, 97–109. Leipzig: Reclam, 1992. Originally published in *Die Grenzboten* 72, no. 39 (1913): 612–620.

———. *Öffentliches Lichtspielrecht.* M. Gladbach [Mönchengladbach]: Volksvereins, 1921.

———. *Rechtsquellen des öffentlichen Kinematographenrechts.* M. Gladbach [Mönchengladbach]: Volksvereins, 1913.

———. *Schundfilms: Ihr Wesen, ihre Gefahren und ihre Bekämpfung.* Halle an der Saale, Germany: Waisenhaus, 1911.

———. "Über die schädliche Suggestivkraft kinematographischer Vorführung." *Aerztliche Sachverständigen-Zeitung* 20, no. 6 (15 March 1914): 122.

———. "Zur Psychologie kinematographischer Vorführungen." *Zeitschrift für Psychotherapie und medizinische Psychologie* 6 (1916): 88–120.

Helmholtz, Hermann von. *Über die Erhaltung der Kraft.* Berlin: Reimer, 1847.

Hennes, Hans. "Die Kinematographie der Bewegungsstörungen." *Die Umschau* 15, no. 29 (1911): 605–606.

———. "Die Kinematographie im Dienste der Neurologie und Psychiatrie, nebst Beschreibung einiger selteneren Bewegungsstörungen." *Medizinische Klinik* 6, no. 51 (18 December 1910): 2010–2014.

Henri, Victor. "Études cinématographique des mouvements browniens." *Comptes rendus hebdomadaires des séances de l'Academie des Sciences* 146 (18 May 1908): 1024–1026.

———. "Influence du milieu sur les mouvements browniens." *Comptes rendus hebdomadaires des séances de l'Academie des Sciences* 147 (6 July 1908): 62–65.

Herbart, Johann Friedrich. *Kurze Encyklopädie der Philosophie.* Hamburg: Voss, 1884.

———. *Lehrbuch zur Einleitung in die Philosophie,* third edition (Königsberg: Unzer, 1834).

Hildebrand, Adolf. *The Problem of Form in the Fine Arts* (1893). In *Empathy, Form, and Space: Problems in German Aesthetics, 1873–1893,* ed. and trans. Harry Francis Mallgrave and Eleftherios Ikonomou, 227–279. Santa Monica, Calif.: Getty Center for the History of Art and the Humanities, 1994.

Jacobj, Carl. "Anschauungsunterricht und Projektion." *Zeitschrift für wissenschaftliche Mikroskopie und mikroskopische Technik* 36, no. 4 (1919): 273–314.

Jendrassik, Ernst. "Klinische Beiträge zum Studium der normalen und pathologischen Gangarten." *Deutsche Archiv für klinische Medizin* 70 (1901): 81–132.

Kalbus, Oskar. "Abriß einer Geschichte der deutschen Lehrfilmbewegung." In *Das Kulturfilmbuch*, ed. Edgar Beyfuß and Alexander Kossowsky, 1–13. Berlin: Chryselius'scher, 1924.

———. *Der Deutsche Lehrfilm in der Wissenschaft und im Unterricht*. Berlin: Heymanns, 1922.

Kandel, I. L. "Germany." In *Comparative Education: Studies of the Educational Systems of Six Modern Nations*, ed. Peter Sandiford, 121–130. London and Toronto: Dent, 1918.

Kant, Immanuel. *Critique of Judgement*. Translated by James Creed Meredith. Oxford: Oxford University Press, 1952.

———. "What Is Enlightenment?" In *German Aesthetic and Literary Criticism*, ed. David Simpson, 29–34. Cambridge: Cambridge University Press, 1984.

Kästle, C., H. Rieder, and J. Rosenthal. "The Bioroentgenography of the Internal Organs." *Archives of the Roentgen Ray* 15, no. 1 (June 1910): 3–12.

———. "Ueber kinematographisch aufgenommene Röntgenogramme (Bio-Röntgenographie) der inneren Organe des Menschen." *Münchener medizinische Wochenschrift* 56, no. 6 (9 February 1909): 280–283.

Key, Ellen. "Erziehung," *Das Jahrhundert des Kindes* (Berlin, 1905). In *Die deutsche Reformpädagogik*, ed. Wilhelm Flitner and Gerhard Kudritzki, 52–54. Düsseldorf and Munich: Küpper, 1961.

Kienzl, Hermann. "Theater und Kinematograph." In *Prolog vor dem Film: Nachdenken über ein neues Medium, 1909–1914*, ed. Jörg Schweinitz, 230–234. Leipzig: Reclam, 1992. Originally published in *Der Strom* 1, no. 7 (October 1911): 219–221.

"Kinematograph als Krankheitsstifter." *Fortschritte der Medizin* 30 (1912): 302.

"Kinematographische Reformvereinigung." *Der Kinematograph* no. 43 (23 October 1907).

"Kino und Buchhandel." In *Prolog vor dem Film: Nachdenken über ein neues Medium, 1909–1914*, ed. Jörg Schweinitz, 272–289. Leipzig: Reclam, 1992.

"Kino und Wissenschaft." *Bild und Film* 1, no. 2 (1912): 55.

Kleibömer, Georg. "Kinematograph und Schuljugend." *Der Kinematograph* no. 124 (12 May 1909).

Knospe, Paul. *Der Kinematograph im Dienste der Schule. Unter besonderer Berücksichtigung des erdkundlichen Unterrichts*. Halle an der Saale, Germany: Waisenhaus, 1913.

Kracauer, Siegfried. "Cult of Distraction: On Berlin's Picture Palaces." In *The Mass Ornament: Weimar Essays*, trans. and ed. Thomas Y. Levin, 323–328. Cambridge, Mass.: Harvard University Press, 1995.

Kraepelin, Emil. "Demonstration von Kinematogrammen." *Centralblatt für Nervenheilkunde und Psychiatrie* 32 (1909): 689.

———. *Memoirs*. Edited by H. Hippius, G. Peters, and D. Ploog in collaboration with P. Hoff and A. Kreuter. Translated by Cheryl Wooding-Deane. Berlin and New York: Springer-Verlag, 1987.

———. "Zur Entartungsfrage." *Zentralblatt für Nervenheilkunde und Psychiatrie* 31 (1908): 745–751. Translated as "On the Question of Degeneration," *History of Psychiatry* 18, no. 3 (2007): 399–404.

Kretz, Richard. "Die Anwendung der Photographie in der Medicin." *Wiener klinische Wochenschrift* 7, no. 44 (1 November 1894): 832.

Kutner, Robert. "Die Bedeutung der Kinematographie für medizinische Forschung und Unterricht sowie für die volkshygienische Belehrung." *Zeitschrift für Ärztliche Fortbildung* 8, no. 8 (15 April 1911): 249–251.

Landau, Joseph. "Mechanisierte Unsterblichkeit." In *Der Deutsche Kaiser im Film. Zum 25jährigen Regierungs-Jubiläum Seiner Majestät des Deutschen Kaisers Königs von Preußen Wilhelm II*, ed. Paul Klebinder, 18–22. Berlin: Klebinder, 1912.

Langbehn, August Julius. "Rembrandt als Erzieher." In *Die Kunsterziehungsbewegung*, ed. Hermann Lorenzen, 7–17. Bad Heilbrunn, Germany: Klinkhardt, 1966.

Lange, Konrad. "Das Wesen der künstlerischen Erziehung." In *Die Kunsterziehungsbewegung*, ed. Hermann Lorenzen, 21–26. Bad Heilbrunn, Germany: Klinkhardt, 1966.

———. *Die künstlerische Erziehung der deutschen Jugend*. Darmstadt: Bergstraeßer, 1893.

Laquer, Leopold. "Über die Schädlichkeit kinematographischer Veranstaltungen für die Psyche des Kindesalters." *Aerztliche Sachverständigen-Zeitung* 27, no. 11 (1 June 1911): 221–222.

Laycock, Thomas. *Lectures on the Principles and Methods of Medical Observation and Research*. Philadelphia: Blanchard and Lea, 1857.

Le Bon, Gustave. *The Crowd: A Study of the Popular Mind*. Atlanta, Ga.: Cherokee, 1982. Originally published as *Psychologie des foules* (Paris: Alcan, 1895). Translated and published in German as *Psychologie der Massen* (Leipzig: Klinkhardt, 1908).

Learned, William S. *An American Teacher's Year in a Prussian Gymnasium*. New York: Educational Review, 1911.

Lecomte du Noüy, P[ierre]. *Biological Time*. With a foreword by Alexis Carrel. New York: Macmillan, 1937.

Lee, Vernon. "Weiteres über Einfühlung und ästhetisches Miterleben." *Zeitschrift für Ästhetik und allgemeine Kunstwissenschaft* 5 (1910): 145–190.

———, and C. Anstruther-Thomson. *Beauty and Ugliness and Other Studies in Psychological Aesthetics*. London and New York: Lane, 1912.

Lehmann, Hans. *Die Kinematographie: ihren Grundlagen und ihre Anwendungen*. Leipzig: Teubner, 1911.

Leipziger Lehrerverein, ed. *Bildbetrachtungen: Arbeiten aus der Abteilung für Kunstpflege des Leipziger Lehrervereins*. Leipzig: Teubner, 1906.

Lemke, Hermann. *Die Kinematographie der Gegenwart, Vergangenheit und Zukunft*. Leipzig: Demme, 1911.

———. "Die kinematographische Reformpartei, ihre Aufgaben und Ziele." *Der Kinematograph* no. 42 (16 October 1907).

———. *Die kinematographische Unterrichtsstunde*. Leipzig: Demme, 1911.

———. "Die Verwertung und Nutzbarmachung neuer Film-Ideen—Künstlerische Films." *Der Kinematograph* no. 57 (29 January 1908).

———. *Durch die Technik zur Schulreform. Zwei modern-technische Lehrmethoden und Veranschaulichungsmittel in der Schule der Zukunft*. Leipzig: Demme, 1911.

———. *Praktische Forderungen für die Verwertung der Kinematographie im Unterricht*. Friedenau: Schule und Technik, 1909.

———. "Volkstümliche Reisebeschreibungen." *Der Kinematograph* no. 34 (21 August 1907).
Lichtwark, Alfred. "Der Deutsche der Zukunft." In *Die deutsche Reformpädagogik*, ed. Wilhelm Flitner and Gerhard Kudritzki, 99–110. Düsseldorf and Munich: Küpper, 1961.
———. "Die Aufgaben der Kunsthalle: Antrittsrede den 9. December 1886." In *Drei Programme*, 2d ed., 11–31. Berlin: Cassirer, 1902.
———. *Die Bedeutung der Amateur-Photographie*. Halle an der Saale, Germany: Knapp, 1894.
———. "Museen als Bildungsstätten." *Der Deutsche der Zukunft*, 89–107. Berlin: Cassirer, 1905.
———. *Übungen in der Betrachtung von Kunstwerken*. Dresden: Kühtmann, 1900.
Liesegang, F. Paul. *Wissenschaftliche Kinematographie*. Düsseldorf: Liesegang, 1920.
Lipps, Theodor. *Ästhetik. Psychologie des Schönen und der Kunst. Erster Teil: Grundlegung der Ästhetik*. Hamburg and Leipzig: Voss, 1903.
———. "Einfühlung, innere Nachahmung, und Organempfindungen." *Archiv für die gesamte Psychologie* 1, nos. 2/3 (1903): 185–204. Translated as "Empathy, Inner Imitation, and Sense-Feeling," in *A Modern Book of Aesthetics*, ed. Melvin M. Rader, 291–304 (New York: Holt, 1935).
———. "Einfühlung und ästhetischer Genuß." *Die Zukunft* 54, no. 14 (20 January 1906): 100–114.
Locke, John. *An Essay Concerning Human Understanding*. Edited and with an introduction by Peter H. Nidditch. Oxford: Oxford University Press, 1975.
Lomon, André, and Jean Comandon. "Radiocinématographie par la photographie des écrans intensificateurs." *La Presse Médicale* 35 (3 May 1911): 359.
Londe, Albert. *Notice sur les titres et travaux scientifique*. Paris: Masson, 1911.
———. *Nouvelle iconographie de la Salpêtrière*. Paris: Masson, 1888–1918.
———. *Photographie médicale*. Paris: Gauthiers-Villars, 1893.
Lorenz, Richard, and W. Eitel. "Über die örtliche Verteilung von Rauchteilchen." *Zeitschrift für anorganische Chemie* 87, no. 1 (12 May 1914): 357–374.
Lukács, Georg. *History and Class Consciousness*. Translated by Rodney Livingstone. Cambridge, Mass.: MIT Press, 1971.
———. *The Theory of the Novel*. Translated by Anna Bostock. Cambridge, Mass.: MIT Press, 1971.
———. "Thoughts Toward an Aesthetic of the Cinema." Translated by Janelle Blankenship. *Polygraph* 13 (2001): 13–18. Originally published as "Gedanken zu einer Aesthetik des 'Kino,'" *Frankfurter Zeitung* 251 (10 September 1913): 1–2.
———. "Über den Dostojewski-Nachlass." *Moskauer Rundschau* 17 (22 March 1931): 4.
Mach, Ernst. *Die Principien der Wärmlehre: Historisch-kritisch Entwickelt*. Leipzig: Barth, 1896.
Macintyre, John. "X-Ray Records for the Cinematograph." *Archives of Skiagraphy* 1, no. 2 (April 1897): 37.
Magnus-Levy, Adolf. "Ueber Organ-Therapie beim endemischen Kretinismus." *Verhandlungen der Berliner medicinischen Gesellschaft* 34, part 2 (1903): 350–357; 34, part 1 (1903): 246–249.

Marey, Étienne-Jules. *Animal Mechanism: A Treatise on Terrestrial and Aerial Locomotion*. New York: Appleton, 1874.

——. "Études sur la marche de l'homme." *Revue Militaire de Médecine et de Chirurgie* 1 (1880): 244–246.

——. *La chronophotographie*. Paris: Gauthier-Villars, 1899.

——. *La methode graphique dans les sciences éxperimentales et principlement en physiologie et en médicine*. Paris: Masson, 1885.

——. *Movement*. Translated by Eric Pritchard. London: Heinemann, 1895. Originally published as *Le movement*. Paris: Masson, 1894.

Marinescu, Gheorghe. "Les troubles de la marche dans l'hémiplégie organique étudiés à l'aide du cinématographe." *La semaine Médicale* (1899): 225–228.

May, Bruno. *Das Recht des Kinematographen*. Berlin: Falk, 1912.

"Medical News." *Lancet* 177 (27 May 1911): 1470.

"Medical News." *Lancet* 182 (11 October 1913): 1083–1084.

"Medizinisch-Naturwissenschaftlicher Verein Tübingen." *Münchener medizinische Wochenschrift* 56, no. 3 (19 January 1909): 154–155.

Meier, Konrad. "Vorschriften über Lüftung von Kinotheatern." *Gesundheits-Ingenieur* 36, no. 26 (28 June 1913): 483–484.

Mellini, Arthur. "Die ganze Richtung passt uns nicht!" *Lichtbild-Bühne* 5 (4 February 1911): 3–4.

Mendel, Georg Victor. *Kinematographie und Schule: Plan zur Gründung eines rein wissenschaftlichen Theaters für Kinematographie und Projektion*. Berlin: privately printed, 1909.

Meumann, Ernst. "Wilhelm Wundt. Zu seinem achtzigsten Geburtstag." *Deutsche Rundschau* 38, no. 11 (August 1912): 193–224.

Moll, Albert. *Hypnotism*. London: Scott, 1890. Originally published as *Der Hypnotismus* (Berlin: Fischer's Medicinische, 1889).

Mosso, Angelo. *Fatigue*. Translated by Margaret Drummond and W. B. Drummond. New York: Putnam, 1904.

Müller-Sanders, Hans. "Die Kinematographenzensur in Preußen." PhD diss., Badischen Ruprecht-Karls-Universität, Heidelberg, 1912.

Münsterberg, Hugo. *The Photoplay: A Psychological Study*. London and New York: Appleton, 1916.

Murawski, Friedrich. *Die Kinematographie und ihre Beziehungen zu Schule und Unterricht*. Dresden: Bieyl and Kaemmerer, 1914.

Muybridge, Eadweard. *Animal Locomotion: An Electro-Photographic Investigation of Consecutive Phases of Animal Movement*. Philadelphia: University of Pennsylvania, 1887.

——. *Muybridge's Complete Human and Animal Locomotion*. New York: Dover, 1979.

"Neuland für Kinematographentheater." In *Kino-Debatte: Texte zum Verhältnis von Literatur und Film 1909–1929*, ed. Anton Kaes, 41. Tübingen: Niemeyer, 1978. Originally published in *Lichtbild-Bühne* 3 (September 1910): 3.

Nietzsche, Friedrich. *Beyond Good and Evil*. In *Basic Writings of Nietzsche*, trans. Walter Kaufmann, 181–435. New York: Modern Library, 1966.

Nitze, Max. *Kystophotographischer Atlas.* Wiesbaden: Bergmann, 1894.
Nordau, Max. *Degeneration.* London: Appleton, 1895. Originally published as *Entartung*, 2 vols. (Berlin: Duncker, 1892–1893).
"Novel Uses for Moving Pictures." *Moving Picture World* 1, no. 3 (23 March 1907): 39–40.
Paget, Sir James. "An Address on the Utility of Scientific Work in Practice." *British Medical Journal* (15 October 1887): 811–814.
Patterson, Frances Taylor. *Scenario and Screen.* New York: Harcourt, Brace, 1928.
Perrin, Jean. "La realité des molecules." *Revue Scientifique* 49, no. 2 (1911): 774–784.
———. *Les atoms.* 4th ed. rev. Paris: Librarie Félix Alcan, 1914. Translated by D. L. Hammick as *Atoms*, 2d English ed. rev. (London: Constable, 1923).
———. "Mouvement brownien et molécules." *Annales de chimie et de physique* 18 (September 1909): 1–114. Translated by F. Soddy as "Brownian Movement and Molecular Reality," in *The Question of the Atom*, ed. Mary Jo Nye, 507–601 (Los Angeles, Calif.: Tomash, 1984).
Pestalozzi, Johann Heinrich. *How Gertrude Teaches Her Children.* Edited by Ebenezer Cooke. Translated by Lucy E. Holland and Frances C. Turner. 2d ed. Syracuse, N.Y.: Bardeen, 1898. Originally published as *Wie Gertrud ihre Kinder lehrt, ein Versuch den Müttern Anleitung zu geben, ihre Kinder selbst zu unterrichten, in Briefen* (Bern and Zürich: Geßner, 1801).
Pfeffer, Wilhelm. "Die Anwendung des Projectionsapparates zur Demonstration von Lebensvorgängen." *Jahrbücher wissenschaftliche Botanik* 35 (1900): 711–745.
Pfemfert, Franz. "Kino als Erzieher." In *Prolog vor dem Film: Nachdenken über ein neues Medium, 1909–1914*, ed. Jörg Schweinitz, 165–169. Leipzig: Reclam, 1992. Originally published in *Die Aktion* 1, no. 18 (19 June 1911): 560–563.
Philippson, Martin. *L'autonomie et la centralisation dans le système nerveux des animaux: étude de physiologie expérimentale et compare.* Brussels: Falk, 1905.
Pieper, Lorenz. "Kino und Drama." *Bild und Film* 1, no. 1 (1912): 5.
Pinthus, Kurt. *Das Kinobuch.* Frankfurt: Fischer Taschenbuch, 1983.
Polgar, Alfred. "Das Drama im Kinematographen." In *Kein Tag ohne Kino: Schriftsteller über den Stummfilm*, ed. Fritz Güttinger, 56–61. Frankfurt: Deutsches Filmmuseum, 1984. Originally published in *Der Strom* 1, no. 2 (May 1911).
Polimanti, Oswald. "Der Kinematograph in der biologischen und medizinischen Wissenschaft." *Naturwissenschaftliche Wochenschrift* 10, no. 49 (3 December 1911): 769–774.
———. "Die Anwendung der Kinematographie in den Naturwissenschaften, der Medizin und im Unterricht." In *Wissenschaftliche Kinematographie*, by Franz Paul Liesegang, with Karl Kieser and Oswald Polimanti, 257–310. Düsseldorf: Liesegang, 1920.
———. "Über Ataxie cerebralen und cerebellaren Ursprungs." *Archiv für Physiologie* (1909): 123–134.
———. "Zur Physiologie der Stirnlappen." *Archiv für Physiologie* (1912): 337–342.
Powers, Samuel A. *Variola: A Series of Twenty-One Heliotype Plates Illustrating the Progressive Stages of the Eruption.* Boston: Samuel A. Powers, 1882.
Quensel, Paul. *Meisterbilder und Schule: Anregungen zu praktischen Versuchen.* Munich: Kunstwart-Verl, 1905.

Ramón y Cajal, Santiago. "New Observations on the Development of Neuroblasts, with Comments on the Neurogenetic Hypothesis of Hensen-Held [1908]." In *Studies on Vertebrate Neurogenesis*, trans. Lloyd Guth, 71–76. Springfield, Ill.: Thomas, 1960.

Rath, Willy. *Kino und Bühne*. M. Gladbach [Mönchengladbach]: Volksvereins, 1913.

Rauscher, Ulrich. "Die Kino-Ballade." In *Kein Tag ohne Kino: Schriftsteller über den Stummfilm*, ed. Fritz Güttinger, 143–149. Frankfurt: Deutsches Filmmuseum, 1984. Originally published in *Der Kunstwart* 26, no. 13 (1 April 1913): 1–6.

Reicher, Karl. "Kinematographie in der Neurologie." *Verhandlungen der Gesellschaft deutscher Naturforscher und Ärzte* 79, part 2 (1907): 235–236.

Rein, Wilhelm. *Pädagogik im Grundriß* (1890). 4th ed. Leipzig: Göschen, 1907.

Rennert, Malwine. "Die Zaungäste des Lebens im Kino." *Bild und Film* 4, no. 11 (1914/1915): 217–218.

Richter, Wilhelm. "Der Kinematograph als naturwissenschaftliches Anschauungsmittel." *Naturwissenschaftliche Wochenschrift* 12, no. 52 (28 December 1913): 817–820.

——. "Hochschulkinematographie." *Bild und Film* 2, nos. 11/12 (1912/1913): 253–257.

Ries, Julius. "Kinematographie der Befruchtung und Zellteilung." *Archiv für Mikroskopische Anatomie und Entwicklungsgeschichte* 74 (1909): 1–31.

Risak, Erwin. *Der klinische Blick*. 7th and 8th eds. Vienna: Springer, 1943.

"The Royal Society Conversazione." *Lancet* 149 (19 June 1897): 1706.

Rüswald, K. "Der Film im Erdkundlichen und Naturwissenschaftlichen Unterricht." In *Dokumente zur Geschichte der Schulfilmbewegung in Deutschland*, ed. Fritz Terveen, 43–44. Emsdetten, Germany: Lechte, 1959.

Samuleit, Paul, and Emil Borm. *Der Kinematograph als Volks- und Jugendbildungsmittel*. Berlin: Gesellschaft für Verbreitung von Volksbildung, 1912.

Schenk, Paul. "Der Kinematograph und die Schule." *Aerztliche Sachverständigen-Zeitung* 14, no. 15 (1 August 1908): 312–313.

Schiller, Friedrich. *On the Aesthetic Education of Man*. Translated by Elizabeth M. Wilkinson and L. A. Willoughby. Oxford: Oxford University Press, 1967.

Schmarsow, August. "The Essence of Architectural Creation (1893)." In *Empathy, Form, and Space: Problems in German Aesthetics, 1873–1893*, ed. and trans. Harry Francis Mallgrave and Eleftherios Ikonomou, 281–297. Santa Monica, Calif.: Getty Center for the History of Art and the Humanities, 1994.

Schmid, Bastian. "Kinematographie und Schule." *Die Naturwissenschaften* 1, no. 6 (7 February 1913): 145–146.

Schopenhauer, Arthur. "On the Metaphysics of Music." In *The World as Will and Representation*. Volume 2. Translated by E. F. J. Payne, 447–457. New York: Dover, 1966.

——. "On the Pure Subject of Knowing." In *The World as Will and Representation*. Volume 2. Translated by E. F. J. Payne, 367–375. New York: Dover, 1966.

——. *The World as Will and Representation*. Volume 1. Translated by E. F. J. Payne. New York: Dover, 1966.

Schultze, Ernst. *Der Kinematograph als Bildungsmittel*. Halle an der Saale, Germany: Waisenhaus, 1911.

Schuster, Paul. "Vorführung pathologischer Bewegungscomplexe mittelst des Kinematographen und Erläuterung derselben." *Verhandlungen der Gesellschaft deutscher Naturforscher und Ärzte* 69, part 1 (1898): 196–199.

Schweisheimer, Waldemar. *Die Bedeutung des Films für soziale Hygiene und Medizin.* Munich: Müller, 1920.

Seddig, Max. "Exacte Messung des Zeitintervalles bei kinematographischen Aufnahmen." *Jahrbuch für Photographie und Reproduktionstechnik* 26 (1912): 654–657.

———. "Messung der Temperatur-Abhängigkeit der Brown'schen Molekularbewegung." Habilitationsschrift, Akademie in Frankfurt a. M., 1909.

———. "Messung der Temperatur-Abhängigkeit der Brown-Zsigmondyschen Bewegung." *Zeitschrift für Anorganische Chemie* 73–74 (1912): 360–384.

———. "Über die Messung der Temperaturabhängigkeit der Brownschen Molekularbewegung." *Physikalische Zeitschrift* 9, no. 14 (15 July 1908): 465–468.

———. "Ueber Abhängigkeit der Brown'schen Molekularbewegung von der Temperatur." *Sitzungsberichte der Gesellschaft zur Beförderung der gesammten Naturwissenschaften zu Marburg* 18 (1907): 182–188.

Segmiller, L. "Das Skizzieren nach Lichtbildern bei Tageslicht und künstlicher Beleuchtung." *Film und Lichtbild* 1, no. 4 (1912): 35–39.

Sellmann, Adolf. "Das Geheimnis des Kinos." *Bild und Film* 1, nos. 3–4 (1912): 65–67.

———. *Der Kinematograph als Volkserzieher?* Langensalza, Germany: Beyer, 1912.

———. *Kino und Schule.* M. Gladbach [Mönchengladbach]: Volksvereins, 1914.

———. *Kino und Volksbildung.* M. Gladbach [Mönchengladbach]: Volksvereins, 1914.

Serner, Walter. "Kino und Schaulust." In *Prolog vor dem Film: Nachdenken über ein neues Medium, 1909–1914,* ed. Jörg Schweinitz, 208–214. Leipzig: Reclam, 1992. Originally published in *Die Schaubühne* 9, nos. 34/35 (1913): 807–811.

Siedentopf, Henry. "Über ultramikroskopische Abbildung." *Zeitschrift für wissenschaftliche Mikroskopie und mikroskopische Technik* 26 (1909): 391–410.

———, and E. Sommerfeldt, "Über die Anfertigung kinematographischer Mikrophotographien der Kristallisationserscheinungen." *Zeitschrift für Elektrochemie* 13, no. 24 (14 June 1907): 325–326.

Simmel, Georg. "The Metropolis and Mental Life." In *The Sociology of Georg Simmel,* ed. and trans. Kurt H. Wolff, 409–424. New York: Free Press, 1950.

Smoluchowski, Maryan. "Essai d'une thèorie cinètique du movement brownien et des milieux troubles." *Krakau Anzeiger* 7 (1906): 585–586.

Solvay, Ernest. *Notes sur le productivisme et le comptabilisme.* Brussels: Misch and Thron, 1900.

Sommer, Ph. "Zur Psychologie des Kinematographen." *Der Kinematograph* no. 227 (3 May 1911).

Sommerfeldt, Ernst. "Über flüssige und scheinbar lebende Kristalle; mit kinematographischen Projektionen." *Verhandlungen der Gesellschaft deutscher Naturforscher und Ärzte* 79, part 2 (1907): 202.

"Special Correspondence: Berlin." *British Medical Journal* (5 March 1910): 598.

Spencer, Herbert. *The Principles of Psychology*. 2 vols. London: Williams and Norgate, 1855.
Spier, Ike. "Die sexuelle Gefahr des Kinos." *Die neue Generation* 8 (1912): 192–198.
Stapel, Wilhelm. "Der homo cinematicus." *Deutsches Volkstum* 21 (October 1919): 319–320.
Stein, Albert E. "Ueber medizinisch-photographische und -kinematographische Aufnahmen." *Deutsche medizinische Wochenschrift* 38 (20 June 1912): 1184–1186.
Stein, Sigmund Theodor. *Das Licht im Dienste wissenschaftlicher Forschung: Handbuch der Anwendung des Lichtes und der Photographie in der Natur- und Heilkunde*. Leipzig: Spamer, 1877.
——. *Die optische Projektionskunst im dienste der exakten Wissenschaften: ein Lehr- und Hilfsbuch zur unterstützung des naturwissenschaftlichen Unterrichts*. Halle an der Saale, Germany: Knapp, 1887.
Stigler, R. "Über das Flimmern der Kinematographen." *Archiv für die gesamte Physiologie des Menschen und der Tiere* (Bonn) 123 (1908): 224–232.
Stratz, Carl Heinrich. *Die Frauenkleidung und ihre natürliche Entwicklung*. Stuttgart: Enke, 1900.
Strobl, Karl Hans. "Der Kinematograph." In *Kein Tag ohne Kino: Schriftsteller über den Stummfilm*, ed. Fritz Güttinger, 50–54. Frankfurt: Deutsches Filmmuseum, 1984. Originally published in *Die Hilfe* 17, no. 9 (2 March 1911).
Sully, James. "Pleasure of Visual Form." *Mind: A Quarterly Review of Psychology and Philosophy* 5, no. 18 (April 1880): 181–201.
Svedberg, Theodor. *Studien zur Lehre von der Kolloiden Lösungen*. Uppsala: Akademische Buchdruckerei Edv. Berling, 1907.
Taylor, Frederick Winslow. *The Principles of Scientific Management* [1911]. New York: Norton, 1947.
Thielemann, W. "Kinematographie und biologische Forschung." *Bild und Film* 3, no. 7 (1913/1914): 171–172.
Tönnies, Ferdinand. *Gemeinschaft und Gesellschaft*. Leipzig: Fues, 1887. Translated by Charles Loomis as *Community and Society* (East Lansing: Michigan State University Press, 1957).
Universum-Film A. G. *Das medizinische Filmarchiv bei der Kulturabteilung der Universum-Film A.G.* Berlin: Gahl, 1919.
Van Gehuchten, Arthur. "Coup de couteau dans la moelle lombaire. Essai de physiologie pathologique." *Le Névraxe* 9 (1907): 208–232.
"Verwandte Gebiete." *Zentralblatt für Röntgenstrahlen, Radium und verwandte Gebiete* 1, no. 2 (1910): 78–80.
"Verzeichnis wissenschaftlich und technisch wervoller Films." *Film und Lichtbild* 1, no. 2 (1912): 16.
Virchow, Rudolph. *Post-mortem Examinations*. Translated by T. P. Smith. 3d ed. Philadelphia: Blakiston, 1895.
Vischer, Robert. *On the Optical Sense of Form* (1873). In *Empathy, Form, and Space: Problems in German Aesthetics, 1873–1893*, ed. and trans. Harry Francis Mallgrave

and Eleftherios Ikonomou, 89–124. Santa Monica, Calif.: Getty Center for the History of Art and the Humanities, 1994.

Vogt, Walther. "Hermann Braus." *Münchener medizinische Wochenschrift* 72, no. 8 (20 February 1925): 304–305.

Volkelt, Johannes. *System der Ästhetik, Erster Band: Grundlegung der Ästhetik.* 2d ed. Munich: Beck, 1927.

von Hanstein, Otfrid. *Kinematographie und Schule. Ein Vorschlag zur Reform des Anschauungs-Unterrichts.* Berlin: Lichtspiele Mozartsaal, 1911.

von Molo, Walter. "Im Kino." In *Prolog vor dem Film: Nachdenken über ein neues Medium, 1909–1914,* ed. Jörg Schweinitz, 28–39. Leipzig: Reclam, 1992. Originally published in *Velhagen & Klasings Monatshefte* 26, no. 8 (April 1912): 618–627.

Warstat, Willi, and Franz Bergmann. *Kino und Gemeinde.* M. Gladbach [Mönchengladbach]: Volksvereins, 1913.

Wassermann, A. "Die medizinische Fakultät." In *Die Universitäten im Deutschen Reich,* ed. W. Lexis, 146–147. Berlin: Asher, 1904.

Weber, Wilhelm. *Mechanik der menschlichen Gehwerkzeuge: Eine anatomisch-physiologische Untersuchung.* In *Wilhelm Weber's Werke,* vol. 6, ed. Der königlichen Gesellschaft der Wissenschaften zu Göttingen, 1–305. Berlin: Springer, 1894.

Weichardt, Wilhelm. "Ermüdungsbekämpfung durch Antikenotoxin." *Deutsche militärärztliche Zeitschrift* 42, no. 1 (5 January 1913): 12–13.

Weisenburg, T. H. "Moving Picture Illustrations in Medicine, with Special Reference to Nervous and Mental Diseases." *Journal of the American Medical Association* 59, no. 26 (28 December 1912): 2310–2312.

Weiser, Martin. *Medizinische Kinematographie.* Dresden and Leipzig: Steinkopff, 1919.

"Wissenschaftliche Abende." *Film und Lichtbild* 1, no. 3 (1912): 30–31.

"Wissenschaft und Lichtspiele." *Bild und Film* 1, no. 2 (1912): 49–50.

Wolf-Czapek, Karl Wilhelm, ed. *Angewandte Photographie in Wissenschaft und Technik.* Berlin: Union Deutsche Verlagsgesellschaft, 1911.

———. "Die Kinematographie im medizinische Unterricht." *Jahrbuch für Photographie und Reproduktionstechnik* 22 (1908): 58–59.

———. *Die Kinematographie: Wesen, Entstehung und Ziele des lebenden Bildes.* Berlin: Union Deutsche Verlagsgesellschaft, 1908; 2d enl. ed., 1911.

Wolgast, Heinrich. "Die Bedeutung der Kunst für die Erziehung." In *Die Kunsterziehungsbewegung,* ed. Hermann Lorenzen, 17–20. Bad Heilbrunn, Germany: Klinkhardt, 1966.

Wundt, Wilhelm. *Grundzüge der physiologischen Psychologie,* 5th ed. Leipzig: Engelmann, 1902–1903.

Wychgram, Engelhard. "Aus optischen und mechanischen Werkstätten IV." *Zeitschrift für wissenschaftliche Mikroskopie und für mikroskopishe Technik* 28 (1911): 337–361.

Zimmermann, Robert. *Allgemeine Aesthetik als Formwissenschaft.* Vienna: Braumüller, 1865.

Zuntz, Nathan, and Wilhelm Schumberg. *Studien zu einer Physiologie des Marsches.* Berlin: Hirschwald, 1901.

"Zur Eröffnung des Ernemann-Kino in Dresden." *Der Kinematograph* no. 134 (21 July 1909).

SECONDARY SOURCES

Abel, Richard, ed. *French Film Theory and Criticism*, vols. 1 and 2. Princeton, N.J.: Princeton University Press, 1988.

——. *The Ciné Goes to Town: French Cinema, 1896–1914*. Berkeley: University of California Press, 1994.

Acland, Charles R., and Haidee Wasson, eds. *Useful Cinema*. Durham, N.C.: Duke University Press, 2011.

Albert, David Z. *Time and Chance*. Cambridge, Mass.: Harvard University Press, 2000.

Albisetti, James C. *Secondary School Reform in Imperial Germany*. Princeton, N.J.: Princeton University Press, 1983.

Alexander, Jennifer Karns. *The Mantra of Efficiency: From Waterwheel to Social Control*. Baltimore: Johns Hopkins University Press, 2008.

Allen, Ann Taylor. *Feminism and Motherhood in Germany, 1800–1914*. New Brunswick, N.J.: Rutgers University Press, 1991.

Allesch, Christian G. *Geschichte der psychologischen Ästhetik*. Göttingen: Verlag für Psychologie, 1987.

Alovisio, Silvio. *L'occhio sensibile. Cinema e scienze della mente nell'Italia del primo Novecento. Con una antologia di testi d'epoca*. Turin: Edizioni Kaplan, 2013.

Amad, Paula. "Archiving the Everyday: A Topos in French Film History, 1895–1931." PhD diss., University of Chicago, 2002.

——. *Counter-Archive: Film, The Everyday, and Albert Kahn's Archives de la Planète*. New York: Columbia University Press, 2010.

Andrew, Dudley. "Film and Society: Public Rituals and Private Space." In *Exhibition: The Film Reader*, ed. Ina Rae Hark, 161–172. New York: Routledge, 2001. Originally published in *East-West Film Journal* 1, no. 1 (1986): 7–22.

Andriopoulos, Stefan. *Possessed: Hypnotic Crimes, Corporate Fiction, and the Invention of Cinema*. Chicago: University of Chicago Press, 2008.

——. "Spellbound in Darkness: Hypnosis as an Allegory of Early Cinema." *Germanic Review* 77, no. 2 (Spring 2002): 102–116.

Anz, Thomas. "Gesundheit, Krankheit und literarische Norm: Max Nordaus 'Entartung' als Paradigma pathologisierender Kunstkritik." In *Gesund oder Krank?: Medizin, Moral und Ästhetik in der deutschen Gegenwartsliteratur*, 33–52. Stuttgart: Metzler, 1989.

Apfelbaum, Erika, and Gregory R. McGuire. "Models of Suggestive Influence and the Disqualification of the Social Crowd." In *Changing Conceptions of Crowd Mind and Behavior*, ed. C. F. Graumann and S. Moscovici, 27–50. New York and Berlin: Springer, 1986.

Arabatzis, Theodore. "Experiment." In *The Routledge Companion to Philosophy of Science*, ed. Stathis Psillos and Martin Curd, 159–170. London and New York: Routledge, 2008.

Arendt, Hannah. *The Human Condition*. Chicago: University of Chicago Press, 1958.

Arndt, Margarete, and Barbara Bigelow. "Toward the Creation of an Institutional Logic for the Management of Hospitals: Efficiency in the Early Nineteen Hundreds." *Medical Care Research and Review* 63, no. 3 (June 2006): 369–394.

Asche, Susanne. "Fürsorge, Partizipation und Gleichberechtigung—die Leistungen der Karlsruherinnen für die Entwicklung zur Großstadt (1859–1914)." In *Karlsruher Frauen, 1715–1945: Eine Stadtgeschichte*, 171–256. Karlsruhe: Badenia, 1992.

Ashwin, Clive. "Pestalozzi and the Origins of Pedagogical Drawing." *British Journal of Educational Studies* 29, no. 2 (June 1981): 138–151.

Askari, Kaveh. *Making Movies Into Art: Picture Craft from the Magic Lantern to Early Hollywood*. London: British Film Institute, 2014.

Aubert, Geneviève. "Arthur Van Gehuchten Takes Neurology to the Movies." *Neurology* 59 (November 2002): 1612–1618.

———. "From Photography to Cinematography: Recording Movement and Gait in a Neurological Context." *Journal of the History of the Neurosciences* 11, no. 3 (2002): 255–264.

Aubin, David. "'The Memory of Life Itself': Bénard's Cells and the Cinematography of Self-Organization." *Studies in History and Philosophy of Science* 39, no. 3 (2008): 359–369.

Bachelard, Gaston. *The New Scientific Spirit*. Translated by Arthur Goldhammer. Boston: Beacon, 1984.

Baird, Davis. *Thing Knowledge: A Philosophy of Scientific Instruments*. Berkeley: University of California Press, 2004.

Baptista, Tiago. "'Il faut voir le maître': A Recent Restoration of Surgical Films by E.-L. Doyen (1859–1916)." *Journal of Film Preservation* 70 (November 2005): 42–50.

Barboi, Alexandru C., Christopher G. Goetz, and Radu Musetoiu. "The Origins of Scientific Cinematography and Early Medical Applications." *Neurology* 62 (June 2004): 2082–2086.

Barkin, Kenneth D. "The Crisis of Modernity, 1887–1902." In *Imagining Modern German Culture, 1889–1910*, ed. Françoise Forster-Hahn, 19–35. Washington, D.C.: National Gallery of Art, 1996.

Barrows, Susanna. *Distorting Mirrors: Visions of the Crowd in Late Nineteenth-Century France*. New Haven: Yale University Press, 1981.

Bauer, Herbert. *Zur Theorie und Praxis der ersten deutschen Landerziehungsheime: Erfahrungen zur Internats- und Ganztagserziehung aus den Hermann-Lietz-Schulen*. Berlin: Volk und Wissen, 1961.

Berg, Christa, ed. *Handbuch der deutschen Bildungsgeschichte*. Munich: Beck, 1991.

Bernard, Claude. *Introduction à l'étude de la médecine expérimentale*. Paris: Baillière, 1865. Translated by Henry Copley Greene as *An Introduction to the Study of Experimental Medicine* (New York: Macmillan, 1927).

Bernheimer, Charles. "Decadent Diagnostics." In Charles Bernheimer, *Decadent Subjects: The Idea of Decadence in Art, Literature, Philosophy, and Culture of the Fin de Siècle in Europe*, ed. T. Jefferson Kline and Naomi Schor, 139–162. Baltimore: Johns Hopkins University Press, 2002.

Berrios, G. E., and M. Gili. "Will and Its Disorders: A Conceptual History." *History of Psychiatry* 6 (1995): 87–104.

Bettelheim, Bruno. *Freud and Man's Soul*. New York: Knopf, 1983.

Bigg, Charlotte. "Evident Atoms: Visuality in Jean Perrin's Brownian Motion Research." *Studies in History and Philosophy of Science* 39 (2008): 312–322.

———. "A Visual History of Jean Perrin's Brownian Motion Curves." In *Histories of Scientific Observation*, ed. Lorraine Daston and Elizabeth Lunbeck, 156–180. Chicago: University of Chicago Press, 2011.

Billings, Susan M. "Concepts of Nerve Fiber Development, 1839–1930." *Journal of the History of Biology* 4, no. 2 (Fall 1971): 275–305.

Blachut, Teodor J., and Rudolf Burkhardt. *Historical Development of Photogrammetric Methods and Instruments*. Falls Church, Va.: American Society for Photogrammetry and Remote Sensing, 1989.

Blackbourn, David. *History of Germany, 1780–1918: The Long Nineteenth Century* 2d ed. Malden, Mass.: Blackwell, 2003.

Blackmore, John T. *Ernst Mach: His Work, Life, and Influence*. Berkeley: University of California Press, 1972.

Blankenship, Janelle. "Futurist Fantasies: Lukács's Early Essay 'Thoughts Toward an Aesthetic of the Cinema.'" *Polygraph* 13 (2001): 21–36.

Block, Ed Jr. "T. H. Huxley's Rhetoric and the Popularization of Victorian Scientific Ideas, 1854–1874." In *Energy & Entropy: Science and Culture in Victorian Britain*, ed. Patrick Brantlinger, 205–228. Bloomington, Ind.: Indiana University Press, 1988.

Bokhove, Niels W., and Karl Schuhmann. "Bibliographie der Schriften von Theodor Lipps." *Zeitschrift für philosophische Forschung* 45, no. 1 (January–March 1991): 112–130.

Bolz, Norbert. *Am Ende der Gutenberg-Galaxis. Die neuen Kommunikationsverhältnisse*. Munich: Fink, 1993.

Boon, Timothy. *Films of Fact: A History of Science in Documentary Films and Television*. New York: Wallflower, 2008.

Booth, Jeremy. "A Short History of Blood Pressure Measurement." *Proceedings of the Royal Society of Medicine* 70, no. 11 (November 1977): 793–799.

Bordwell, David. *On the History of Film Style*. Cambridge, Mass.: Harvard University Press, 1997.

Borell, Merriley. "Extending the Senses: The Graphic Method." *Medical Heritage* 2, no. 2 (March/April 1986): 114–121.

Bourdieu, Pierre. *Distinction*. Cambridge, Mass.: Harvard University Press, 1984.

Bowser, Eileen. *The Transformation of Cinema, 1907–1915*. New York: Scribner, 1990.

Brain, Robert M. "Representation on the Line: Graphic Recording Instruments and Scientific Modernism." In *From Energy to Information: Representation in Science and Technology, Art, and Literature*, ed. Bruce Clarke and Linda Dalrymple Henderson, 155–177. Stanford, Calif.: Stanford University Press, 2002.

Brain, Robert Michael. "The Pulse of Modernism: Experimental Physiology and Aesthetic Avant-Gardes Circa 1900." *Studies in History and Philosophy of Science* 39, no. 3 (2008): 393–417.

———. "Self-Projection: Hugo Münsterberg on Empathy and Oscillation in Cinema Spectatorship." *Science in Context* 25, no. 3 (September 2012): 329–353.

Braun, Marta. *Picturing Time: The Work of Étienne-Jules Marey*. Chicago: University of Chicago Press, 1992.

Breidbach, Olaf. "Representation of the Microcosm: The Claim for Objectivity in 19th Century Scientific Microphotography." *Journal of the History of Biology* 35 (2002): 221–250.

Brown, Elspeth H. *The Corporate Eye: Photography and the Rationalization of American Commercial Culture, 1884–1929*. Baltimore: Johns Hopkins University Press, 2005.

Brush, Stephen G. "A History of Random Processes: I. Brownian Movement from Brown to Perrin." *Archive for History of Exact Sciences* 5, no. 1 (1968): 1–36.

———. *The Kind of Motion We Call Heat: A History of the Kinetic Theory of Gases in the 19th Century*. Amsterdam and New York: North-Holland and American Elsevier, 1976.

———. *Statistical Physics and the Atomic Theory of Matter from Boyle and Newton to Landau and Onsager*. Princeton, N.J.: Princeton University Press, 1983.

Bryson, Norman. *Vision and Painting: The Logic of the Gaze*. New Haven: Yale University Press, 1983.

Bucchi, Massimiano. "Images of Science in the Classroom: Wallcharts and Science Education, 1850–1920." *British Journal for the History of Science* 31, no. 2 (1998): 161–184.

Buroker, Jill Vance. "Descartes on Sensible Qualities." *Journal of the History of Philosophy* 29, no. 4 (October 1991): 585–611.

Burwick, Frederick, and Paul Douglass, eds. *The Crisis in Modernism: Bergson and the Vitalist Controversy*. Cambridge: Cambridge University Press, 1992.

Bynum, W. F. *Science and the Practice of Medicine in the Nineteenth Century*. Cambridge: Cambridge University Press, 1994.

Cahan, David, ed. *Hermann von Helmholtz and the Foundations of Nineteenth-Century Science*. Berkeley and Los Angeles: University of California Press, 1993.

———. "The Zeiss Werke and the Ultramicroscope: The Creation of a Scientific Instrument in Context." In *Scientific Credibility and Technical Standards in 19th and Early 20th Century German and Britain*, ed. Jed Z. Buchwald, 67–115. Dordrecht, Netherlands: Kluwer Academic, 1996.

Canales, Jimena. "Desired Machines: Cinema and the World in Its Own Image." *Science in Context* 24, no. 3 (September 2011): 329–359.

———. "Einstein, Bergson, and the Experiment That Failed: Intellectual Cooperation at the League of Nations." *Modern Language Notes* 120, no. 5 (2006): 1168–1191.

———. "Movement Before Cinematography: The High-Speed Qualities of Sentiment." *Journal of Visual Culture* 5, no. 3 (December 2006): 275–294.

———. "Photogenic Venus: The 'Cinematographic Turn' in Science and Its Alternatives." *Isis* 93 (2002): 585–613.

———. "Sensational Differences: The Case of the Transit of Venus." *Cahiers François Viète* 1, nos. 11/12 (September 2007): 15–40.

———. *A Tenth of a Second: A History*. Chicago: University of Chicago Press, 2009.

Canguilhem, Georges. *The Normal and the Pathological*. Translated by Carolyn R. Fawcett in collaboration with Robert S. Cohen. New York: Zone, 1991.

Canning, Kathleen. *Languages of Labor and Gender: Female Factory Work in Germany, 1850–1914*. Ithaca, N.Y.: Cornell University Press, 1996.

Carleton, H. M. "Tissue Culture: A Critical Summary." *British Journal of Experimental Biology* 1, no. 1 (October 1923): 131–151.

Carroll, Noel. "Modernity and the Plasticity of Perception." *Journal of Aesthetics and Art Criticism* 59, no. 1 (Winter 2001): 11–18.
Cartwright, Lisa. "'Experiments of Destruction': Cinematic Inscriptions of Physiology." *Representations* 40 (Fall 1992): 129–152.
———. *Screening the Body: Tracing Medicine's Visual Culture*. Minneapolis: University of Minnesota Press, 1995.
Chamberlin, J. Edward, and Sander L. Gilman, eds. *Degeneration: The Dark Side of Progress*. New York: Columbia University Press, 1985.
Chanan, Michael. *The Dream That Kicks: The Prehistory and Early Years of Cinema in Britain*. London and Boston: Routledge & Kegan Paul, 1980.
Charney, Leo, and Vanessa Schwartz, eds. *Cinema and the Invention of Modern Life*. Berkeley: University of California Press, 1995.
Cherchi Usai, Paolo, and Lorenzo Codelli, eds. *Before Caligari: German Cinema, 1895–1920*. Madison: University of Wisconsin Press, 1990.
Chertok, Léon, and Isabelle Stengers. *A Critique of Psychoanalytic Reason: Hypnosis as a Scientific Problem from Lavoisier to Lacan*. Translated by Martha Noel Evans. Stanford, Calif.: Stanford University Press, 1992.
Chytry, Josef. *The Aesthetic State: A Quest in Modern German Thought*. Berkeley and Los Angeles: University of California Press, 1989.
Clark, Peter. "Atomism Versus Thermodynamics." In *Method and Appraisal in the Physical Sciences*, ed. Colin Howson, 41–106. Cambridge: Cambridge University Press, 1976.
Cobley, Evelyn. *Modernism and the Culture of Efficiency: Ideology and Fiction*. Toronto and Buffalo, N.Y.: University of Toronto Press, 2009.
Coopmans, Catelijne, Janet Vertesi, Michael E. Lynch, and Steve Woolgar, eds. *Representation in Scientific Practice Revisited*. Cambridge, Mass.: MIT Press, 2014.
Cooter, Roger, and Stephen Pumfrey. "Separate Spheres and Public Places: Reflections on the History of Science Popularization and Science in Popular Culture." *History of Science* 32, no. 97 (1994): 237–267.
Corrigan, Timothy. *A Cinema Without Walls: Movies and Culture After Vietnam*. New Brunswick, N.J.: Rutgers University Press, 1991.
Corzo-Duchardt, Beth. "Primal Screen: Primitivism and American Silent Film Spectatorship." PhD diss., Northwestern University, 2013.
Cowan, Michael. *Cult of the Will: Nervousness and German Modernity*. University Park: Pennsylvania State University Press, 2008.
Crary, Jonathan. *Suspensions of Perception: Attention, Spectacle, and Modern Culture*. Cambridge, Mass.: MIT Press, 1999.
———. *Techniques of the Observer: On Vision and Modernity in the Nineteenth Century*. Cambridge, Mass.: MIT Press, 1990.
———. "Unbinding Vision." *October* 68 (Spring 1994): 21–44.
Curtis, Robin. "*Einfühlung* and Abstraction in the Moving Image: Historical and Contemporary Reflections." *Science in Context* 25, no. 3 (September 2012): 425–446.

Curtis, Scott. "Between Observation and Spectatorship: Medicine, Movies, and Mass Culture in Imperial Germany." In *Film 1900: Technology, Perception, Culture*, ed. Annemone Ligensa and Klaus Kreimeier, 87–98. New Barnet, U.K.: Libbey, 2009.

———. "Die kinematographische Methode. Das 'Bewegte Bild' und die Brownsche Bewegung." *montage/AV: Zeitschrift für Theorie & Geschichte audiovisueller Kommunikation* 14, no. 2 (2005): 23–43.

———. "Dissecting the Medical Training Film." In *Beyond the Screen: Institutions, Networks and Publics of Early Cinema*, ed. Marta Braun, Charlie Keil, Rob King, Paul Moore, and Louis Pelletier, 161–167. New Barnet, U.K.: Libbey, 2012.

———. "The Efficiency of Images: Educational Effectiveness and the Modernity of Motion Pictures." In *The Visual Culture of Modernism*, SPELL: Swiss Papers in English Language and Literature 26, ed. Deborah L. Madsen and Mario Klarer, 41–59. Tübingen: Narr, 2011.

———. "Einfühlung und frühe deutsche Filmtheorie." In *Einfühlung. Zur Geschichte und Gegenwart eines ästhetischen Konzepts*, ed. Robin Curtis and Gertrud Koch, 61–84. Paderborn, Germany: Fink, 2009.

———. "Images of Efficiency: The Films of Frank B. Gilbreth." In *Films That Work: Industrial Film and the Productivity of Media*, ed. Vinzenz Hediger and Patrick Vonderau, 85–99. Amsterdam: Amsterdam University Press, 2009.

———. "Photography and Medical Observation." In *The Educated Eye: Visual Pedagogy in the Life Sciences*, ed. Nancy Anderson and Michael R. Dietrich, 68–93. Hanover, N.H.: Dartmouth College Press, 2012.

———. "Science Lessons." *Film History* 25, nos. 1–2 (2013): 45–54.

———. "Still/Moving: Digital Imaging and Medical Hermeneutics." In *Memory Bytes: History, Technology, and Digital Culture*, ed. Lauren Rabinovitz and Abraham Geil, 218–254. Durham, N.C.: Duke University Press, 2004.

———. "'Tangible as Tissue': Arnold Gesell, Infant Behavior, and Film Analysis." *Science in Context* 24, no. 3 (September 2011): 417–442.

———. "The Taste of a Nation: Training the Senses and Sensibility of Cinema Audiences in Imperial Germany." *Film History* 6, no. 4 (Winter 1994): 445–469.

———. "Vergrösserung und das mikroskopische Erhabene." *Zeitschrift für Medienwissenschaft* 5 (2011): 96–110.

Dagognet, Francois. *Étienne-Jules Marey: A Passion for the Trace*. Translated by Robert Galeta with Jeanine Herman. New York: Zone, 1992.

Dalton, Margaret Stieg. *Catholicism, Popular Culture, and the Arts in Germany, 1880–1933*. Notre Dame, Ind.: University of Notre Dame Press, 2005.

Daston, Lorraine. "On Scientific Observation." *Isis* 99, no. 1 (2008): 97–110.

———, and Elizabeth Lunbeck, eds. *Histories of Scientific Observation*. Chicago: University of Chicago Press, 2011.

———, and Peter Galison. "The Image of Objectivity." *Representations* 40 (Fall 1992): 81–128.

———. *Objectivity*. New York: Zone, 2007.

de Broglie, Louis. "The Concepts of Contemporary Physics and Bergson's Ideas on Time and Motion." In *Bergson and the Evolution of Physics*, ed. and trans. P. A. Y. Gunter, 45–62. Knoxville: University of Tennessee Press, 1969.

de Chadarevian, Soraya. "Graphical Method and Discipline: Self-Recording in Nineteenth-Century Physiology." *Studies in History and Philosophy of Science* 24, no. 2 (June 1993): 267–291.
de Duve, Thierry. "Time Exposure and Snapshot: The Photograph as Paradox." *October* 5 (Summer 1978): 113–125.
de Pastre, Béatrice and Thierry Lefebvre, eds. *Filmer la science, comprendre la vie: Le cinema de Jean Comandon.* Paris: Centre national de la cinématographie, 2012.
de Rijcke, Sarah. "Drawing Into Abstraction. Practices of Observation and Visualisation in the Work of Santiago Ramón y Cajal." *Interdisciplinary Science Reviews* 33, no. 4 (2008): 287–311.
Deleuze, Gilles. *Cinema 1: The Movement-Image.* Translated by Hugh Tomlinson and Barbara Habberjam. Minneapolis: University of Minnesota, 1986.
———. *Cinema 2: The Time-Image.* Translated by Hugh Tomlinson and Barbara Habberjam. Minneapolis: University of Minnesota, 1989.
Deussing, Gottlieb Gustav. "Der Anschauungsunterricht in der deutschen Schule von Comenius bis zur Gegenwart." PhD diss., Universität Jena, 1884.
Didier, Robert. *Le Docteur Doyen: Chirurgien de la Belle Époque.* Paris: Librairie Maloine, 1962.
Didi-Huberman, Georges. *Invention of Hysteria: Charcot and the Photographic Iconography of the Salpêtrière.* Translated by Alisa Hartz. Cambridge, Mass.: MIT Press, 2003.
Diederichs, Helmut H. *Anfänge deutscher Filmkritik.* Stuttgart: Fischer, 1986.
———. "Frühgeschicht deutscher Filmtheorie. Ihre Entstehung und Entwicklung bis zum Ersten Weltkrieg." Unpublished Habilitationsschrift, J. W. Goethe-Universität Frankfurt am Main, 1996.
———. "Hermann Häfker." In *Cinegraph*, ed. Hans-Michael Bock, Munich: edition text + kritik, 1984.
———. "Kino und die Wortkünste: Zur Diskussionen der deutschen literarischen Intelligenz 1910 bis 1915." *KINtop* 13 (2004): 9–23.
———. "Naturfilm als Gesamtkunstwerk: Hermann Häfker und sein 'Kinetographie'-Konzept." *Augenblick* 8 (1990): 37–60.
———. "The Origins of the *Autorenfilm*." In *Before Caligari: German Cinema, 1895–1920*, ed. Paolo Cherchi Usai and Lorenzo Codelli, 380–401. Madison: University of Wisconsin Press, 1990.
Diederichs, Helmut H., ed. *Geschichte der Filmtheorie: Kunsttheoretische Texte von Méliès bis Arnheim.* Frankfurt: Suhrkamp, 2004.
Dilthey, Wilhelm. "The Construction of the Historical World in the Human Studies." In *Selected Writings*, ed., trans., and with an introduction by H. P. Rickman, 168–245. Cambridge: Cambridge University Press, 1976.
Doane, Mary Ann. *The Emergence of Cinematic Time: Modernity, Contingency, the Archive.* Cambridge, Mass.: Harvard University Press, 2002.
Dommann, Monika, *Durchsicht, Einsicht, Vorsicht: Eine Geschichte der Röntgenstrahlen, 1896–1963.* Zürich: Chronos, 2003.
do O'Gomes, Isabelle. "L'oeuvre de Jean Comandon." In *Le cinéma et la science*, ed. Alexis Martinet, 78–85. Paris: CNRS Éditions, 1994.

Douglas, Ann. *The Feminization of American Culture*. New York: Knopf, 1977.
Dräger, Horst. *Die Gesellschaft für Verbreitung von Volksbildung: Eine historisch-problemgeschichtliche Darstellung von 1871–1914*. Stuttgart: Klett, 1975.
Duckwitz, Amelie, Martin Loiperdinger, and Susanne Theisen. "'Kampf dem Schundfilm!': Kinoreform and Jugendschutz in Trier." *KINtop* 9 (2000): 53–63.
Dunkel, Harold B. *Herbart and Education*. New York: Random House, 1969.
Durst, David C. *Weimar Modernism: Philosophy, Politics, and Culture in Germany 1918–1933*. Lanham, Md.: Lexington, 2004.
Duttlinger, Carolin. "Between Contemplation and Distraction: Configurations of Attention in Walter Benjamin." *German Studies Review* 30, no. 1 (February 2007): 33–54.
Eagleton, Terry. *The Ideology of the Aesthetic*. Cambridge, Mass.: Blackwell, 1990.
Ebel, Gerhard, and Otto Lührs. "Urania—eine Idee, eine Bewegung, eine Institution wird 100 Jahre alt." In *100 Jahre Urania: Wissenschaft heute für morgen*, 15–74. Berlin: Urania Berlin, 1988.
Egginton, William. "Intimacy and Anonymity, or How the Audience Became a Crowd." In *Crowds*, ed. Jeffrey T. Schnapp and Matthew Tiews, 97–110. Stanford, Calif.: Stanford University Press, 2006.
Elias, Norbert. *The Civilizing Process*. Vol. 1, *The History of Manners*. Translated by Edmund Jephcott. New York: Pantheon, 1978.
Ellis, John. *Visible Fictions: Cinema, Television, Video*. London: Routledge, 1982.
Elsaesser, Thomas. "Die Stadt von Morgen: Filme zum Bauen und Wohnen in der Weimarer Republik." In *Geschichte des dokumentarischen Film in Deutschland, Band 2: Weimarer Republik (1918–1933)*, ed. Klaus Kreimeier, Antje Ehmann, and Jeanpaul Goergen, 381–410. Stuttgart: Reclam, 2005.
——, and Michael Wedel, eds. *Kino der Kaiserzeit: Zwischen Tradition und Moderne*. Munich: edition text + kritik, 2002.
——, with Michael Wedel, eds. *A Second Life: German Cinema's First Decades*. Amsterdam: Amsterdam University Press, 1996.
Engstrom, Eric J. *Clinical Psychiatry in Imperial Germany: A History of Psychiatric Practice*. Ithaca, N.Y.: Cornell University Press, 2003.
——. "Emil Kraepelin: Psychiatry and Public Affairs in Wilhelmine Germany." *History of Psychiatry* 2, no. 6 (June 1991): 111–132.
Ermarth, Michael. *Wilhelm Dilthey: The Critique of Historical Reason*. Chicago: University of Chicago Press, 1978.
Evans, Hughes. "Losing Touch: The Controversy Over the Introduction of Blood Pressure Instruments Into Medicine." *Technology and Culture* 34, no. 4 (October 1993): 784–807.
Fisch, Jörg. "Zivilisation, Kultur." In *Geschichtliche Grundbegriffe. Historisches Lexikon zur politisch-sozialen Sprache in Deutschland*, vol. 7, ed. Otto Brunner, Werner Conze, and Reinhardt Koselleck, 679–774. Stuttgart: Klett, 1992.
Fitzi, Gregor. *Soziale Erfahrung und Lebensphilosophie. Georg Simmels Beziehung zu Henri Bergson*. Konstanz, Germany: UVK Verlagsgesellschaft, 2002.
Flanagan, Maureen A. *America Reformed: Progressives and Progressivisms, 1890s–1920s*. New York: Oxford University Press, 2007.

Fleck, Ludwik. *Cognition and Fact: Materials on Ludwik Fleck*. Edited by Robert S. Cohen and Thomas Schnelle. Dordrecht, Netherlands, and Boston: Reidel, 1986.
———. *Genesis and Development of a Scientific Fact*. Edited by Thaddeus J. Trenn and Robert K. Merton. Translated by Fred Bradley and Thaddeus J. Trenn. Chicago: University of Chicago Press, 1979.
———. "Scientific Observation and Perception in General [1935]." In *Cognition and Fact: Materials on Ludwik Fleck*, ed. Robert S. Cohen and Thomas Schnelle, 59–78. Dordrecht, Netherlands, and Boston: Reidel, 1986.
———. "Some Specific Features of the Medical Way of Thinking [1927]." In *Cognition and Fact: Materials on Ludwik Fleck*, ed. Robert S. Cohen and Thomas Schnelle, 39–46. Dordrecht, Netherlands, and Boston: Reidel, 1986.
———. "To Look, To See, To Know [1947]." In *Cognition and Fact: Materials on Ludwik Fleck*, ed. Robert S. Cohen and Thomas Schnelle, 129–151. Dordrecht, Netherlands, and Boston: Reidel, 1986.
Fleckenstein, Karen J. "The Mosso Plethysmograph in 19th-Century Physiology." *Medical Instrumentation* 18, no. 6 (November–December 1984): 330–331.
Foucault, Michel. *The Birth of the Clinic: An Archaeology of Medical Perception*. Translated by A. M. Sheridan Smith. New York: Vintage, 1973.
———. *Discipline and Punish*. Translated by Alan Sheridan. New York: Vintage, 1979.
———. *The History of Sexuality*. Translated by Robert Hurley. New York: Vintage, 1990.
Fox, Daniel M., and Christopher Lawrence. *Photographing Medicine: Images and Power in Britain and America Since 1840*. New York: Greenwood, 1988.
Frank, Robert G. Jr. "The Telltale Heart: Physiological Instruments, Graphic Methods, and Clinical Hopes, 1854–1914." In *The Investigative Enterprise: Experimental Physiology in Nineteenth-Century Medicine*, ed. William Coleman and Frederic L. Holmes, 211–290. Berkeley: University of California Press, 1988.
Fraser, Hilary. "Women and the Ends of Art History: Vision and Corporeality in Nineteenth-Century Critical Discourse." *Victorian Studies* 42, no. 1 (Autumn 1998): 77–100.
Frevert, Ute. *Krankheit als politisches Problem 1770–1880. Soziale Unterschichten in Preußen zwischen medizinischer Polizei und staatlicher Sozialversicherung*. Göttingen: Vandenhoeck & Ruprecht, 1984.
Frey, Erwin, and Klaus Kroy. "Brownian Motion: A Paradigm of Soft Matter and Biological Physics." *Annalen der Physik* 14, nos. 1–3 (February 2005): 20–50.
Fried, Michael. *Absorption and Theatricality: Painting and Beholder in the Age of Diderot*. Chicago: University of Chicago Press, 1980.
Fuller, P. W. W. "Carl Cranz, His Contemporaries, and High-Speed Photography." *Proceedings of SPIE*, no. 5580, 26th International Congress on High-Speed Photography and Photonics (25 March 2005): 250–260.
Fye, W. Bruce. "Franz M. Groedel." *Clinical Cardiology* 23, no. 2 (February 2000): 133–134.
Gailus, Andreas. "Of Beautiful and Dismembered Bodies: Art as Social Discipline in Schiller's *On the Aesthetic Education of Man*." In *Impure Reason: Dialectic of Enlightenment in Germany*, ed. W. Daniel Wilson and Robert C. Holub, 146–165. Detroit, Mich.: Wayne State University Press, 1993.

Garfinkel, Harold, Michael Lynch, and Eric Livingston. "The Work of a Discovering Science Construed with Materials from the Optically Discovered Pulsar." *Philosophy of the Social Sciences* 11, no. 2 (June 1981): 131–158.
Gaudreault, André. *Film and Attraction: From Kinematography to Cinema*. Translated by Timothy Barnard. Urbana: University of Illinois Press, 2011.
Gay, Frederick P. "Medical Logic." *Bulletin of the History of Medicine* 7 (1939): 6–27.
Gay, Peter. *The Education of the Senses*, vol. 1. New York: Oxford University Press, 1984.
Gaycken, Oliver. *Devices of Curiosity: Early Cinema and Popular Science*. Oxford and New York: Oxford University Press, 2015.
———. "'A Drama Unites Them in a Fight to the Death': Some Remarks on the Flourishing of a Cinema of Scientific Vernacularization in France, 1909–1914." *Historical Journal of Film, Radio and Television* 22, no. 3 (2002): 353–374.
———. "The Secret Life of Plants: Visualizing Vegetative Movement, 1880–1903." *Early Popular Visual Culture* 10, no. 1 (2012): 51–69.
———. "'The Swarming of Life': Moving Images, Education, and Views Through the Microscope." *Science in Context* 24, no. 3 (September 2011): 361–380.
Gebhard, Julius. *Alfred Lichtwark und die Kunsterziehungsbewegung in Hamburg*. Hamburg: Hoffmann und Campe, 1947.
Geimer, Peter. "Living and Non-living Pictures." In *Undead: Relations Between the Living and the Lifeless*, ed. Peter Geimer, 39–51. Berlin: Max-Planck-Institut für Wissenschaftsgeschichte, 2003.
Gernsheim, Alison. "Medical Photography in the Nineteenth Century." *Medical and Biological Illustration* (London) 11, no. 2 (April 1961): 85–92.
Geuss, Raymond. "Kultur, Bildung, Geist." *History and Theory* 35, no. 2 (May 1996): 151–164.
Ginzburg, Carlo. "Clues: Roots of an Evidential Paradigm." In *Clues, Myths, and the Historical Method*, trans. John and Ann C. Tedeschi, 96–125. Baltimore: Johns Hopkins University Press, 1989.
Glasser, Otto. *Wilhelm Conrad Roentgen and the Early History of the Roentgen Rays*. Springfield, Ill.: Thomas, 1934.
Gordon, Rae Beth. *Why the French Love Jerry Lewis: From Cabaret to Early Cinema*. Stanford, Calif.: Stanford University Press, 2001.
Greve, Ludwig, Margot Pehle, and Heidi Westhoff, eds. *Hätte ich das Kino! Die Schriftsteller und der Stummfilm*. Munich: Kösel, 1976.
Grieveson, Lee. *Policing Cinema: Movies and Censorship in Early-Twentieth-Century America*. Berkeley: University of California Press, 2004.
Gross, David L. "*Kultur* and Its Discontents: The Origins of a 'Critique of Everyday Life' in Germany, 1880–1925." In *Essays on Culture and Society in Modern Germany*, ed. Gary D. Stark and Bede Karl Lackner, 70–97. College Station: Texas A&M University Press, 1982.
Guerlac, Suzanne. *Thinking in Time: An Introduction to Henri Bergson*. Ithaca, N.Y.: Cornell University Press, 2006.
Gunning, Tom. "An Aesthetic of Astonishment: Early Film and the (In)Credulous Spectator." *Art and Text* 34 (Spring 1989): 31–45.

———. "The Cinema of Attractions: Early Film, Its Spectator and the Avant-Garde." In *Early Cinema: Space, Frame, Narrative*, ed. Thomas Elsaesser, with Adam Barker, 56–62. London: British Film Institute, 1990.

———. "Film History and Film Analysis: The Individual Film in the Course of Time." *Wide Angle* 12, no. 3 (July 1990): 4–19.

———. "Modernity and Cinema: A Culture of Shocks and Flows." In *Cinema and Modernity*, ed. Murray Pomerance, 297–315. New Brunswick, N.J.: Rutgers University Press, 2006.

———. "Systematizing the Electric Message: Narrative Form, Gender, and Modernity in *The Lonedale Operator*." In *American Cinema's Transitional Era: Audiences, Institutions, Practices*, ed. Charlie Keil and Shelley Stamp, 15–50. Berkeley: University of California Press, 2004.

Güttinger, Fritz, ed. *Kein Tag ohne Kino: Schriftsteller über den Stummfilm*. Frankfurt: Deutsches Filmmuseum, 1984.

Hacking, Ian. *Representing and Intervening: Introductory Topics in the Philosophy of Natural Science*. Cambridge: Cambridge University Press, 1983.

Hake, Sabine. *The Cinema's Third Machine: Writing on Film in Germany, 1907–1933*. Lincoln: University of Nebraska Press, 1993.

Hankins, Thomas L., and Robert J. Silverman. "The Magic Lantern and the Art of Demonstration." In *Instruments and the Imagination*, 37–71. Princeton, N.J.: Princeton University Press, 1995.

Hansen, Miriam. *Babel and Babylon: Spectatorship in American Silent Film*. Cambridge, Mass.: Harvard University Press, 1991.

———. "Early Silent Cinema: Whose Public Sphere?" *New German Critique* 29 (Spring/Summer 1983): 147–184.

Hansen, Miriam Bratu. *Cinema and Experience: Siegfried Kracauer, Walter Benjamin, and Theodor W. Adorno*. Berkeley: University of California Press, 2012.

Harrington, Anne. *Reenchanted Science: Holism in German Culture from Wilhelm II to Hitler*. Princeton, N.J.: Princeton University Press, 1996.

Harris, Stefanie. *Mediating Modernity: German Literature and the "New" Media, 1895–1930*. University Park: Pennsylvania State University Press, 2009.

Hau, Michael. "The Holistic Gaze in German Medicine, 1890–1930." *Bulletin of the History of Medicine* 74, no. 3 (Fall 2000): 495–524.

Hediger, Vinzenz, and Patrick Vonderau, eds. *Films That Work: Industrial Film and the Productivity of Media*. Amsterdam: Amsterdam University Press, 2009.

———. "Record, Rhetoric, Rationalization: Industrial Organization and Film." In *Films That Work: Industrial Film and the Productivity of Media*, ed. Vinzenz Hediger and Patrick Vonderau, 35–49. Amsterdam: Amsterdam University Press, 2009.

Heller, Heinz-B. *Literarische Intelligenz und Film: Zu Veränderungen der ästhetischen Theorie und Praxis unter dem Eindruck des Films 1910–1930 in Deutschland*. Tübingen: Niemeyer, 1985.

Hepp, Corona. *Avantgarde: Moderne Kunst, Kulturkritik und Reformbewegungen nach der Jahrhundertwende*. Munich: Deutscher Taschenbuch, 1987.

Herf, Jeffrey. *Reactionary Modernism: Technology, Culture, and Politics in Weimar and the Third Reich*. Cambridge and New York: Cambridge University Press, 1984.

Herr, Harry W. "Max Nitze, the Cystoscope and Urology." *Journal of Urology* 176, no. 4 (October 2006): 1313–1316.

Höfler, Günther A. "La naissance de la 'nervosité' issue de l'esprit de la modernité technologique. Dégénérescence et nomadisation chez Max Nordau et Adolph Wahrmund." In *Max Nordau (1849–1923): Critique de la Dégénérescence, Médiateur Franco-Allemand, Père Fondateur du Sionisme*, ed. Delphine Bechtel, Dominique Bourel, and Jacques Le Rider, 149–160. Paris: Cerf, 1996.

Holdorff, Bernd. "Die privaten Polikliniken für Nervenkranke vor und nach 1900." In *Geschichte der Neurologie in Berlin*, ed. Bernd Holdorff and Rolf Winau, 127–137. Berlin and New York: de Gruyter, 2001.

———. "Zwischen Hirnforschung, Neuropsychiatrie und Emanzipation zur klinischen Neurologie bis 1933." In *Geschichte der Neurologie in Berlin*, ed. Bernd Holdorff and Rolf Winau, 157–174. Berlin and New York: de Gruyter, 2001.

Hoof, Florian. " 'The One Best Way': Bildgebende Verfahren der Ökonomie und die Innovation der Managementtheorie Nach 1860." *montage/AV: Zeitschrift für Theorie und Geschichte audiovisueller Kommunikation* 15, no. 1 (2006): 123–138.

Hopwood, Nick. "Producing Development: The Anatomy of Human Embryos and the Norms of Wilhelm His." *Bulletin of the History of Medicine* 74 (2000): 29–79.

Howell, Joel D. *Technology in the Hospital: Transforming Patient Care in the Early Twentieth Century*. Baltimore: Johns Hopkins University Press, 1995.

Huerkamp, Claudia. *Der Aufstieg der Ärzte im 19. Jahrhundert: Vom gelehrten Stand zum professionellen Experten: Das Beispiel Preußens*. Göttingen: Vandenhoeck & Ruprecht, 1985.

———. "The Making of the Modern Medical Profession, 1800–1914: Prussian Doctors in the Nineteenth Century." In *German Professions, 1800–1950*, ed. Geoffrey Cocks and Konrad H. Jarausch, 66–84. New York and Oxford: Oxford University Press, 1990.

Husserl, Edmund. *The Crisis of European Sciences and Transcendental Phenomenology*. Translated by David Carr. Evanston, Ill.: Northwestern University Press, 1970.

Ilgner, Christian, and Dietmar Linke, "Filmtechnik: Vom Malteserkreuz zum Panzerkino." In *Oskar Messter: Filmpioneer der Kaiserzeit*, ed. Martin Loiperdinger, 93–134. Basel and Frankfurt: Stroemfeld/Roter Stern, 1994.

Imorde, Joseph. "Einfühlung in der Kunstgeschichte." In *Einfühlung. Zur Geschichte und Gegenwart eines ästhetischen Konzepts*, ed. Robin Curtis and Gertrud Koch, 127–142. Paderborn, Germany: Fink, 2009.

Jahoda, Gustav. "Theodor Lipps and the Shift from 'Sympathy' to 'Empathy.' " *Journal of the History of the Behavioral Sciences* 4, no. 2 (2005): 151–163.

Jarre, Hans A. "Roentgen Cinematography." In *The Science of Radiology*, ed. Otto Glasser, 198–209. Springfield, Ill.: Thomas, 1933.

Jay, Martin. *Downcast Eyes: The Denigration of Vision in Twentieth-Century French Thought*. Berkeley and Los Angeles: University of California Press, 1993.

Jelavich, Peter. " 'Am I Allowed to Amuse Myself Here?': The German Bourgeoisie Confronts Early Film." In *Germany at the Fin de Siècle: Culture, Politics, and Ideas*, ed. Suzanne Marchand and David Lindenfeld, 227–249. Baton Rouge: Louisiana State University Press, 2004.

Jenkins, Jennifer. *Provincial Modernity: Local Culture and Liberal Politics in Fin-de-Siècle Hamburg*. Ithaca, N.Y.: Cornell University Press, 2003.
Joerissen, Peter. *Kunsterziehung und Kunstwissenschaft im wilhelminischen Deutschland, 1871–1918*. Cologne and Vienna: Böhlau, 1979.
Jones, Caroline A., and Peter Galison, eds. *Picturing Science, Producing Art*. New York: Routledge, 1998.
Jülich, Solveig. "Seeing in the Dark: Early X-Ray Imaging and Cinema." In *Moving Images: From Edison to the Webcam*, ed. John Fullerton and Astrid Söderberg Widding, 47–58. Sydney: Libbey, 2000.
Jung, Uli. "Film für Lehre und Bildung." In *Geschichte des dokumentarischen Films in Deutschland*, vol. 1, ed. Uli Jung and Martin Loiperdinger, 333–340. Stuttgart: Reclam, 2005.
Kaes, Anton, ed. *Kino-Debatte: Texte zum Verhältnis von Literatur und Film 1909–1929*. Tübingen: Niemeyer, 1978.
———. "Literary Intellectuals and the Cinema: Charting a Controversy (1909–1929)." *New German Critique* 40 (Winter 1987): 7–34.
Kaiser, Céline. *Rhetorik der Entartung: Max Nordau und die Sprache der Verletzung*. Bielefeld, Germany: Transcript, 2007.
Kaiser, David Aram. *Romanticism, Aesthetics, and Nationalism*. Cambridge: Cambridge University Press, 1999.
Kalbus, Oskar. *Pionere des Kulturfilms: Ein Beitrag zur Geschichte des Kulturfilmschaffens in Deutschland*. Karlsruhe: Neue Verlags-Gesellschaft, 1956.
Kater, Michael H. "Professionalization and Socialization of Physicians in Wilhelmine and Weimar Germany." *Journal of Contemporary History* 20 (1985): 677–701.
Kaufmann, Doris. "'Pushing the Limits of Understanding': The Discourse on Primitivism in German *Kulturwissenschaften*, 1880–1930." *Studies in History and Philosophy of Science* 39 (2008): 434–443.
Kay, Carolyn. *Art and the German Bourgeoisie: Alfred Lichtwark and Modern Painting in Hamburg, 1886–1914*. Toronto and Buffalo, N.Y.: University of Toronto Press, 2002.
Keathley, Christian. *Cinephilia and History; or, The Wind in the Trees*. Bloomington: Indiana University Press, 2006.
Keele, Kenneth D. *The Evolution of Clinical Methods in Medicine*. London: Pitman, 1963.
Keene, Melanie Judith. "Object Lessons: Sensory Science Education, 1830–1870." PhD diss., University of Cambridge, 2008.
Keil, Charlie. "'To Here from Modernity': Style, Historiography, and Transitional Cinema." In *American Cinema's Transitional Era: Audiences, Institutions, Practices*, ed. Charlie Keil and Shelley Stamp, 51–65. Berkeley: University of California Press, 2004.
Kemp, Martin. "Temples of the Body and Temples of the Cosmos: Vision and Visualization in the Vesalian and Copernican Revolutions." In *Picturing Knowledge: Historical and Philosophical Problems Concerning the Use of Art in Science*, ed. Brian S. Baigrie, 40–84. Toronto and Buffalo, N.Y.: University of Toronto Press, 1996.
Kenkel, Karen J. "The Adult Children of Early Cinema." *Women in German Yearbook* (2000): 137–160.

———. "The Nationalisation of the Mass Spectator in Early German Film." In *Celebrating 1895: The Centenary of Cinema*, ed. John Fullerton, 155–162. Sydney: Libbey, 1998.
Kerker, Milton. "Brownian Movement and Molecular Reality Prior to 1900." *Journal of Chemical Education* 51, no. 12 (December 1974): 764–768.
Kern, Stephen. *Anatomy and Destiny: A Cultural History of the Human Body*. Indianapolis: Bobbs-Merrill, 1975.
———. "Freud and the Emergence of Child Psychology, 1880–1910." PhD diss., Columbia University, 1970.
Kerschensteiner, Georg. "Begriff der Arbeitsschule." In *Die deutsche Reformpädagogik*, ed. Wilhelm Flitner and Gerhard Kudritzki, 222–238. Düsseldorf and Munich: Küpper, 1961.
Kessler, Frank. "The Cinema of Attractions as *Dispositif*." In *The Cinema of Attractions Reloaded*, ed. Wanda Strauven, 57–69. Amsterdam: Amsterdam University Press, 2006.
———. "Henri Bergson und die Kinematographie." *KINtop* 12 (2003): 12–16.
———. "La cinématographie comme dispositif (du) spectaculaire." *CiNéMAS* 14, no. 1 (2003): 21–34.
———. "Viewing Change, Changing Views: The 'History of Vision' Debate." In *Film 1900: Technology, Perception, Culture*, ed. Annemone Ligensa and Klaus Kreimeier, 23–35. New Barnet, U.K.: Libbey, 2009.
Kessler, Frank, Sabine Lenk, and Martin Loiperdinger, eds. *Oskar Messter—Erfinder und Geschäftsmann*. KINtop Schriften 3. Basel and Frankfurt: Stoemfeld/Roter Stern, 1994.
Killen, Andreas. *Berlin Electropolis: Shock, Nerves, and German Modernity*. Berkeley: University of California Press, 2006.
———. "Psychiatry, Cinema, and Urban Youth in Early-Twentieth-Century Germany." *Harvard Review of Psychiatry* 14, no. 1 (2006): 38–43.
———. "The Scene of the Crime: Psychiatric Discourses on the Film Audience in Early Twentieth Century Germany." In *Film 1900: Technology, Perception, Culture*, ed. Annemone Ligensa and Klaus Kreimeier, 99–111. New Barnet, U.K.: Libbey, 2009.
King, Lester S. "Medical Logic." *Journal of the History of Medicine and Allied Sciences* 33, no. 3 (July 1978): 377–385.
———. *Medical Thinking: A Historical Preface*. Princeton, N.J.: Princeton University Press, 1982.
Kipp, Rudolf W. *Bilddokumente zur Geschichte des Unterrichtsfilms*. Grünwald, Germany: Institut für Film und Bild in Wissenschaft und Unterricht, 1975.
Kittler, Friedrich A. *Gramophone, Film, Typewriter*. Translated and with an introduction by Geoffrey Winthrop-Young and Michael Wutz. Stanford, Calif.: Stanford University Press, 1999.
Klein, Martin J. "Fluctuations and Statistical Physics in Einstein's Early Work." In *Albert Einstein: Historical and Cultural Perspectives*, ed. Gerald Holton and Yehuda Elkana, 39–58. Princeton, N.J.: Princeton University Press, 1982.
Knorr-Cetina, Karin. *The Manufacture of Knowledge: An Essay on the Constructivist and Contextual Nature of Science*. Oxford: Pergamon, 1981.

Koebner, Thomas. "Der Film als neue Kunst: Reaktionen der literarischen Intelligenz: Zur Theorie des Stummfilms (1911–1924)." In *Literaturwissenschaft-Medienwissenschaft*, ed. Volker Canaris and Helmut Kreuzer, 1–31. Heidelberg: Quelle und Meyer, 1977.
Koenigsberger, Leo. *Hermann von Helmholtz*. Braunschweig, Germany: Viewig, 1911.
Kommer, Helmut. *Früher Film und späte Folgen: Zur Geschichte der Film- und Fernseherziehung*. Berlin: Basis, 1979.
Koss, Juliet. "On the Limits of Empathy." *Art Bulletin* 88, no. 1 (March 2006): 139–157.
———. *Modernism After Wagner*. Minneapolis: University of Minnesota Press, 2010.
Kracauer, Siegfried. *From Caligari to Hitler*. Princeton, N.J.: Princeton University Press, 1947.
Kubler, George. *The Shape of Time: Remarks on the History of Things*. New Haven: Yale University Press, 1962.
Labisch, Alfons. *Homo Hygienicus: Gesundheit und Medizin in der Neuzeit*. Frankfurt and New York: Campus, 1992.
Landecker, Hannah. "Cellular Features: Microcinematography and Film Theory." *Critical Inquiry* 31, no. 4 (2005): 903–937.
———. "Creeping, Drinking, Dying: The Cinematic Portal and the Microscopic World of the Twentieth-Century Cell." *Science in Context* 24, no. 3 (September 2011): 381–416.
———. *Culturing Life: How Cells Became Technologies*. Cambridge, Mass.: Harvard University Press, 2007.
———. "The Life of Movement: From Microcinematography to Live-Cell Imaging." *Journal of Visual Culture* 11, no. 3 (December 2012): 378–399.
———. "Microcinematography and the History of Science and Film." *Isis* 97, no. 1 (2006): 121–132.
———. "New Times for Biology: Nerve Cultures and the Advent of Cellular Life in Vitro." *Studies in the History and Philosophy of Biological and Biomedical Sciences* 33, no. 4 (2002): 667–694.
———. "Technologies of Living Substance: Tissue Culture and Cellular Life in Twentieth Century Biomedicine." PhD diss., Massachusetts Institute of Technology, 1999.
Lanzoni, Susan. "Empathy in Translation: Movement and Image in the Psychological Laboratory." *Science in Context* 25, no. 3 (September 2012): 301–327.
Latour, Bruno. *Science in Action*. Cambridge, Mass.: Harvard University Press, 1987.
———. "Visualization and Cognition: Thinking with Eyes and Hands." *Knowledge and Society: Studies in the Sociology of Culture Past and Present* 6 (1986): 1–40.
———, and Steve Woolgar. *Laboratory Life: The Social Construction of Scientific Facts*. London and Beverly Hills, Calif.: Sage, 1979.
Lees, Andrew. *Cities, Sin, and Social Reform in Imperial Germany*. Ann Arbor: University of Michigan Press, 2002.
———, and Lynn Lees, eds. *The Urbanization of European Society in the Nineteenth Century*. Lexington, Mass.: Heath, 1976.
Lefebvre, Thierry. "Contribution à l'histoire de la microcinématographie: De François-Franck à Comandon." *1895* 14 (June 1993): 35–43.
———. "Die Trennung der Siamesischen Zwillinge Doodica und Radica durch Dr. Doyen." *KINtop* 6 (1997): 97–101.

———. "Flimmerndes Licht: Zur Geschichte der Filmwahrnehmung im frühen Kino." *KINtop* 5 (1996): 71–80.
———. *La chair et le celluloid: Le cinéma chirurgical du Docteur Doyen*. Brionne: Jean Doyen éditeur, 2004.
———. "La collection des films du Dr Doyen." *1895* 17 (December 1994): 100–114.
———. "Le cas étrange du Dr Doyen, 1859–1916." *Archives* 29 (February 1990): 1–12.
———. "Le Dr Doyen, un précurseur." In *Le cinéma et la science*, ed. Alexis Martinet, 70–77. Paris: CNRS Éditions, 1994.
———. "The Scientia Production (1911–1914): Scientific Popularization Through Pictures." *Griffithiana* no. 47 (May 1993): 137–153.
Lenk, Sabine, and Frank Kessler. "The Institutionalization of Educational Cinema: The Case of the *Kinoreformbewegung* in Germany." In *The Institutionalization of Educational Cinema: Educational Cinemas in North America and Europe in the 1910s and 1920s*, ed. Marina Dahlquist and Joel Frykholm. Bloomington: Indiana University Press, forthcoming.
Levin, Tom. "From Dialectical to Normative Specificity: Reading Lukács on Film." *New German Critique* 40 (Winter 1987): 35–61.
Lewis, Beth Irwin. "*Lustmord*: Inside the Windows of the Metropolis." In *Berlin: Culture and Metropolis*, ed. Charles W. Haxthausen and Heidrun Suhr, 111–140. Minneapolis: University of Minnesota Press, 1990.
Lindstrom, J. A. "'Getting a Hold Deeper in the Life of the City': Chicago Nickelodeons, 1905–1914." PhD diss., Northwestern University, 1998.
Lindstrom, Richard. "'They All Believe They Are Undiscovered Mary Pickfords': Workers, Photography, and Scientific Management." *Technology and Culture* 41, no. 4 (2000): 725–751.
Lucien Bull. Brussels: Hayez, 1967.
Lutz, Tom. *American Nervousness, 1903: An Anecdotal History*. Ithaca, N.Y.: Cornell University Press, 1991.
Lynch, Michael. "Discipline and the Material Form of Images: An Analysis of Scientific Visuality." *Social Studies of Science* 15, no. 1 (February 1985): 43–44.
———. "The Externalized Retina: Selection and Mathematization in the Visual Documentation of Objects in the Life Sciences." In *Representation in Scientific Practice*, ed. Michael Lynch and Steve Woolgar, 153–186. Cambridge, Mass.: MIT Press, 1990.
———, and Steve Woolgar, eds. *Representation in Scientific Practice*. Cambridge, Mass.: MIT Press, 1990.
Maase, Kaspar. "Krisenbewußtsein und Reformorientierung: Zum Deutungshorizont der Gegener der modernen Populärkünste 1880–1918." In *Schund und Schönheit: Populäre Kultur um 1900*, ed. Kaspar Maase and Wolfgang Kaschuba, 290–342. Cologne: Böhlau, 2001.
———. "Massenkunst und Volkserziehung: Die Regulierung von Film und Kino im deutschen Kaiserreich." *Archiv für Sozialgeschichte* 41 (2001): 39–77.
———. "Struggling About 'Filth and Trash': Educationalists and Children's Culture in Germany Before the First World War." *Paedagogica Historica* 34, no. 1 (1998): 8–28.

Maehle, Andreas-Holger. "The Search for Objective Communication: Medical Photography in the Nineteenth Century." In *Non-verbal Communication in Science Prior to 1900*, ed. Renato G. Mazzolini, 563–586. Florence: Olschki, 1993.

Maiocchi, Roberto. "The Case of Brownian Motion." *British Journal for the History of Science* 23 (September 1990): 257–283.

Mallgrave, Harry Francis, and Eleftherios Ikonomou. "Introduction." In *Empathy, Form, and Space: Problems in German Aesthetics, 1873–1893*, ed. and trans. Harry Francis Mallgrave and Eleftherios Ikonomou, 1–85. Santa Monica, Calif.: Getty Center for the History of Art and the Humanities, 1994.

Marchand, Suzanne L. *Down from Olympus: Archaeology and Philhellenism in Germany, 1750–1970*. Princeton, N.J.: Princeton University Press, 1996.

Marcus, Laura. *The Tenth Muse: Writing About Cinema in the Modernist Period*. Oxford and New York: Oxford University Press, 2007.

Márkus, György. "Life and the Soul: The Young Lukács and the Problem of Culture." In *Lukács Revalued*, ed. Agnes Heller, 1–26. London: Blackwell, 1983.

Maulitz, Russell C. "Physician Versus Bacteriologist: The Ideology of Science in Clinical Medicine." In *The Therapeutic Revolution: Essays in the Social History of American Medicine*, ed. Morris J. Vogel and Charles E. Rosenberg, 91–107. Philadelphia: University of Pennsylvania Press, 1979

Mayer, Andreas. *Sites of the Unconscious: Hypnosis and the Emergence of the Psychoanalytical Setting*. Chicago: University of Chicago Press, 2013.

——. *Wissenschaft vom Gehen. Die Erforschung der Bewegung im 19. Jahrhundert*. Frankfurt: Fischer, 2013.

Mayer, William. "Robert Gaupp." *American Journal of Psychiatry* 108, no. 10 (April 1952): 724–725.

McClelland, Charles E. "Modern German Doctors: A Failure of Professionalization?" In *Medicine and Modernity: Public Health and Medical Care in Nineteenth- and Twentieth-Century Germany*, ed. Manfred Berg and Geoffrey Cocks, 81–97. Cambridge and New York: Cambridge University Press, 1997.

——. "The Wise Man's Burden: The Role of Academicians in Imperial German Culture." In *Essays on Culture and Society in Modern Germany*, ed. Gary D. Stark and Bede Karl Lackner, 45–69. College Station, Tex.: Texas A & M University Press, 1982.

McCormick, Richard W., and Alison Guenther-Pal, eds. *German Essays on Film*. New York: Continuum, 2004.

McCormmach, Russell. "On Academic Scientists in Wilhelmian Germany." In *Science and Its Public: The Changing Relationship*, ed. Gerald Horton and William A. Blanpied, 157–171. Dordrecht, Netherlands, and Boston: Reidel, 1976.

Meyer, Rudolf W. "Bergson in Deutschland. Unter besonderer Berücksichtigung seiner Zeitauffassung." In *Studien zum Zeitproblem in der Philosophie des 20. Jahrhunderts*, ed. Rudolf W. Meyer, Ernst Wolfgang Orth, Rudolf Boehm, and Wolfgang Krewani, 10–64. Munich: Alber, 1982.

Michaelis, Anthony R. *Research Films in Biology, Anthropology, Psychology, and Medicine*. New York: Academic, 1955.

Mosse, George L. *The Crisis of German Ideology: Intellectual Origins of the Third Reich*. New York: Grosset & Dunlap, 1964.

———. "Max Nordau and His Degeneration." In Max Nordau, *Degeneration*, xiii–xxxvi. New York: Fertig, 1968.

Müller, Corinna. "Der frühe Film, das frühe Kino und seine Gegner und Befürworter." In *Schund und Schönheit: Populäre Kultur um 1900*, ed. Kaspar Maase und Wolfgang Kaschuba, 62–91. Cologne, Weimar, and Vienna: Böhlau, 2001.

———. "The Emergence of the Feature Film in Germany Between 1910 and 1911." In *Before Caligari: German Cinema, 1895–1920*, ed. Paolo Cherchi Usai and Lorenzo Codelli, 94–113. Madison: University of Wisconsin Press, 1990.

———. *Frühe deutsche Kinematographie: formale, wirtschaftliche und kulturelle Entwicklungen 1907–1912*. Stuttgart and Weimar: Metzler, 1994.

Musser, Charles. "A Cinema of Contemplation, A Cinema of Discernment: Spectatorship, Intertextuality and Attractions in the 1890s." In *The Cinema of Attractions Reloaded*, ed. Wanda Strauven, 159–179. Amsterdam: Amsterdam University Press, 2006.

———. *The Emergence of Cinema: The American Screen to 1907*. New York: Scribner, 1990.

Myers, Greg. "Nineteenth-Century Popularizations of Thermodynamics and the Rhetoric of Social Prophecy." In *Energy & Entropy: Science and Culture in Victorian Britain*, ed. Patrick Brantlinger, 307–338. Bloomington: Indiana University Press, 1988.

Nead, Lynda. *The Haunted Gallery: Painting, Photography, Film c. 1900*. New Haven and London: Yale University Press, 2007.

Newman, Lex. "Ideas, Pictures, and the Directness of Perception in Descartes and Locke." *Philosophy Compass* 4, no. 1 (2009): 134–154.

Nicholas, J. S. "Ross Granville Harrison, 1870–1959." *Yale Journal of Biology and Medicine* 32, no. 6 (June 1960): 407–412.

Nichtenhauser, Adolf. "History of Motion Pictures in Medicine." Unpublished manuscript. MS C 380, History of Medicine Division, National Library of Medicine, Bethesda, Md.

North, Paul. *The Problem of Distraction*. Stanford, Calif.: Stanford University Press, 2012.

Nye, Mary Jo. *Molecular Reality: A Perspective on the Scientific Work of Jean Perrin*. New York: American Elsevier, 1972.

———. "The Nineteenth-Century Atomic Debates and the Dilemma of an 'Indifferent Hypothesis.'" *Studies in History and Philosophy of Science* 7, no. 3 (1976): 245–268.

Nye, Robert A. *Crime, Madness, and Politics in Modern France: The Medical Concept of National Decline*. Princeton, N.J.: Princeton University Press, 1984.

———. *The Origins of Crowd Psychology: Gustave Le Bon and the Crisis of Mass Democracy in the Third Republic*. London and Beverly Hills, Calif.: Sage, 1975.

Nyhart, Lynn K. "Learning from History: Morphology's Challenges in Germany ca. 1900." *Journal of Morphology* 252 (April 2002): 2–14.

———. *Modern Nature: The Rise of the Biological Perspective in Germany*. Chicago: University of Chicago Press, 2009.

———. "Science, Art, and Authenticity in Natural History Displays." In *Models: The Third Dimension of Science*, ed. Soraya De Chadarevian and Nick Hopwood, 307–335. Stanford, Calif.: Stanford University Press, 2004.

Odin, Steve. *Artistic Detachment in Japan and the West: Psychic Distance in Comparative Aesthetics*. Honolulu: University of Hawaii Press, 2001.

Oelkers, Jürgen. *Reformpädagogik. Eine kritische Dogmengeschichte*. Weinheim and Munich: Juventa, 1989.

Oesterlen, Friedrich. *Medical Logic*. Translated by G. Whitley. London: Sydenham Society, 1855.

Oksiloff, Assenka. *Picturing the Primitive: Visual Culture, Ethnography, and Early German Cinema*. New York: Palgrave, 2001.

Orgeron, Devin, Marsha Orgeron, and Dan Streible, eds. *Learning with the Lights Out*. New York: Oxford University Press, 2012.

Ostherr, Kirsten. *Cinematic Prophylaxis: Globalization and Contagion in the Discourse of World Health*. Durham, N.C.: Duke University Press, 2005.

———. *Medical Visions: Producing the Patient Through Film, Television, and Imaging Technologies*. New York: Oxford University Press, 2013.

Pasveer, Bernike. "Knowledge of Shadows: The Introduction of X-Ray Images in Medicine." *Sociology of Health and Illness* 11, no. 4 (December 1989): 361–381.

———. "Representing or Mediating: A History and Philosophy of X-Ray Images in Medicine." In *Visual Cultures of Science: Rethinking Representational Practices in Knowledge Building and Science Communication*, ed. Luc Pauwels, 41–62. Lebanon, N.H.: Dartmouth College Press/University Press of New England, 2006.

Pernick, Martin. *The Black Stork: Eugenics and the Death of "Defective" Babies in American Medicine and Motion Pictures Since 1915*. New York: Oxford University Press, 1999.

Peterson, Jennifer Lynn. *Education in the School of Dreams: Travelogues and Early Nonfiction Film*. Durham, N.C.: Duke University Press, 2013.

Petro, Patrice. *Joyless Streets: Women and Melodramatic Representation in Weimar Germany*. Princeton, N.J.: Princeton University Press, 1989.

Pflug, Günther. "Die Bergson-Rezeption in Deutschland." *Zeitschrift für philosophische Forschung* 45, no. 2 (April–June 1991): 257–266.

Podoll, K., and J. Lüning. "Geschichte des wissenschaftlichen Films in der Nervenheilkunde in Deutschland 1895–1929." *Fortschritte der Neurologie, Psychiatrie* 66 (1998): 122–132.

Pollmann, Inga. "Cinematic Vitalism: Theories of Life and the Moving Image." PhD diss., University of Chicago, 2011.

Praehauser, Ludwig. *Erfassen und Gestalten: Die Kunsterziehung als Pflege formender Kräfte*. Salzburg: Müller, 1950.

Präffcke, Hans. *Der Kunstbegriff Alfred Lichtwarks*. Hildesheim, Zürich, and New York: Olms, 1986.

Prigogine, Ilya. *From Being to Becoming: Time and Complexity in the Physical Sciences*. San Francisco: Freeman, 1980.

———, and Isabelle Stengers. *The End of Certainty: Time, Chaos, and the New Laws of Nature*. New York: Free Press, 1997.

———. *Order Out of Chaos: Man's New Dialogue with Nature.* Boulder, Colo.: New Science Library, 1984.
Quaresima, Leonardo. "*Dichter, Heraus!* The *Autorenfilm* and German Cinema of the 1910s." *Griffithiana* 38/39 (October 1990): 101–120.
Rabinbach, Anson. *The Human Motor: Energy, Fatigue, and the Origins of Modernity.* Berkeley and Los Angeles: University of California Press, 1990.
Radder, Hans. *In and About the World: Philosophical Studies of Science and Technology.* Albany: State University of New York, 1996.
———. *The Material Realization of Science.* Assen and Maastricht: Van Gorcum, 1988.
———, ed. *The Philosophy of Scientific Experimentation.* Pittsburgh, Pa.: University of Pittsburgh Press, 2003.
Radkau, Joachim. *Das Zeitalter der Nervosität: Deutschland zwischen Bismarck und Hitler.* Munich: Hanser, 1998.
Ramsey, L. J. "Early Cineradiography and Cinefluorography." *History of Photography* 7, no. 4 (October–December 1983): 311–322.
Rancière, Jacques. *The Emancipated Spectator.* Translated by Gregory Elliott. New York: Verso, 2009.
Rasmussen, Nicolas. *Picture Control: The Electron Microscope and the Transformation of Biology in America, 1940–1960.* Stanford, Calif.: Stanford University Press, 1997.
Reichert, Ramón. "Der Arbeitstudienfilm: Eine verborgene Geschichte des Stummfilms." *medien & zeit: Kommunikation in Vergangenheit und Gegenwart* 5 (2002): 46–57.
Reill, Peter Hanns. *Vitalizing Nature in the Enlightenment.* Berkeley: University of California Press, 2005.
Reiser, Stanley Joel. *Medicine and the Reign of Technology.* Cambridge: Cambridge University Press, 1978.
Renn, Jürgen. "Einstein's Invention of Brownian Motion." *Annalen der Physik* 14, supplement (2005): 23–37.
Repp, Kevin. *Reformers, Critics, and the Paths of German Modernity: Anti-Politics and the Search for Alternatives, 1890–1914.* Cambridge, Mass.: Harvard University Press, 2000.
Reulecke, Jürgen. *Geschichte der Urbanisierung in Deutschland.* Frankfurt: Suhrkamp, 1985.
———. *Sozialer Frieden durch soziale Reform: Der Centralverein für das Wohl der Arbeitenden Klassen in der Frühindustrialisierung.* Wuppertal: Hammer, 1983.
Rheinberger, Hans-Jörg. *Toward a History of Epistemic Things: Synthesizing Proteins in the Test Tube.* Stanford, Calif.: Stanford University Press, 1997.
Rich, Norman. *The Age of Nationalism and Reform, 1850–1890.* New York: Norton, 1970.
Ringer, Fritz K. *The Decline of the German Mandarins: The German Academic Community, 1890–1933.* Cambridge, Mass.: Harvard University Press, 1969.
Ritter, Christopher Owen. "Re-presenting Science: Visual and Didactic Practice in Nineteenth-Century Chemistry." PhD diss., University of California, Berkeley, 2001.
Ritzheimer, Kara L. "Protecting Youth from 'Trash': Anti-*Schund* Campaigns in Baden, 1900–1933." PhD diss., State University of New York–Binghamton, 2007.

Rosa, Hartmut, and William E. Scheuerman, eds. *High-Speed Society: Social Acceleration, Power, and Modernity.* University Park: Pennsylvania State University Press, 2009.
Rosen, George. "The Efficiency Criterion in Medical Care, 1900–1920." *Bulletin of the History of Medicine* 50, no. 1 (Spring 1976): 28–44.
Ross, Corey. *Media and the Making of Modern Germany: Mass Communications, Society, and Politics from the Empire to the Third Reich.* New York: Oxford University Press, 2008.
Rossell, Deac. "Beyond Messter: Aspects of Early Cinema in Berlin." *Film History* 10 (1998): 52–69.
———. *Faszination der Bewegung: Ottomar Anschütz zwischen Photographie und Kino.* KINtop Schriften 6. Frankfurt: Stroemfeld/Roter Stern, 2001.
———. *Living Pictures: The Origins of the Movies.* Albany: State University of New York Press, 1998.
Rubin, Lewis Philip. "Leo Loeb's Role in the Development of Tissue Culture." *Clio Medica* 12, no. 1 (1977): 33–56.
Sachße, Christoph. *Mütterlichkeit als Beruf: Sozialarbeit, Sozialreform und Frauenbewegung, 1871–1929.* Frankfurt: Suhrkamp, 1986.
Sarasin, Philipp. "Die Rationalisierung des Körpers: Über 'Scientific Management' und 'biologische Rationalisierung.'" In *Geschichtswissenschaft und Diskursanalyse*, 61–99. Frankfurt: Suhrkamp, 2003.
Sayre, Robert, and Michael Löwy. "Figures of Romantic Anti-capitalism." *New German Critique* 32 (Spring/Summer 1984): 42–92.
Schaffer, Simon. "Transport Phenomena: Space and Visibility in Victorian Physics." *Early Popular Visual Culture* 10, no. 1 (February 2012): 71–91.
Scheibe, Wolfgang. *Die Reformpädagogische Bewegung, 1900–1932: Eine einführende Darstellung.* 9th ed. Weinheim and Basel: Beltz, 1984.
Schenda, Rudolf. *Die Lesestoffe der kleinen Leute: Studien zur populären Literatur im 19. und 20. Jahrhundert.* Munich: Beck, 1976.
Schlich, Thomas. "'Wichtiger als der Gegenstand selbst': Die Bedeutung des fotografischen Bildes in der Begründung der bakteriologischen Krankheitsauffassung durch Robert Koch." In *Neue Wege in der Seuchengeschichte*, ed. Martin Dinges and Thomas Schlich, 143–174. Stuttgart: Steiner, 1995.
Schlüpmann, Heide. *Unheimlichkeit des Blicks: Das Drama des frühen deutschen Kinos.* Frankfurt: Stroemfeld/Roter Stern, 1990.
Schmidgen, Henning. "1900—The Spectatorium: On Biology's Audiovisual Archive." *Grey Room* 43 (2011): 42–65.
———. "Pictures, Preparations, and Living Processes: The Production of Immediate Visual Perception (*Anschauung*) in Late-19th-Century Physiology." *Journal of the History of Biology* 37, no. 3 (October 2004): 477–513.
Schmidt, Gunnar. *Anamorphotische Körper: Medizinische Bilder vom Menschen im 19. Jahrhundert.* Cologne: Böhlau, 2001.
Schmitt, Heiner. *Kirche und Film: Kirchliche Filmarbeit in Deutschland von ihren Anfängen bis 1945.* Boppard: Boldt, 1979.

Schorr, Thomas. "Die Film- und Kinoreformbewegung und die Deutsche Filmwirtschaft. Eine Analyse des Fachblatts *Der Kinematograph* (1907–1935) unter pädagogischen und publizistischen Aspekten." PhD diss., Universität der Bundeswehr, Munich, 1990.

Schulte, Christoph. *Psychopathologie des Fin de Siècle: Der Kulturkritiker, Arzt und Zionist Max Nordau.* Frankfurt: Fischer Taschenbuch, 1997.

Schulze, Volker. "Frühe kommunale Kinos und die Kinoreformbewegung in Deutschland bis zum Ende des ersten Weltkriegs." *Publizistik* 22, no. 1 (January–March 1977): 61–71.

Schweinitz, Jörg, ed. *Prolog vor dem Film: Nachdenken über ein neues Medium, 1909–1914.* Leipzig: Reclam, 1992.

Secord, Anne. "Botany on a Plate: Pleasure and the Power of Pictures in Promoting Early Nineteenth-Century Scientific Knowledge." *Isis* 93, no. 1 (March 2002): 28–57.

Shapin, Steven. "Pump and Circumstance: Robert Boyle's Literary Technology." *Social Studies of Science* 14, no. 4 (November 1984): 481–520.

———, and Simon Schaffer. *Leviathan and the Air-Pump: Hobbes, Boyle, and the Experimental Life.* Princeton, N.J.: Princeton University Press, 1985.

Sicard, Monique, Robert Pujade, and Daniel Wallach. *À corps et à raison: Photographies médicales, 1840–1920.* Paris: Marval, 1995.

Singer, Ben. "The Ambimodernity of Early Cinema: Problems and Paradoxes in the Film-and-Modernity Discourse." In *Film 1900: Technology, Perception, Culture*, ed. Annemone Ligensa and Klaus Kreimeier, 37–51. New Barnet, U.K.: Libbey, 2009.

———. *Melodrama and Modernity: Early Sensational Cinema and Its Contexts.* New York: Columbia University Press, 2001.

Snyder, Joel. "*Res Ipsa Loquitur.*" In *Things That Talk: Object Lessons from Art and Science*, ed. Lorraine Daston, 195–221. New York: Zone, 2004.

Sobchack, Vivian. *The Address of the Eye: A Phenomenology of Film Experience.* Princeton, N.J.: Princeton University Press, 1992.

Söder, Hans-Peter. "Disease and Health as Contexts of Modernity: Max Nordau as a Critic of Fin-de-Siècle Modernism." *German Studies Review* 14, no. 3 (October 1991): 473–487.

Spengler, Oswald. *The Decline of the West.* Translated by Charles Francis Atkinson. New York: Knopf, 1926–1928.

Springman, Luke. "Poisoned Hearts, Diseased Minds, and American Pimps: The Language of Censorship in the *Schund und Schmutz* Debates." *German Quarterly* 68, no. 4 (Autumn 1995): 408–429.

Stachel, John. "Einstein on Brownian Motion." In *The Collected Papers of Albert Einstein*, vol. 2, *The Swiss Years: Writings, 1900–1909*, ed. John Stachel, 206–222. Princeton, N.J.: Princeton University Press, 1989.

Stark, Gary D. "Cinema, Society, and the State." In *Essays on Culture and Society in Modern Germany*, ed. Gary D. Stark and Bede Karl Lackner, 122–166. College Station: Texas A & M University Press, 1982.

Starr, Paul. *The Social Transformation of American Medicine.* New York: Basic, 1982.

Stern, Fritz. *The Politics of Cultural Despair: A Study in the Rise of the Germanic Ideology.* Berkeley: University of California Press, 1961.

Stone, Judith F. *The Search for Social Peace: Reform Legislation in France, 1890–1914.* Albany, N.Y.: State University of New York Press, 1985.

Storim, Mirjam. "'Einer, der besser ist, als sein Ruf': Kolportageroman und Kolportagebuchhandel um 1900 und die Haltung der Buchbranche." In *Schund und Schönheit: Populäre Kultur um 1900*, ed. Kaspar Maase and Wolfgang Kaschuba, 252–282. Cologne: Böhlau, 2001.

Sweeney, Dennis. "Reconsidering the Modernity Paradigm: Reform Movements, the Social and the State in Wilhelmine Germany." *Social History* 31, no. 4 (November 2006): 405–434.

Szczepaniak-Gillece, Jocelyn. "Machines for Seeing: Cinema, Architecture, and Midcentury American Spectatorship." PhD diss., Northwestern University, 2013.

Szczepanik, Petr. "Modernism, Industry, Film: A Network of Media in the Bat'a Corporation and the Town of Zlín in the 1930s." In *Films That Work: Industrial Film and the Productivity of Media*, ed. Vinzenz Hediger and Patrick Vonderau, 349–376. Amsterdam: Amsterdam University Press, 2009.

Szöllösi-Janze, Margit. "Science and Social Space: Transformations in the Institutions of *Wissenschaft* from the Wilhelmine Empire to the Weimar Republic." *Minerva* 43 (2005): 339–360.

Takaya, Keiichi. "The Method of *Anschauung*: From Johann H. Pestalozzi to Herbert Spencer." *Journal of Educational Thought* 37, no. 1 (2003): 77–99.

Taurek, Renata. *Die Bedeutung der Photographie für die medizinische Abbildung im 19. Jahrhundert*. Cologne: Hansen, 1980.

Taylor, Peter J., and Ann S. Blum. "Pictorial Representation in Biology." *Biology and Philosophy* 6, no. 2 (April 1991): 125–134.

Teuteberg, Hans Jürgen, ed. *Urbanisierung im 19. und 20. Jahrhundert: historische und geographische Aspekte*. Cologne: Böhlau, 1983.

Tosi, Virgilio. *Cinema Before Cinema: The Origins of Scientific Cinematography*. Translated by Sergio Angelini. London: British Universities Film & Video Council, 2005.

Tsivian, Yuri. *Early Cinema in Russia and Its Cultural Reception*. Edited by Richard Taylor. Translated by Alan Bodger. London and New York: Routledge, 1994.

Tucker, Jennifer. *Nature Exposed: Photography as Eyewitness in Victorian Science*. Baltimore: Johns Hopkins University Press, 2005.

———. "Photography as Witness, Detective, and Imposter: Visual Representation in Victorian Science." In *Victorian Science in Context*, ed. Bernard Lightman, 378–408. Chicago: University of Chicago Press, 1997.

Ulich, Robert. "Pestalozzi, Johann Heinrich." In *The Encyclopedia of Philosophy*, vol. 6, ed. Paul Edwards, 121–122. New York: Macmillan, 1967.

Valéry, Paul. *Idée Fixe*. Translated by David Paul. New York: Pantheon, 1965.

Veatch, Robert M. "Generalization of Expertise." *Hastings Center Studies* 1, no. 2 (1973): 29–40.

Virilio, Paul. *War and Cinema: The Logistics of Perception*. Translated by Patrick Camiller. London: Verso, 1989.

vom Bruch, Rüdiger, ed. *Weder Kommunismus noch Kapitalismus: Bürgerliche Sozialreform in Deutschland vom Vormärz bis zur Ära Adenauer*. Munich: Beck, 1985.

———. *Wissenschaft, Politik und öffentliche Meinung: Gelehrtenpolitik im Wilhelminischen Deutschland, 1890–1914*. Husum, Germany: Matthiesen, 1980.

Vondung, Klaus. "Zur Lage der Gebildeten in der wilhelminischen Zeit." In *Das wilhelminische Bildungsbürgertum Zur Sozialgeschichte seiner Ideen*, ed. Klaus Vondung, 20–33. Göttingen: Vandenhoeck & Ruprecht, 1976.
Wagner, Juliet Clare. "Twisted Bodies, Broken Minds: Film and Neuropsychiatry in the First World War." PhD diss., Harvard University, 2009.
Warner, John Harley. *Against the Spirit of System: The French Impulse in Nineteenth-Century American Medicine*. Princeton, N.J.: Princeton University Press, 1998.
——. "Ideals of Science and Their Discontents in Late Nineteenth-Century American Medicine." *Isis* 82, no. 3 (September 1991): 454–478.
Warnotte, Daniel. *Ernest Solvay et l'Institut de Sociologie: Contribution à l'histoire de l'énergétique sociale*. Brussels: Bruylant, 1946.
Warwick, Andrew. "X-Rays as Evidence in German Orthopedic Surgery, 1895–1900." *Isis* 96, no. 1 (March 2005): 1–24.
Weber, Max. "Science as a Vocation." In *The Vocation Lectures*, ed. and with an introduction by David Owen and Tracy B. Strong, trans. Rodney Livingstone, 1–31. Indianapolis: Hackett, 2004.
Wehler, Hans-Ulrich. *The German Empire, 1871–1918*. Leamington Spa, U.K.: Berg, 1985.
Weindling, Paul. "Bourgeois Values, Doctors and the State: The Professionalization of Medicine in Germany 1848–1933." In *The German Bourgeoisie*, ed. David Blackbourn and Richard J. Evans, 198–223. New York: Routledge, 1991.
——. *Health, Race, and German Politics Between National Unification and Nazism, 1870–1945*. Cambridge and New York: Cambridge University Press, 1989.
——. "Public Health in Germany." In *The History of Public Health and the Modern State*, ed. Dorothy Porter, 119–131. Atlanta, Ga.: Rodopi, 1994.
Wesfreid, José Eduardo. "Scientific Biography of Henri Bénard (1874–1939)." In *Dynamics of Spatio-Temporal Cellular Structures: Henri Bénard Centenary Review*, ed. Innocent Mutabazi, José Eduardo Wesfreid, and Étienne Guyon, 9–40. New York: Springer, 2006.
Whitbeck, Caroline. "What Is Diagnosis? Some Critical Reflections." *Metamedicine* 2, no. 3 (October 1981): 319–329.
Wilder, Kelly. *Photography and Science*. London: Reaktion, 2009.
Wilkending, Gisela. *Volksbildung und Pädagogik "vom Kinde aus": Eine Untersuchung zur Geschichte der Literaturpädagogik in den Anfängen der Kunsterziehungsbewegung*. Weinheim, Germany: Beltz, 1980.
Winston, Brian. "The Documentary Film as Scientific Inscription." In *Theorizing Documentary*, ed. Michael Renov, 37–57. London and New York: Routledge, 1993.
Wohlring, Herman J. Review of *The Human Gait*. *Human Movement Science* 8, no. 1 (February 1989): 79–83.
Yoxen, Edward. "Seeing with Sound: A Study of the Development of Medical Images." In *The Social Construction of Technological Systems: New Directions in the Sociology and History of Technology*, ed. Wiebe E. Bijker, Thomas P. Hughes, and Trevor J. Pinch, 281–303. Cambridge, Mass.: MIT Press, 1987.

INDEX

ABC der Anschauung (Pestalozzi), 173
abstraction: of movement, 34, 37–38, 49–63, 74–75; of spectators, 11–12
active and passive viewers, 200, 232. *See also* expert/lay distinctions
Adorno, Theodor, 199
adult education and motion pictures, 184–90
aesthetic contemplation, 8, 10, 17, 211, 251, 197–202; agency or free will and, 201, 213, 214; alertness (*Geistesgegenwart*) and, 213; Benjamin and, 235, 240–41; distraction and, 201, 233–35; as expert vision or observation, 10, 135, 195; gender and, 231–32; as ground for aesthetic experience, 219, 230; indeterminacy and, 208; Kracauer and, 234; Lukács and, 235–40; mass reception and, 201, 216, 217; moral significance of, 241; motion pictures and, 17, 189, 195–96, 230–31; movement and, 226, 230–31; as observational training, 170–72; passivity and, 201; political implications of, 232; Schiller and, 190; Schopenhauer and, 207, 214. *See also* aesthetic experience
aesthetic education, 143–47; film reform and, 172; as German tradition, 143; moral renewal and, 176, 169–70; museums and, 170; nationalism and, 168–71; observational training and, 163, 168–71; Schiller and, 167–68, 171; suggestibility and, 167

aesthetic experience, 5, 7, 8, *198*; aesthetic education and, 143; agency within, 201, 202–14; descriptions of, 196–201; detachment and, 196–98, 202, 209, 220, 241; embodied vision and, 214–30; as emotional projection, 217, 220–21; identity and, 214–16; moral significance of, 197–201, 207; movement and, 224–30; as renewal of perception, 213; repose and, 209; Schiller and, 205–07; Schopenhauer and, 207, 214; synesthesia and, 221–22; as system of dichotomies, 196–201; taste and, 191; temporality and, 204–7. *See also* aesthetic contemplation
aesthetics, 1, 4–5; *Einfühlung* as solution to problems in, 215–16; formalist and scientific approaches to, 215; reception and form in, 195; "traditional," 17, 190, 195, 207, 214, 216–17, 237
agency, 5, 17, 197–99, 201, 202–14. *See also* free will; volition; will
Die Aktion, 239
alertness or presence of mind (*Geistesgegenwart*), 212–14, 240
all-at-onceness, 120. *See also* observation: gaze and glance
alignment of image, object, and technology, 38, 43–62
Allgemeines Krankenhaus, 103
The Alps (film), 243
Altenloh, Emilie, 165, 200
alternative public sphere, early film exhibition as, 184
American College of Surgeons, 109
American Hospital Association, 109
analysis and synthesis, 11, 25, 36–37, 42, 112, 117–18, 124; medical observation and, 120–25; motion pictures and, 32, 36; scientific cinematography's relationship to, 87, 117–18; spectatorship and, 165
analysis (close reading), 4–5, 12, 247–49

analysis of motion, 20, 38, 40, 44, 77, 115–16; Bergson's critique of, 32–33, 87; frame-by-frame, 60, 98, 117, 125
Andreas-Salomé, Lou, 209–10
Andrew, Dudley, 201
Andriopolous, Stefan, 135
anesthesia, 138
animal locomotion, 41, 45
Anschaulichkeit (vividness), 107–8, 179
Anschauung (sense impression), 172–76
Anschauungsunterricht (visual means of instruction), 9, 10, 16, 145–46, 172–84; as apperception, 174–75; correlation and, 174–75; detail and, 173–74; elementary and adult education and, 176–77; Herbartian principles of, 182–83; Lichtwark and, 163; motion pictures and, 146, 182–83; natural sciences and, 175; as object lesson, 173–75; as observational training, 10, 145, 162, 173–75; Pestalozzi and, 172, 174–75; relationship to images and words, 175–76; self-cultivation (*Bildung*) and, 145; social acceleration and, 173–74. *See also* object lesson
Anschütz, Ottomar, 45, 265n73
apperception, 173–74, 181–82, 226–27
Arbeitsschule (works schools), 149
Arbeitswissenschaft (the science of work), 42
architecture, 190, 224–27, 241
archives, of images, 97, 103–5, 114–16
Aristotelian science, 35
Arnheim, Rudolf, 244
art education movement (*Kunsterziehungsbewegung*), 16, 149, 169–71, 176
Ashwin, Clive, 173, 176
atomic-kinetic model of heat, 20, 65, 67
atomic-kinetic theory of matter, 64–65
attention (*Aufmerksamkeit*), 126, 132–33, 139–40, 174, 178, 189, 198–99, 201, 220
aura, 190

automatic writing, 45
autopsies, 119
Autorenfilm (author's film), 194

Bachelard, Gaston, 23
Balàzs, Béla, 244, 250, 279n70
bearers of culture (*Kulturträger*), 127. See also physicians
beautiful, theory of the, 214, 304n10
Bellevue Hospital, 104–5
Bellour, Raymond, 136
Bénard, Henri, 76
Benjamin, Walter, 193, 202, 218, 236–37, 244, 250; alertness and, 213; aura and, 190; Bergson and, 36; contemplation and, 201, 208, 235, 240–41; distraction and, 201, 233, 235; Duhamel and, 212; expert/lay distinctions and, 12; mode of perception and, 13; optical unconscious and, 86; traditional aesthetics and, 195, 214
Bergson, Henri, 21, 25, 38, 39, 43, 60, 73, 116, 117, 235; analysis and, 87–88; Benjamin and, 36; cinematographical thinking and, 15, 22, 32–37; critique of science and, 32–37, 62–64; *eidos* and, 59; German counterparts to, 36–37; movement and, 33–34; synthesis and, 36; time and, 74–76; United States and, 22, 32, 36
Bernheim, Hippolyte, 135
Bildbetrachtung (image viewing), 10, 171, 189
Bild und Film (trade journal), 152, 160, 184
Bild und Wort (Image and Word society) (film exhibitions), 184
Bildung (self-cultivation), 145; visual training and, 141
Bildungsbürgertum (educated middle-class), 126, 134
Billings, Susan, 81
Billroth, Theodor, 119
Binet, Alfred, 164

biology, 14, 22, 32, 97; cell, 6, 21, 25, 37, 76–88, 250
Der Blaue Engel (The Blue Angel) (film) 142
Bloch, Ernst, 194, 222–24
The Blue Angel (film), 142
Borbacher Knabenmord (child murder of Borbach), 287n151
Botryllus (sea squirts), 76
Bourdieu, Pierre, 143
Braun, Ludwig, 98–100, 111–21, 124, 248, 250
Braune, Wilhelm, 19–20, 24–26, 38, 41–44, 62–63, 88, 91, 98, 116. See also *The Human Gait*
Braus, Hermann, 20, 24, 25, 29, 78; nerve fibers and, 77, 80, 82–86, 88
Brecht, Bertolt, 240, 241
British Medical Association, 106
Brod, Max, 194, 212
Brown, Robert, 65
Brownian motion, 20, 24–25, 27, 62–76, 71, 72, 77, 81, 87, 266n97; Einstein's theory of, 64–69, 266n94, 266n98; importance of displacement to Einstein's theory, 68–69
Brunner, Karl, 152–53, 157–58, 292n34
Brush, Stephen, 69
Bull, Lucien, 27, 76
Burrows, Montrose, 83

The Cabinet of Dr. Caligari (film), 244
capitalism, 143, 144, 147, 161
cardiac dynamics and mechanics, 98, 99, 100
Carlet, H. M., 45
Carrel, Alexis, 83
Cartwright, Lisa, 104, 113
Carvallo, Joachim-Léon, 27, 100, 107
cell biology, 6, 21, 25, 37, 76–88, 250
censorship, 153, 158, 186
Century of the Child (Key), 163
Charcot, Jean-Martin, 115, 137, 164
Charité Hospital, 105

Charles Urban Trading Company, 156
Chase, Walter, 103
Chevroton, Lucienne, 27, 76
child psychology, 127, 146, 162–64. See also children and crowds: analogies between; crowd psychology
children and crowds: analogies between, 143, 162; aesthetic sensibilities of, 143; as models for descriptions of film spectatorship, 162, See also crowd psychology
chronophotography, 15, 21, 23, 25, 28, 60–61, 62, 116; Brownian motion and, 20, 69, 73; ergonomics and, 41; for *The Human Gait*, 44–62, 52; human locomotion and, 19–20; two-sided, 46
cinema: theater compared to, 236–39
Cinema and Art (*Kino und Kunst*) (Häfker), 184
"Cinema and Schoolchildren" (Kleibömer), 202
Cinema Reform Association, 155, 162
The Cinematic Lesson Plan (*Die kinematographische Unterrichtsstunde*) (Lemke), 182, 183
cinematographical thinking, 15, 15, 22, 32–37. See also Bergson, Henri
cinematography, 21–22; adaptation of for biology and physics, 76–77; as aid to correlation, 116, 123; analysis and synthesis and, 117; Bergson's analogy of, 32–37; Braun and, 111–14; Braus and, 20, 84–87; Brownian motion and, 66, 69–70; as confirmation of biological theory, 85; high-speed, 27, 30, 76; medical, 92–93, 98 114, 243; microcinematography, 27, 76, 92–93, 98; science and, 32–37; slow-motion, 27, 86, 123; as substitute for object of study, 84; time-lapse, 23, 30, 84–85, 86, 98, 123, 250; as optical unconscious, 86; as virtual experiment, 84, 100; as virtual witness, 84, 100; X-rays and, 27, 100–102, *101*. See also chronophotography; *The Human Gait*; medical filmmaking; research film
class: educated middle-class, 126, 134; ideals, 245; identity, 13, 122; warfare, 292n34; working, 232
class-based distinctions, 12, 95, 245
classical physics, 63, 65, 68
Clausius, Rudolf, 40
clinical gaze, 115, 118–19, 122. See also observation: gaze and glance
clothing reform, 149–50
Cole, Lewis Gregory, 100
Collège de France, 27, 76
Comandon, Jean, 28, 29, 76, 102, 104
communal sense (*sensus communis*), 191
continuity and discontinuity, 5, 25, 32–34, 36, 61, 64, 75, 82, 83, 86–87, 117, 178
correlation, 8, 111, 124, 139, 163, 173, 175; definition of, 114; medical observation and, 115–17; film as aid to, 122, 123, 246
country boarding schools (*Landerziehungsheime*), 149
Cranz, Carl, 28
Crary, Jonathan, 132–33, 215
Creative Evolution (Bergson), 22, 25, 32, 37
crime, hypnotism and, 135–36
critical method (*kritischer Methode*), 130. See also observation
Critique of Judgment (Kant), 197
Critiques (Kant), 172, 197
Cros, Charles, 19
crowd psychology, 137, 146; theories of, 164–65; suggestibility and, 165–66; will or volition and, 166. See also children and crowds: analogies between
"The Cult of Distraction" (Kracauer), 233
cultural capital, 122
cultural pessimism, 128

cultural series, 13
The Culture of the Female Body as a Foundation for Women's Clothing (Schultze-Naumberg), 149–50
curricular integration, 181

Dadaists, 232
Darwin, Charles, 128, 137, 164, 296n80
Daston, Lorraine, 120
de Broglie, Louis, 74
Degeneration (Nordau), 128
degeneration (concept), 40 128, 131–32
de Lagarde, Paul, 148, 168
Deleuze, Gilles, 37, 75–76
Descartes, René, 63, 119, 210–11
The Descent of Man (Darwin), 164
Dessoir, Max, 307n51, 308n55
detachment: aesthetic, 191, 196–98, 202, 209, 220, 241; expert, 11, 12, 245. *See also* observation: aesthetic experience
detail: as fidelity to nature, 5, 93, 107–08, 111–13, 179; as ground for authenticity, 173–75, 179; as ground for observational practice, 119–20, 139, 204, 246; scientific management of, 25, 46, 52, 60, 125; as symptom of modernity, 174, 189, 211
Dewey, John, 171
Diderot, Denis, 232
Didi-Huberman, Georges, 115
Diederichs, Helmut, 194
differential reproduction, 250, 257n9
Dilthey, Wilhelm, 36
direct perception (immediate or sensual), 7, 108, 110, 122, 145, 172, 173, 221, 250, 279n70
disciplinary agendas, 2–3, 4, 88, 89, 245–46, 249; technology and, 3–4, 14–15, 249
disciplinary logic, 4–6, 9–10, 14–15, 23–24, 109, 248–49; film form and, 5–6, 23–24, 111–25. *See also* medical logic; patterns of use: film form and

discontinuity. *See* continuity and discontinuity
disease, 91, 101, 105, 119, 127–28; cinematography and, 140; neurological, 103; series photography and, 113–15
disinterest. *See* detachment; aesthetic
displacement, of particles, 64, 68–70, 72–73. *See also* Brownian motion
dispositif, 246; as experimental material, 250–51
dissemination of media technologies, 6, 10
distraction, 7, 9, 195, 240, 244; attention and, 133; Benjamin and, 201, 233, 235; Kracauer and, 233; proximity and, 190
Döblin, Alfred, 194
docile bodies (*aka* working objects), 42–44
doctor–patient relationship, 95, 126
documentary function, of medical filmmaking, 102–5
domestication of the image, 42–44, 51–62; of the body, 42–44. *See also* docile bodies
Don Quixote (novel), 236
Doodica (conjoined twin), 106
Doyen, Eugène Louis, 106–10, 279n63
"Drama in the Film Theaters" (Polgar), 220
Driesch, Hans, 32, 39
Duenschmann, Hermann, 210
Duhamel, Georges, 212
duration, 5, 60, 133, 246; cinematography and, 24, 86, 250. *See also durée*
Duration and Simultaneity (Bergson), 32
durée (duration), 32–33, 43, 74; form as interruption of, 62–64
Duttlinger, Carolin, 201, 240

Eclipse (film company), 155–56
educated middle-class (*Bildungsbürgertum*), 126, 134

education, 1, 5, 141; art education movement, 16, 149, 169–71, 176. *See also* adult education and motion pictures; aesthetic education, *Anschauungsunterricht*, elementary education and motion pictures
educational function, of medical filmmaking, 105–10
effects of media technologies, 244–45
efficiency, 38; as social goal, 38–42; and the human body, 38–42; in medicine, 109; of the filmic image, 109–10, 175–76, 210–11
eidetic images, 59
eidos, 59
Einfühlung (emotional projection, feeling into), 17, 202, 215–22, 224–29, 237, 307n51; body as model for, 222; contemplation and, 219, 230; *Kino-Debatte* and, 217, 220; movement and, 224–28; physical response and, 228–30; space and, 225–28
Einstein, Albert, 20–21, 24–25, 64, 66–69, 72–74, 87–88
elementary education and motion pictures, 176–84
Elias, Norbert, 144
embodied perception or vision, 214–30, 237. *See also* aesthetic experience; *Einfühlung*
embourgeoisement, 193
emotional projection. *See Einfühlung*
empathy, 216. *See also Einfühlung*
enemies of the cinema (*Kinogegner*), 130
energy, conservation of, 39–40
Enlightenment, 97, 132, 168, 201
entropy, 40
Epstein, Jean, 136, 250, 279n70
ergograph, 40
ergonomics, 41
Ernemann (film company), 28–29, 180
Essay Concerning Human Understanding (Locke), 211

"The Essence of Architectural Creation" (Schmarsow), 225
Europe, 36, 39, 97, 107, 147
evidence, creation of, 24–25
evil, children as, 163
evolutionary theory, 137
Exercises in the Contemplation of Art Works (Lichtwark), 170–72, 189
Exner, Felix, 66, 69–70
experiment: in biology, 82–83; characteristics of, 98–99; differential reproduction in, 250–51; disciplinary logic and, 23–24; historiographic implications of, 250–51; infrastructure for, 27–28; in medicine, 98–102; motion picture equipment manufacturers and, 28; motion pictures and, 22–24, 27, 88; popular audiences for, 28–29; time and, 139–40
experimental systems, 3–4, 10, 23–24, 64, 257n9
expert modes of viewing, 6–10, 24–25. *See also* observation
expert observation. *See* observation
expert training, 60, 95; advantages of cinematography for, 108; and professional identity, 94; and research films, 106, 117
expert viewers. *See* observation
expert vision. *See* observation
expert/lay distinctions, 11–13, 94–95, 245; historiographic implications of, 245
exploratory function, of medical filmmaking, 98–102
Eykman, P. H., 100

faithful to nature (*Naturgetreu*), 179, 181, 182
Fata Morgana (film theater), 28
fatigue, 39–42, 132, 187
Fechner, Gustav, 215
feeling into. *See Einfühlung*
female hysteria, 137

female spectatorship, 165, 231–32
film drama (Kino-drama), 157, 161, 164, 192
film exhibition, legal restrictions on, 159
film form, 4–5, 6, 14, 246, 249; correspondences between form and logic, 23–24; disciplinary logic and, 248; historiography and, 247
film frame, as spatial and temporal boundary, 116
filmic realism, 5, 157. *See also* Naturgetreu (faithful to nature)
Film-Idea-Central, 156
"Film in the Service of Medicine" (demonstration), 107
film reform, 152–62; distribution and, 159–60; exhibition and, 160, 184–90; foreign films and, 157; goals of, 155; Kino-Kommissions and, 155; Kinoreform and, 155; legacy of, 160–62; narrative and, 157; objections to motion pictures, 154; positive and negative approaches to, 153; production and, 156–59; trade journals and 152–53; uplift and, 154
film strip as line, 74–76
film studies, 3
Film und Lichtbild (periodical), 28
first law of thermodynamics, 40, 65
Fischer, Otto, 19–20, 24–26, 38, 41–44, 63. *See also The Human Gait*
Fleck, Ludwik, 5–7, 113, 249
flicker effect, 130–31, 154, 186, 192
flip books, 44
folk medicine, 126
Forel, Auguste, 135
form: geometry inherent in, 62–63, 74; as temporal interruption in duration, 74. *See also* abstraction: of movement
form drive, 205, 206
Foucault, Michel, 8, 42–43, 91, 115, 118–22
frame-by-frame analysis, 60, 98, 117, 125

France, 27–28, 90, 92, 147, 148
François-Franck, Charles Émile, 27, 76
Fränkel, James, 107
Frankfurter Zeitung, 236
free play of associations, 8
free will, 133, 231; aesthetic contemplation and, 201, 213, 214; aesthetic experience and, 201, 202–14; crowd psychology and, 166; identity and, 217. *See also* agency; volition; will
French Academy of Medicine, 106
Freud, Sigmund, 163–64, 210, 212, 309n72
From Caligari to Hitler (Kracauer), 301n145

Der Gang des Menschen (Braune and Fischer), 19. *See also The Human Gait*
Gaudreault, André, 13, 134
Gaumont (film company), 155
Gaumont, Charles, 76
Gaupp, Robert, 127–28, 136–37, 208
gaze: clinical or medical, 112, 115, 118–19; glance and, 7, 120–22; holistic, 7, 118, 120–21, 165. *See also* aesthetic contemplation, all-at-onceness, observation, Sehen, Schauen
"Gedanken zu einer Aesthetik des 'Kino' (Thoughts Toward an Aesthetic of the Cinema)" (Lukács), 202, 236
Geissler tubes, 46, 48–49, 51–52, 58
Geistesgegenwart (alertness or presence of mind), 212–14, 240
gender. *See* women
gender dynamics, 231
geometry, 58; inherent in form, 62–63, 74. *See also* abstraction: of movement
German High Ministry of War, 44
Germany, reaction to modernity, 13–14
Gesamtkunstwerk (Wagner), 32
Gesamtsinnlichkeit (sensuous totality), 224
Gesamtvorführung (total presentation), 188

Gesellschaft der Freunde des
vaterländischen Schul- und
Erziehungswesens (Society of Friends
of the Schools and Instruction for the
Fatherland), 154
Gesellschaft deutscher Naturforscher und
Ärzte (Society of German Natural
Scientists and Physicians), 83, 111
Gesellschaft zur Verbreitung von
Volksbildung (Society for the
Dissemination of Popular Education)
(GVV), 152, 159
Geuss, Raymond, 144
Gewohnheit (habit), 235, 240
*Giovanna d'Arco (The Maid from
Orleans)* (film), 160
glance, and gaze, 7, 120–22
Goerke, Franz, 105, 106
Goethe, Johann Wolfgang von, 203, 236
Gordon, Rae Beth, 136
Götze, O., 131
Gouy, Louis-Georges, 66
Grau, Alexander, 243
Great Exhibition (1851), 174
Great Lisbon Earthquake, 203
Griffith, D. W., 248
Groedel, Franz, 101–2
Guerlac, Suzanne, 74
Gunning, Tom, 229, 247
gutta-percha, 51
GVV. *See* Gesellschaft zur Verbreitung von
Volksbildung
Gymnasium (high school), 148, 152

habit (*Gewohnheit*), 235, 240
Hacking, Ian, 25, 43
Häfker, Hermann, 16, 152–53, 156,
172, 176–77, 243; contemplation
and, 231, 233; *Kinetographie* and,
188; Lichtwark and, 189; model
presentations and, 147, 184;
photograph of, *185*; taste and, 186;
World War I and, 301n145
Hake, Sabine, 157, 161, 194

Hamburg commission, 152, 154, 165
Hansen, Miriam, 184, 231–32
Harrington, Anne, 32
Harris, Stefanie, 194
Harrison, Ross Granville, 77, 79, 80,
82–83, 85, 88
Hau, Michael, 118, 120, 122
health campaigns, 127
health insurance, 126
heat transference, 40, 65
Hegel, Georg Wilhelm Friedrich, 215
Heimat, 169
Heinrich von Ofterdingen (Novalis), 236
Held, Hans, 81, 83
Heller, Heinz-B., 194
Hellwig, Albert, 130, 154, 158, 166
Hennes, Hans, 103–6
Henri, Victor, 27, 69–70, 73
Herbart, Johann, 174, 176, 181–82, 215,
247
heterogeneity, 245; of early cinema, 2,
12, 13
high school (*Gymnasium*), 148, 152
high-speed cinematography, 27, 30, 76
Hildebrand, Adolf, 199, 221, 224–25
historiography, 4, 247; analogy and
homology in, 249; disciplinary or
expert use of film and, 14, 17, 246–47;
Kino-Debatte and, 195; "tactile,"
246–50
History and Class Consciousness
(Lukács), 235–36, 239–40
holistic gaze, 7, 118, 120–21, 165
Hollywood, mode of production and
reception, 11–12
Huelsenbeck, Richard, 232
Huerkamp, Claudia, 126
The Human Gait (Braune and Fischer),
19, 44–62; camera placement
for, *50*; coordinates graph for,
58, *59*; coordinates table for, *57*;
determination of coordinates for,
53; instrument for coordinate
measurements, *54*; measurement of

coordinate, 55; military recruit for, 47; resulting chronophotograph for, 52; subject at rest, 48; tridimensional model for, 61, 62
human locomotion, 19–20, 26, 38–39, 41–44. See also The Human Gait
human sciences, 36
Husserl, Edmund, 58–59
hypnotism, 138; cinema and, 136–37; as model for observation, 138–40; as model for spectatorship, 135–37
hysteria, 128, 137

Ibsen, Henrik, 128
identity: aesthetic experience and, 197, 214–30; class, 13, 122; free will and, 217; professional, 94, 95, 122, 124, 125; self-identity, 11, 13, 144, 174, 236
Image and Word (*Bild und Wort* society) (film exhibitions), 184
image archives, 97, 103–5, 114–16
image viewing (*Bildbetrachtung*), 10, 171, 189
imagination, 166, 211, 225; architecture and, 226; repose and, 197
imaginative faculty, 209–10
immutability, 31
improper viewing, 11, 94–95. See also spectatorship
incremental exposures, 248. See also series photography
Industrial Revolution, 39, 147–48
"Infantile Sexuality" (Freud), 163
infantilization, of audience, 245. See also children and crowds: analogies between
inner child, 162
Innerlichkeit (inwardness), 239–40
inner movement, 218. See also movement and: aesthetic contemplation; movement and: aesthetic experience
inscription devices, 10, 30–31
Institut de Sociologie, 40
Institut Marey, 27, 100

interdisciplinary research, 3
inwardness (*Innerlichkeit*), 239–40
Die Irrfahrten des Odysseus (*The Wanderings of Odysseus*) (film), 157, 158

Jacobj, Carl, 175
Janssen, Jules, 26
Jelavich, Peter, 194
journals, trade 152–53

Kade, August, 180
Kaes, Anton, 16, 193, 194
Kaiserin-Friedrich-Haus, 107
Kammer-Lichtspiele Theater, 234
Kant, Immanuel, 8, 172, 214–15; aesthetic contemplation and, 201; aesthetic experience and, 7, 195–99; citizenship and, 140–41; taste and, 191
Kästle, C., 100
Keene, Melanie, 174
Kelvin, Lord. See Thomson, William
Key, Ellen, 163
Kienzl, Hermann, 209–10
Kiesler, Frederick, 241
Killen, Andreas, 284n122
Der Kinematograph (trade journal), 152–53, 155, 184, 203
Kinematographie und Schule (Mendel), 180
Die kinematographische Unterrichtsstunde (*The Cinematic Lesson Plan*) (Lemke), 182, 183
Kinetographie, 188
Kinobuch (Pinthus), 239
Kino-Debatte, 16–17, 193–95, 201–2, 236, 241; aesthetic experience and, 208, 217; alertness and, 212; *Einfühlung* and, 217, 220; historiography and, 195; literary emphasis and, 194; somnambulism and, 212; traditional aesthetics and, 214
Kino-drama (film drama), 157, 161, 164, 192

Kinogegner (enemies of the cinema), 130
Kino-Kommissions, 155
Kinoreform, 153–55, 159, 160, 238. See also film reform
Kino und Kunst (Cinema and Art) (Häfker), 184
"Kino und Schaulust" (Serner), 228
Kitsch, 168
Kleibömer, Georg, 202–3
Koch, Robert, 26
Kosmographia, 180
Kracauer, Siegfried, 161, 202, 233–34, 240, 244, 301n145
Kraepelin, Emil, 103, 127
Kretz, Richard, 179
Krieger, Ernst, 243
kritischer Methode (critical method), 130. See also observation
Kultur and *Zivilisation*, 144–45
Kulturträger (bearers of culture), 127. See also physicians
Kunsterziehungsbewegung (art education movement), 16, 149, 169–71, 176
Der Kunstwart, 204
Kutner, Robert, 107, 109
kymograph, 41

laboratories, 19, 77, 81; motion pictures in, 16, 21, 22, 27–30, 92, 125
Ladenkinos (storefront cinemas), 126, 153
Landecker, Hannah, 81
Landerziehungsheime (country boarding schools), 149
Langbehn, August Julius, 168–69
Lange, Konrad, 134, 152, 154, 169–70, 186; *Kino-drama* and, 157; taste and, 143
Latour, Bruno, 30–31, 87
Laycock, Thomas, 119
lay viewers, 232, 245. See also spectatorship
Lebensphilosophie, 39
Le Bon, Gustave, 137, 164

legibility, 31, 221, 224, 249
legitimacy, 61, 100, 124, 244, 249; of motion pictures within a discipline, 1–2, 4–5, 9, 14, 24, 95, 122–23, 195, 217
Lemke, Hermann, 16, 146, 154, 158, 166, 178, 189, 247; Cinema Reform Association and, 155, 162; Film-Idea-Central and, 156; Herbart and, 181–82; Society for the Dissemination of Adult Education and, 159; Society for the Dissemination of Popular Education and, 152
Levin, Tom, 239
Lichtbild-Bühne (trade periodical), 167, 167
Lichtbilderei (film institute), 159–60
Lichtwark, Alfred, 163, 169–72, 189
Lipps, Theodor, 227–28, 230
local exhibition rights, 295n61
Locke, John, 210–11
locomotion: animal, 41, 45; human, 19–20, 26, 38–39, 41–44. See also *The Human Gait*
logic: disciplinary, 4–6, 9–10, 14–15, 23–24, 109, 248–49; film form and, 5–6, 23–24, 111–25; medical, 5, 16, 90–91, 111–25. See also patterns of use: film form and
Lombroso, Cesare, 128
Lomon, André, 102
Londe, Albert, 45
The Lonedale Operator (Griffith), 248
Luisen-Kino, *180*
Lukács, Georg, 9, 194, 202, 240, 288n9, 311n98; child psychology and, 162; contemplation and, 235–37, 239; Romanticism and, 236; time and, 238
Lumière (film company), 28
Lynch, Michael, 43, 49, 59

Mach, Ernst, 65
Macintyre, John, 100

magazines, film 152–53
The Maid from Orleans (Giovanna d'Arco) (film), 160
Maiocchi, Roberto, 66, 68
Mantell, Gideon Algernon, 174
Marey, Étienne-Jules, 26, 40–42, 76, 117, 121; graphic method of, 19, 27, 40; movement and, 27; single-camera system of, 45
Marinescu, Gheorghe, 103
Marital Hygiene (film), 243
Marxism, 236
masculinity, 140, 150–51, 231
mathematical immanent in matter. *See* geometry
Matter and Memory (Bergson), 32
mean square displacement, 68. *See also* Brownian motion
"Measurement of the Temperature Dependency of Brownian Motion" (Seddig), 20
media ensemble *(Medienverbund)*, 3, 178
media technology, 1–4, 9, 21, 189
medical demonstrations, 103–4
medical filmmaking, 92–93, 90–96, 98, 114, 243; documentary function of, 102–5; educational function of, 105–10; exploratory function of, 98–102; multiple functions of, 96–110; observation and, 110–25. *See also* cinematography, research film
medical gaze, 112, 115, 118–19. *See also* gaze; medical perception; observation
medical hermeneutics, 91
medicalization, of society, 127
medical logic, 5, 16, 90–91, 111–25. *See also* disciplinary logic, patterns of use: film form and
medical perception, 119–20. *See also* gaze; medical gaze; observation
medical way of thinking, 16. *See also* medical logic; disciplinary logic; patterns of use: film form and

medicine, 1, 7, 15–16, 103, 113, 122, 123; Doyen and, 106–07; folk, 126; photography and, 98, 112; social, 127; spectacle and, 279n63
Medienverbund (media ensemble), 3, 178
"Melody in the Cinema, or Immanent and Transcendental Music" (Bloch), 222
Mendel, Georg Victor, 180
mental development, 163
Messter, Oskar, 28–29
metaphysics, 239
"The Metropolis and Mental Life" (Simmel), 212
Michaelis, Anthony, 60
microcinematography, 27, 76, 92–93; medical filmmaking and, 98
microphotographs, 26
military, 42
mind, comparisons with motion pictures and modernity, 208–12
model presentations *(Mustervorstellungen)*, 147, 184, 187
mode of perception (Benjamin), 12–13
Modern Hospital, 109
modernity, 36, 38, 143, 194, 200, 201, 235; cultural and social, 41; excesses of, 128, 130, 133, 142, 174, 189, 191; German reaction to, 13–14, 149–51; motion pictures as emblem of, 125–26, 134, 149, 186, 204, 208–9, 217; motion pictures as potential haven from, 187; pace of change within, 131–34, 203–4, 208; as series of shocks, 187, 192. *See also* social acceleration
modernity thesis, 13
Moll, Albert, 138
monopoly system, 295n61
moral weakness, 166
Mosso, Angelo, 40, 42
Le mouvement (Marey), 76

movement, 21, 40; aesthetic education and, 178–79; analysis of, 20, 38, 40, 44, 77, 115–16; animal locomotion, 41, 45; architecture and, 226; Bergson and, 33–34, 75, 87; contemplation and, 226, 230–31; education and, 178; *Einfühlung* and, 224–28; human locomotion, 19–20, 26, 38–39, 41–44; immobility and, 37–38; inner, 218; Marey and, 27; pathological, 104. *See also* Brownian motion; *The Human Gait*
Moving Picture World, 42
multicellular theory of nerve development, 77
Münsterberg, Hugo, 136, 209
music, 222–24
Musser, Charles, 231
Mustervorstellungen (model presentations), 147, 184, 187
Muybridge, Eadweard, 26–27, 45

Nature, 190–91
nature films, 187, 191
Naturgetreu (faithful to nature), 179, 181, 182
negative space, spectatorship as, 11, 250
neo-Kantian tradition, 199
nerve fibers, 76–88, 80
nervousness, 128
neurasthenia, 189
neurological diseases, 103
neurology, 92; cinematography and, 103–5
Newton, Isaac, 63, 65
Nielsen, Asta, 150, 193
Nietzsche, Friedrich, 128, 148
Noguès, Pierre, 27
nontheatrical uses of film, 2; disciplinary agendas and, 2
Nordau, Max, 128, 129, 131–34, 139
Novalis, 236
novels, 236; serialized, 151
Nye, Mary Jo, 68–69

objectivity, 9, 95, 123
object lesson, 16, 108, 145, 173–75, 177–79, 181, 183–84. *See also* Anschauungsunterricht
observation: the accommodation of motion pictures to, 5–6, 9–11, 111–12, 190; aesthetic contemplation and, 10, 135, 170–72, 195; aesthetic education and, 163, 168–71; *Anschauungsunterricht* and, 10, 145, 162, 172–76; analysis and synthesis and, 121, 124; attention and, 126; in biology, 80–82; as correlation, 8, 111–17, 175; detail and, 119–20, 139, 204, 246; disciplinary logic and, 6, 9–10, 23–24, 91, 92, 111–25; as engine of progress, 132; experiment and, 25, 123; gaze and glance as, 7, 119–22; gender and, 140, 165, 231–32; hypnotism and, 135, 138–40; as ideology, 8–9, 95; in medicine, 91, 97, 98, 110–25; as method of ordering thought, 123; modernity and, 132–34, 163, 189, 191–92; photography and, 179; as practice, 7, 113, 118–20; spectatorship and, 11, 13, 16, 91, 92, 95, 96, 110, 139–40, 163, 189, 191–92, 250, 251; temporality of, 118–26; theory and, 64, 66–67, 73, 80–81; training and, 6–7, 10–11, 146, 163, 171, 192, 145–46. *See also* all-at-onceness, analysis and synthesis; "critical method"; expert modes of viewing
On the Optical Sense of Form (Vischer), 215
opera, 224
optical unconscious, 86
"ordinary perception" (Bergson), 30, 33–36, 98, 102, 213, 235
organ function, 100
"The Origin of Geometry" (Husserl), 58
Ostwald, Wilhelm, 65

outgrowth theory of nerve development, 80, 81, 85

pace: human liberty or potential and, 208; of modern life. 131, 133, 134, 163, 173, 186, 189, 208, 244; of motion pictures, 5, 14, 118, 125, 126, 131, 139, 181, 200, 202, 210, 244; of observation, 93, 95, 139; superficiality and, 178, 203–4; tastelessness and, 186. See also modernity; observation; social acceleration
Paget, Sir James, 7, 91, 123
PAGU (film company), 157
Parnaland, Ambroise-François, 106
particle displacement, 64, 68–70, 72–73
passive and active viewers, 200, 232. See also expert/lay distinctions
passivity, 201, 232
Pasveer, Bernike, 113
Pathé Frères, 28, 106
pathological movement, 104
pathology, 91
patterns of use: film form and, 20–22, 248–51; historiography and, 20–22, 246, 248–49; medicine and, 111, 121–23; professional identity and, 125
perception: apperception, 173–74, 181–82, 226–27; embodied, 215–17, 220, 231; direct (immediate or sensual), 7, 108, 110, 145, 172, 173, 221, 250; mode of (Benjamin), 12–13; "ordinary" (Bergson), 30, 33–36, 98, 102, 213, 235. See also aesthetic experience; medical gaze; medical perception; observation
Perrin, Jean, 69, 73
Pestalozzi, Johann Heinrich, 7, 148, 172–76, 181
Pfeffer, Wilhelm, 28
Pfemfert, Franz, 239
photogrammetry, 52
photography, 26, 29–30, 170, 193, 211, 216; Brownian motion and, 66–67, 69–70, 71, 73–75; medicine and, 97, 98, 104–05, 112–14, 115; Naturgetreu and, 179; series, 112–15, 116, 248. See also chronophotography
physical sciences, 36
physicians: as bearers of culture (Kulturträger), 127; doctor-patient relationship, 95, 126; as experts, 126–27
physics, 35, 40, 76–77; classical, 63, 65, 68; metaphysics, 239
physiology, 41, 97, 215, 225
Pinthus, Kurt, 194, 239
Pizon, Antoine, 27, 76
Plato, 211
play drive, 206–7
pleasure, visual, 228–29
Polgar, Alfred, 220, 221, 224
Polimanti, Osvaldo, 103
political consciousness, 233
politics, of contemplation, 230–42
Populist movement, 147
Porten, Henny, 193
positivism, 41
presence, 108, 110, 124, 237–38
primitivism, 296n80
The Problem of Form in the Fine Arts (Hildebrand), 221
Proceedings of the Royal Saxon Society for Sciences, 44
professional identity, 94, 95, 122, 124, 125
Progressive movement, 147
projection, 95, 102, 110, 117, 136; emotional. See Einfühlung; slow-motion, 118; speeds of, 117, 123, 125
proper viewing, 11, 94–95. See also observation
protoplasmic bridge theory of nerve development, 80, 81, 83
psychoanalysis, 86, 138
psychology: of children, 127, 146, 162–64; of crowds, 137, 146, 164–66; social, 162, 164

public health, 94, 127
public sphere, 16, 184; women in, 231
pulp fiction (*Schundliteratur*), 127, 151

quantitative frame analysis, 60, 98, 117, 125
Quo Vadis? (film), 159

Rabinbach, Anson, 39, 41, 166, 187
Radica (conjoined twin), 106
radiology, 92, 101
Ramón y Cajal, Santiago, 81
Rancière, Jacques, 199–200
Raumgestaltung (spatial creation), 226
Rauscher, Ulrich, 211–12
readability. *See* legibility
realism. *See* filmic realism
Refining the Cinema (*Veredelung des Kinos*), 156
reform: of clothing, 149–50; of education, 148–49, 151; of film, 147, 149, 152–62, 163, 172; in Germany, 147–52; Kinoreform, 152, 154–55, 160, 184, 238; *Kulturkritik* and, 148; modernity and, 149; spirit of, 147–62
Reformkinos, 180
regulation, of cinemas, 158–59
Reicher, Karl, 107
Reich Health Office, 284n122
Rembrandt as Educator (Langbehn), 168–69
repeatability, 98, 113
repose, 197, 207, 209
research film: 15, 19–26, 118, 187, 243, 248; Bergson and, 32–37; Brownian motion and, 62–76, 71, 72; disciplinary contexts of, 97; distribution of, 104; documentary function of, 102–5; educational function of, 105–10; exploratory function of, 98–102; medical, 92–93, 96–125; nerve fibers and, 76–88, 80; overview of, 26–31; public screening of, 107; science of work and, 37–44; tissue cultures and, 76–88. *See also* chronophotography; cinematography; *The Human Gait*; medical filmmaking
Rheinberger, Hans-Jörg, 24, 250
Richer, Paul, 103
Rieder, H., 100
Ries, Julius, 27, 76
Ritter, Christopher, 174
Ritzheimer, Kara, 151
Rockefeller Institute, 83
Romanticism, 236
Rosenthal, J., 100
Ross, Corey, 151
Rousseau, Jean-Jacques, 148
Runge, Philipp Otto, 170
Russia, 90
Rüswald, K. (teacher), 178

Saint-Louis Hospital, 104
Samuleit, Paul, 157
scanning (*Schauen*), 7, 225
Schaar, Eckard, 170
Schauen (scanning), 7, 225
Schaulust (scopophilia), 165, 228–30, 309n72
Schelling, Friedrich Wilhelm Joseph, 215
Schenk, Paul, 130
Schiller, Friedrich, 147, 191, 213, 215; aesthetic contemplation and, 190, 201; aesthetic education and, 167–68, 171; aesthetic experience and, 8, 14, 195–98, 204–7; form drive and, 205, 206; play drive and, 206–7; sense drive and, 204, 206; temporality and, 204–7; vision and, 190
Schlanger, Benjamin, 241
Schlüpmann, Heide, 231–32
Schmarsow, August, 225–27, 230
Schopenhauer, Arthur, 195, 199, 204, 215, 241; aesthetic experience and, 8, 196, 213, 235; self and, 214; temporality and, 207

Schultze-Naumberg, Paul, 150
Schülervorstellungen (commercial screenings for children), 181
Schulvorstellungen (screenings for classes), 181
Schumberg, Wilhelm, 42
Schund campaigns, 151–52, 154
Schundfilme (trashy films), 126, 157, 158
Schundliteratur (pulp fiction), 127, 151
Schuster, Paul, 103, 104
scientific filmmaking. See research film. See also chronophotography; cinematography; The Human Gait; medical filmmaking
scientific method, 15, 22, 32, 35, 37, 64, 117, 125, 127, 133, 135. See also experiment; observation
scientific observation. See observation
scientific theater, 180
scopophilia (Schaulust), 165, 228–30, 309n72
sea squirts (Botryllus), 76
second law of thermodynamics, 40, 65
Seddig, Max, 20–21, 25–26, 29, 64, 69–75, 87–88
Sehen (seeing), 7, 225
self, loss of, 197, 214
self-awareness, 208, 214
self-cultivation (Bildung), 141, 145
self-identity, 13, 144, 174
self-mastery, 245
Sellmann, Adolf, 1, 172, 178, 208
sense drive, 204, 206
sense impression (Anschauung), 172–73, 176
sensuality, 142–43
sensuous totality (Gesamtsinnlichkeit), 224
sensus communis (communal sense), 191
sequential images, 86
serialized novels, 151

series photography, 112–15, 116, 248. See also chronophotography; medicine; photography
Serner, Walter, 194, 228–29
sexual arousal, 228–29
sexual images, 128, 130
sexuality, 231
Shapin, Steven, 100
shocks, modernity as series of, 187, 192
Siedentopf, Henry, 77
Simmel, Georg, 212
slow-motion cinematography, 27, 86, 123
slow-motion projection, 118
Smoluchowski, Maryan, 73
Sobchack, Vivian, 220
social acceleration, 126, 131, 173, 202, 213, 244; Anschauungsunterricht and, 173–74. See also modernity; pace: human liberty or potential and
Social Democratic Party of Germany (SPD), 292n34
social psychology, 162, 164
Society for the Dissemination of Popular Education (Gesellschaft zur Verbreitung von Volksbildung) (GVV), 152, 159
Society of Friends of the Schools and Instruction for the Fatherland (Gesellschaft der Freunde des vaterländischen Schul- und Erziehungswesens), 154
Society of German Natural Scientists and Physicians (Gesellschaft deutscher Naturforscher und Ärzte), 83, 111
Solvay, Ernest, 39
"Some Specific Features of the Medical Way of Thinking" (Fleck), 113
somnambulism, 212–13
soullessness, 239–40
spatial creation (Raumgestaltung), 226
SPD. See Social Democratic Party of Germany

spectacle, in medicine, 279n63
spectatorship, 146, 191, 200–202, 208; analysis, synthesis and, 165; children, crowds and, 162–65; embodied, 216, 220–25; gender and, 165, 231–32; hypnotism and, 135–37; observation and, 11, 13, 16, 91, 92, 95, 96, 110, 139–40, 163, 189, 191–92, 250, 251; theories of film, 11–12. *See also* improper viewing; lay viewers; negative space
"Spectacles of the Earth" (Häfker), 187
Spencer, Herbert, 175
"The Stage Considered as a Moral Institution" (Schiller), 167
staining techniques, 84–85
Stapel, Wilhelm, 204, 208
statistical aspect, of medical logic, 113
stillness and movement, 11, 122, 124, 226
stimulus shield, 212
storefront cinemas (*Ladenkinos*), 126, 153
Stratz, Carl Heinrich, 150
strobe effect, 51–52
Strobl, Karl Hans, 203, 208, 213
Studien zu einer Physiologie des Marsches (Zuntz and Schumberg), 42
subjectivity, 8, 190, 214, 215, 220; objectivity and, 95, 123
sublime, theory of the, 214
suggestibility, 136, 162, 164–65, 167
surgery of the dead, 108
"Surrealism" (Benjamin), 233
synesthesia, 221–22, 237
synthesis. *See* analysis and synthesis
Szczepaniak-Gillece, Jocelyn, 241

taste, 94, 143, 146, 147, 155, 160, 166, 200; aesthetics; morality and, 167–68; geometry of, 168; Häfker and, 186–89; ideology and, 191; nation and, 170–71; as *sensus communis*, 191; and vision, 145–46, 191
Tausk, Viktor, 210
technological reproducibility, 235

technology, 144; formal relationships in, 4; media, 1–4, 9
temperance movement, 127
temporal discontinuity. *See* continuity and discontinuity
temporal interruption, 64, 74
temporality: of aesthetic experience, 202–14; of observation, 118–26
temporal phenomena, 22
Tews, Johannes, 159
textual analysis. *See* analysis (close reading)
theater: cinema compared to, 236–39; scientific, 180
theory: disciplinary logic and, 24; observation and, 64, 66–67, 73, 80–81
Theory of the Novel (Lukács), 235–36, 240
thermodynamics, 40, 65, 67
Thomson, William (Lord Kelvin), 40
"Thoughts Toward an Aesthetic of the Cinema (Gedanken zu einer Aesthetik des 'Kino')" (Lukács), 202, 236
thought styles, 249, 250. *See also* disciplinary logic
Three Essays on Sexuality (Freud), 163
time, 197, 207; Lukács and, 238; reversibility of, 74–75; spatialization of, 62–64, 74; spectatorship and will and, 125–41. *See also* duration; *durée*; temporality
time-lapse cinematography, 23, 30, 84–85, 86, 87, 98, 123, 250; Braus and, 20; medical filmmaking and, 98
Tirol in Waffen (*Tirol in Arms*) (film), 160
tissue cultures, 76–88
Titchener, Edward B., 216
Tolstoy, Leo, 128
Tönnies, Ferdinand, 288n9
total presentation (*Gesamtvorführung*), 188
trade journals and magazines, 152–53
traditional aesthetics. *See* aesthetics
training film, 108
trashy films (*Schundfilme*), 126, 157, 158

triangulation, 52
Tucker, Jennifer, 26
tutelage, 140
two-sided chronophotography, 46. See also chronophotography; *The Human Gait*

UFA, 243, 244; cultural division of (*Kulturabteilung*), 243
ultramicroscope, 69, 70
Union-Theater, 219
United Kingdom, 90, 92, 106, 147–48
United States, 90, 92, 97, 107, 109, 147–48; Bergson and, 22, 32, 36; object lesson and, 16
University of Tübingen, 127, 134, 143, 152
uplift and educational film, 154
useful cinema, 2

Valéry, Paul, 90
Van Gehuchten, Arthur, 103
Veredelung des Kinos (Refining the Cinema), 156
Vierordt, Hermann, 45
Virchow, Rudolph, 119
virtual experiments, 84, 100
virtual witnessing, 84, 100
Vischer, Robert, 7, 215, 220, 222, 224–25
visual means of instruction. See *Anschauungsunterricht*
visual pleasure, 228–29
vitalism and mechanism, 39–40
vividness (*Anschaulichkeit*), 107–8, 179
Vlès, Fred, 27
volition, 133, 286n143. See also agency; free will
von Bergmann, Ernst, 107
von Helmholtz, Hermann, 40, 215

Wagner, Richard, 32, 128
The Wanderings of Odysseus (*Die Irrfahrten des Odysseus*) (film), 157, 158

Wanderkino, 160
Warstat, Willi, 186
Weber, Max, 3
Weber, Wilhelm, 263n58
Weber brothers, 45
Weichardt, Wilhelm, 42
Weimar Republic, 151
Weisenburg, Theodore, 103
"What Is Enlightenment?" (Kant), 140
wholeness and fragmentation, 36–37
Wilhelm II (kaiser), 148, 218
Wilhelm Meister (Goethe), 236
will 139. See also agency; free will
Wolf-Czapek, K. W., 124
Wolgast, Heinrich, 171
women, 151; clothing reform and, 149–50; elite and, 150; hysteria and, 137; in public sphere, 150–51, 231; spectatorship by, 165, 231–32
Women's Clothing and Its Natural Development (Stratz), 150
working class, 232
"Work of Art" (Benjamin), 195, 201, 218, 233, 240
working objects. See docile bodies
The World as Will and Representation (Schopenhauer), 215
World War I, 13, 159, 161–62, 193, 202, 236, 241; class and, 232; gender dynamics and, 231; Häfker and, 301n145; health campaigns and, 127; *Kulturabteilung* and, 243; X-ray cinematography and, 102
Wundt, Wilhelm, 215

X-rays, 97, 113–14; cinematography, 27, 100–102, *101*

Zeiss, Carl, 28, 77
Zivilisation and *Kultur*, 144–45
Zoological Gardens, 156
Zsigmondy, Richard, 77
Zuntz, Nathan, 42

FILM AND CULTURE

A series of Columbia University Press
Edited by John Belton

What Made Pistachio Nuts? Early Sound Comedy and the Vaudeville Aesthetic,
    Henry Jenkins
Showstoppers: Busby Berkeley and the Tradition of Spectacle, Martin Rubin
Projections of War: Hollywood, American Culture, and World War II, Thomas Doherty
Laughing Screaming: Modern Hollywood Horror and Comedy, William Paul
Laughing Hysterically: American Screen Comedy of the 1950s, Ed Sikov
*Primitive Passions: Visuality, Sexuality, Ethnography, and Contemporary
    Chinese Cinema*, Rey Chow
The Cinema of Max Ophuls: Magisterial Vision and the Figure of Woman,
    Susan M. White
Black Women as Cultural Readers, Jacqueline Bobo
Picturing Japaneseness: Monumental Style, National Identity, Japanese Film,
    Darrell William Davis
*Attack of the Leading Ladies: Gender, Sexuality, and Spectatorship
    in Classic Horror Cinema*, Rhona J. Berenstein
This Mad Masquerade: Stardom and Masculinity in the Jazz Age, Gaylyn Studlar
Sexual Politics and Narrative Film: Hollywood and Beyond, Robin Wood
The Sounds of Commerce: Marketing Popular Film Music, Jeff Smith
Orson Welles, Shakespeare, and Popular Culture, Michael Anderegg
*Pre-Code Hollywood: Sex, Immorality, and Insurrection in American Cinema,
    1930–1934*, Thomas Doherty
Sound Technology and the American Cinema: Perception, Representation, Modernity,
    James Lastra
Melodrama and Modernity: Early Sensational Cinema and Its Contexts, Ben Singer
Wondrous Difference: Cinema, Anthropology, and Turn-of-the-Century Visual Culture,
    Alison Griffiths
Hearst Over Hollywood: Power, Passion, and Propaganda in the Movies, Louis Pizzitola
Masculine Interests: Homoerotics in Hollywood Film, Robert Lang
Special Effects: Still in Search of Wonder, Michele Pierson
Designing Women: Cinema, Art Deco, and the Female Form, Lucy Fischer
Cold War, Cool Medium: Television, McCarthyism, and American Culture,
    Thomas Doherty
Katharine Hepburn: Star as Feminist, Andrew Britton
Silent Film Sound, Rick Altman
Home in Hollywood: The Imaginary Geography of Hollywood, Elisabeth Bronfen
Hollywood and the Culture Elite: How the Movies Became American, Peter Decherney
Taiwan Film Directors: A Treasure Island, Emilie Yueh-yu Yeh and Darrell
    William Davis
*Shocking Representation: Historical Trauma, National Cinema, and the Modern
    Horror Film*, Adam Lowenstein

China on Screen: Cinema and Nation, Chris Berry and Mary Farquhar
The New European Cinema: Redrawing the Map, Rosalind Galt
George Gallup in Hollywood, Susan Ohmer
Electric Sounds: Technological Change and the Rise of Corporate Mass Media,
	Steve J. Wurtzler
The Impossible David Lynch, Todd McGowan
*Sentimental Fabulations, Contemporary Chinese Films: Attachment in
	the Age of Global Visibility*, Rey Chow
Hitchcock's Romantic Irony, Richard Allen
Intelligence Work: The Politics of American Documentary, Jonathan Kahana
Eye of the Century: Film, Experience, Modernity, Francesco Casetti
Shivers Down Your Spine: Cinema, Museums, and the Immersive View, Alison Griffiths
Weimar Cinema: An Essential Guide to Classic Films of the Era, Edited by
	Noah Isenberg
African Film and Literature: Adapting Violence to the Screen, Lindiwe Dovey
Film, A Sound Art, Michel Chion
Film Studies: An Introduction, Ed Sikov
Hollywood Lighting from the Silent Era to Film Noir, Patrick Keating
Levinas and the Cinema of Redemption: Time, Ethics, and the Feminine, Sam B. Girgus
Counter-Archive: Film, the Everyday, and Albert Kahn's Archives de la Planète,
	Paula Amad
Indie: An American Film Culture, Michael Z. Newman
Pretty: Film and the Decorative Image, Rosalind Galt
Film and Stereotype: A Challenge for Cinema and Theory, Jörg Schweinitz
Chinese Women's Cinema: Transnational Contexts, Edited by Lingzhen Wang
Hideous Progeny: Disability, Eugenics, and Classic Horror Cinema, Angela M. Smith
Hollywood's Copyright Wars: From Edison to the Internet, Peter Decherney
Electric Dreamland: Amusement Parks, Movies, and American Modernity,
	Lauren Rabinovitz
*Where Film Meets Philosophy: Godard, Resnais, and Experiments
	in Cinematic Thinking*, Hunter Vaughan
The Utopia of Film: Cinema and Its Futures in Godard, Kluge, and Tahimik,
	Christopher Pavsek
Hollywood and Hitler, 1933–1939, Thomas Doherty
Cinematic Appeals: The Experience of New Movie Technologies, Ariel Rogers
Continental Strangers: German Exile Cinema, 1933–1951, Gerd Gemünden
Deathwatch: American Film, Technology, and the End of Life, C. Scott Combs
After the Silents: Hollywood Film Music in the Early Sound Era, 1926–1934,
	Michael Slowik
*"It's the Pictures That Got Small": Charles Brackett on Billy Wilder and Hollywood's
	Golden Age*, Edited by Anthony Slide
*Plastic Reality: Special Effects, Technology, and the Emergence of 1970s Blockbuster
	Aesthetics*, Julie A. Turnock

Maya Deren: Incomplete Control, Sarah Keller
Dreaming of Cinema: Spectatorship, Surrealism, and the Age of Digital Media, Adam Lowenstein
Motion(less) Pictures: The Cinema of Stasis, Justin Remes
The Lumière Galaxy: Seven Key Words for the Cinema to Come, Francesco Casetti
The End of Cinema? A Medium in Crisis in the Digital Age, André Gaudreault and Philippe Marion
Studios Before the System: Architecture, Technology, and the Emergence of Cinematic Space, Brian R. Jacobson

GPSR Authorized Representative: Easy Access System Europe, Mustamäe tee 50, 10621 Tallinn, Estonia, gpsr.requests@easproject.com

www.ingramcontent.com/pod-product-compliance
Lightning Source LLC
Chambersburg PA
CBHW021929290426
44108CB00012B/778